International
REVIEW OF
Neurobiology
Volume 100

International

REVIEW OF

Neurobiology

Volume 100

SERIES EDITORS

R. ADRON HARRIS

Waggoner Center for Alcohol and Drug Addiction Research
The University of Texas at Austin
Austin, Texas, USA

PETER JENNER

Division of Pharmacology and Therapeutics
GKT School of Biomedical Sciences
King's College, London, UK

EDITORIAL BOARD

Monoamine Oxidases and Their Inhibitors

EDITED BY

MOUSSA B.H. YOUDIM

Department of Molecular Pharmacology
Eve Topf Centre of Excellence for Neurodegenerative Diseases
Technion-Rappaport Faculty of Medicine
Haifa, Israel

PETER RIEDERER

Clinical Neurochemistry
National Parkinson Foundation Centre of Excellence Laboratories
Clinic and Polyclinic for Psychiatry, Psychosomatic, and Psychotherapy
Medical School
University of Würzburg, Würzburg, Germany

ELSEVIER

AMSTERDAM • BOSTON • HEIDELBERG • LONDON
NEW YORK • OXFORD • PARIS • SAN DIEGO
SAN FRANCISCO • SINGAPORE • SYDNEY • TOKYO
Academic Press is an imprint of Elsevier

Academic Press is an imprint of Elsevier
32 Jamestown Road, London NW1 7BY, UK
Radarweg 29, PO Box 211, 1000 AE Amsterdam, The Netherlands
Linacre House, Jordan Hill, Oxford OX2 8DP, UK
225 Wyman Street, Waltham, MA 02451, USA
525 B Street, Suite 1900, San Diego, CA 92101-4495, USA

First edition 2011

ISBN: 978-0-12-386467-3
ISSN: 0074-7742

For information on all Academic Press publications visit our website at elsevierdirect.com

Printed and bound in USA
11 12 13 14 10 9 8 7 6 5 4 3 2 1

CONTENTS

Kinetic Behavior and Reversible Inhibition of Monoamine Oxidases—Enzymes that Many Want Dead

Keith F. Tipton, Gavin P. Davey and Andrew G. McDonald

The Pharmacology of Selegiline

Kálmán Magyar

Type A Monoamine Oxidase Regulates Life and Death of Neurons in Neurodegeneration and Neuroprotection

Makoto Naoi, Wakako Maruyama, Keiko Inaba-Hasegawa and Yukihiro Akao

Multimodal Drugs and their Future for Alzheimer's and Parkinson's Disease

CORNELIS J. VAN DER SCHYF AND WERNER J. GELDENHUYS

Neuroprotective Profile of the Multitarget Drug Rasagiline in Parkinson's Disease

ORLY WEINREB, TAMAR AMIT, PETER RIEDERER, MOUSSA B.H. YOUDIM AND SILVIA A. MANDEL

Rasagiline in Parkinson's Disease

L.M. CHAHINE AND M.B. STERN

Selective Inhibitors of Monoamine Oxidase Type B and the "Cheese Effect"

JOHN P. M. FINBERG AND KEN GILLMAN

A Novel Anti-Alzheimer's Disease Drug, Ladostigil: Neuroprotective, Multimodal Brain-Selective Monoamine Oxidase and Cholinesterase Inhibitor

Orly Weinreb, Tamar Amit, Orit Bar-Am and Moussa B.H. Youdim

Novel MAO-B Inhibitors: Potential Therapeutic Use of the Selective MAO-B Inhibitor PF9601N in Parkinson's Disease

Mercedes Unzeta and Elisenda Sanz

CONTRIBUTORS

Numbers in parentheses indicate the pages on which the authors' contributions begin.

Yukihiro Akao (85), United Graduate School of Drug Discovery and Medical Information Sciences, Gifu University, Gifu, Japan

Tamar Amit (127, 191), Eve Topf Centre of Excellence for Neurodegenerative Diseases and Department of Molecular Pharmacology, Technion-Rappaport Faculty of Medicine, Efron Street, P.O. Box 9697, Haifa, Israel

Orit Bar-Am (191), Eve Topf Centre of Excellence for Neurodegenerative Diseases and Department of Molecular Pharmacology, Technion-Rappaport Faculty of Medicine, Efron Street, P.O. Box 9697, Haifa, Israel

Claudia Binda (1), Department of Genetics and Microbiology, University of Pavia, Pavia, Italy

Marco Bortolato (13), Department of Pharmacology and Pharmaceutical Sciences, School of Pharmacy, University of Southern California, Los Angeles, California, USA

L.M. Chahine (151), Penn Comprehensive Neuroscience Center, University of Pennsylvania, Philadelphia, Pennsylvania, USA

Gavin P. Davey (43), Department of Biochemistry, Trinity College, Dublin, Ireland

Dale E. Edmondson (1), Department of Biochemistry, Emory University, Atlanta, Georgia, USA

John P.M. Finberg (169), Department of Molecular Pharmacology, Rappaport Faculty of Medicine, Technion, Haifa, Israel

Werner J. Geldenhuys (107), Department of Pharmaceutical Sciences, College of Pharmacy, Northeast Ohio Medical University, Rootstown, Ohio, USA

Ken Gillman (169), Psychotropical Research, Bucasia, Queensland, Australia

Keiko Inaba-Hasegawa (85), Department of Neurosciences, Gifu International Institute of Biotechnology, Kakamigahara, Gifu, Japan

Kálmán Magyar (65), Department of Pharmacodynamics, Semmelweis University, Budapest, Hungary; Neurochemical Research Unit, Hungarian Academy of Sciences, Budapest, Hungary

Silvia A. Mandel (127), Eve Topf Centre of Excellence for Neurodegenerative Diseases and Department of Molecular Pharmacology, Technion Rappaport Faculty of Medicine, Efron Street, P.O. Box 9697, Haifa, Israel

Wakako Maruyama (85), Department of Cognitive Brain Science, National Research Center for Geriatrics and Gerontology, Obu, Aichi, Japan

Andrea Mattevi (1), Department of Genetics and Microbiology, University of Pavia, Pavia, Italy

Andrew G. McDonald (43), Department of Biochemistry, Trinity College, Dublin, Ireland

Makoto Naoi (85), Department of Neurosciences, Gifu International Institute of Biotechnology, Kakamigahara, Gifu, Japan

Peter Riederer (127), Clinical Neurochemistry, National Parkinson Foundation Centre of Excellence Laboratories, Clinic and Polyclinic for Psychiatry, Psychosomatic, and Psychotherapy, Medical School, University of Würzburg, Würzburg, Germany

Elisenda Sanz (217), Institut de Neurociències and Departament de Bioquímica i Biologia Molecular, Universitat Autònoma de Barcelona, Cerdanyola del Vallès (Barcelona), Spain

Jean C. Shih (13), Department of Pharmacology and Pharmaceutical Sciences, School of Pharmacy, University of Southern California, Los Angeles, California, USA; Department of Cell and Neurobiology, Keck School of Medicine; University of Southern California, Los Angeles, California, USA

M.B. Stern (151), Penn Comprehensive Neuroscience Center, University of Pennsylvania, Philadelphia, Pennsylvania, USA

Keith F. Tipton (43), Department of Biochemistry, Trinity College, Dublin, Ireland

Mercedes Unzeta (217), Institut de Neurociències and Departament de Bioquímica i Biologia Molecular, Universitat Autònoma de Barcelona, Cerdanyola del Vallès (Barcelona), Spain

Cornelis J. Van der Schyf (107), Department of Pharmaceutical Sciences, College of Pharmacy, Northeast Ohio Medical University, Rootstown, Ohio, USA

Orly Weinreb (127, 191), Eve Topf Centre of Excellence for Neurodegenerative Diseases and Department of Molecular Pharmacology, Technion-Rappaport Faculty of Medicine, Efron Street, P.O. Box 9697, Haifa, Israel

Moussa B.H. Youdim (127, 191), Eve Topf Centre of Excellence for Neurodegenerative Diseases and Department of Molecular Pharmacology, Technion-Rappaport Faculty of Medicine, Efron Street, P.O. Box 9697, Haifa, Israel; Department of Biology, Yonsei University, Seoul, South Korea

PREFACE

This volume will be concerned with an important enzyme that was identified in the liver of rabbits some 90 years ago, which metabolized tyramine by oxidative deamination. In 1934 Blaschko demonstrated that this enzyme metabolized primary, secondary, and tertiary amines, including adrenaline and noradrenaline, and that tyramine oxidase and noradrenaline oxidase were the same enzyme. Zeller gave it its name as monoamine oxidase (MAO) to differentiate it from diamine oxidases. Its relevance to psychiatry was recognized by serendipity in the early 1950s, where iproniazid, a drug for treatment of tuberculosis, was discovered to be its inhibitor and introduced into the clinic as the first antidepressant. The contributions of studies on MAO and its inhibitor to pharmacology and physiological actions of biogenic amine neurotransmitters can not be exaggerated. Numerous MAO inhibitors were developed by the pharmaceutical companies as antidepressants. Subsequent reports of side effects of MAO inhibitor antidepressants, known as the "cheese reaction," resulting in hypertensive crisis led almost to this appearence of these drugs from the clinic. Between 1965 and 1968, evidences were provided that MAO exists in at least two forms. Johnson named these as MAO-A and MAO-B, where the irreversible propargylamine-derived inhibitor, clorgyline, was shown to be a selective inhibitor of MAO-A. The neurotransmitters serotonin, noradrenaline, and adrenaline were identified as substrates of MAO-A, while phenylethylamine and benzylamine are substrates of MAO-B. Dopamine and tyramine were considered substrates for both enzymes. Knoll and Magyar reported that another propargylamine-derived inhibitor, L-deprenyl (later named selegiline), was a selective inhibitor of MAO-B. The impact of these findings was to result in a flurry of studies with these selective inhibitors to determine the distribution, function, and physiological and neuropharmacological roles of the two enzymes in various animal and human tissues and more profoundly in the brain. One important intriguing pharmacological aspect of L-deprenyl was that at its selective MAO-B-inhibitory dosage, it did not induce the "cheese reaction" in animal studies, thus being the first MAO inhibitory devoid of such property. This feature and predominance of MAO-B in human brain extrapyramidal regions was the impetus for Birkmayer, Riederer, and Youdim to initiate a clinical trial with L-deprenyl in Parkinson's disease in

1974–1975, with positive results and confirmed by other groups. In 1983 the same group suggested that L-deprenyl in Parkinson's disease may induce longevity in parkinsonian patients and was attributed to prevention of the degeneration of nigrostriatal dopamine neurons (neuroprotection). This finding was supported by identification of the dopaminergic neurotoxin, MPTP (*N*-methyl-1,2,3,6-tetrahydropyridine), that caused parkinsonism in drug designer addicts. MPTP, a substrate of MAO-B, is an inert substance; however, when injected into animals, it is converted to the neurotoxin, MPP^+ by MAO-B, which then is taken up by nigrostriatal dopamine neurons and causes neurodegeneration of dopamine neurons in mice, cat, dogs, and human and nonhuman primates. Heikkila and colleagues in 1984 published the exciting paper which showed that when mice were preinjected with L-deprenyl, but not with clorgyline, it prevented the MPTP-induced degeneration of the dopamine neurons. Thus, the concept of neuroprotection was born, which initiated numerous neuroprotective studies with MPTP as the model of Parkinson's disease. The demonstration of MAO-A and -B being different proteins, the anti-Parkinson activity of L-deprenyl and its neuroprotective activity, as confirmed by several groups, led to the search and development of numerous irreversible and reversible selective and specific inhibitors of MAO-A, as antidepressants, and MAO-B as anti-Parkinson drugs, devoid of the cheese reaction. Much credit should go to Moshe Da Prada for advancing the MAO inhibitors, who developed successfully the antidepressant reversible MAO-A inhibitor, moclobemide, and the first reversible MAO-B inhibitor, lazabemide, neither of which initiated the cheese reaction.

L-Deprenyl did not reach the United States until 1989, some 15 years after our first description of its anti-Parkinson action. There is a misconception that no drug is a drug until it is approved by FDA in the United States. Nevertheless, L-deprenyl was patented as an orphan drug in the United States and given the name selegiline. Between 1975 and 1982, there was a belief that there was something unique about the pharmacological action of L-deprenyl (selegiline) that differentiated it from other MAO inhibitors. But during this period, Youdim and colleagues identified the second propargylamine-derived MAO-B inhibitor, AGN1135, which was highly significantly more active than L-deprenyl. This drug eventually became the second MAO-B inhibitor anti-Parkinson drug named, rasagiline. Rasagiline is differentiated from selegiline by the observations in which both the parent compound and its aminoindan metabolite are neuroprotective and MAO-B inhibition is not a prerequisite for neuroprotection. Indeed a recent clinical study, ADAGIO, in parkinsonian subjects with rasagiline has indicated it may be the first disease-modifying neuroprotective drug. Much has been learned about the molecular mechanism of the neuroprotective activities of selegiline and rasagiline, which we hope can contribute to the development of even better antidepressants and anti-Parkinson drugs.

This volume is a collection of chapters by many of the leading individuals that have helped to shape and advance the research in MAO and its more recent developments.

The volume starts with the fascinating chapter by Binda, Mattevi, and Edmondson which describe the fundamental differences between MAO-A and MAO-B with regard to their protein structures as shown by X-ray crystal structure analyses. Identical in their FAD-binding sites, they differ in the structures of their active sites opposite the flavin cofactor. The cavity structures differ considerably for substrate entrance. Both enzymes are dimeric in their membrane-bound forms. These findings are essential for the development of new compounds as MAO-inhibitors. They describe the mechanism and structural requirement for binding of the selective MAO inhibitors.

Both isoenzymes are mitochondrial-bound proteins, catalyzing the oxidative deamination of biogenic amines/neurotransmitters. As Bortolato and Shih elaborate here MAO-A and MAO-B derive from a common ancestral progenitor gene, are located at the X-chromosome, and share 70% structural identity. They can be distinguished by their substrate and inhibitor specificity and selectivity. As such, it is not far fetched to assume that any dysregulations may lead to a variety of behavioral/social alterations with particular phenotypes including anxiety, depression, attention-deficit-hyperactivity disorder, impulse-control disorders, autism, psychosis, etc.

It is another highlight of this book that the Doyen of MAO/MAO-inhibitor biochemistry, Tipton *et al.* delivered an extensive essay on the kinetic properties of both enzymes and their inhibition. Perhaps, it is the last of such chapters dealing with "classic biochemistry" for the next decade (and until biochemistry again becomes important in molecular biology when it is fundamental again to translate static measures into "functions").

One of the reasons that MAO-inhibitors of the 50th and 60th of the past century could not be further used and developed has been the liver toxicity of many such compounds based on a hydrazine structure and the development of hypertensive crises called cheese effect (as it was thought that diet composed of cheese, beer, and red wine in patients on MAO-inhibitor therapy would be causal for such blood pressure crisis because of the sympathomimetic amines present in such diets which are substrates of MAO and prevention of their metabolism in MAO treatment can result in the hypertensive crisis).

Finberg and Gillman in an extremely elegant and comprehensive review describe the hard facts about the potential of MAO-inhibitors to release the cheese effect. Importantly, they describe the translation of animal experimental work and the patient response when using such medication under clinical dose regime. They describe the mechanism why selective inhibitors of MAO-B in contrast to MAO-A inhibitors do not induce the cheese reaction.

Magyar delineates the development of the first selective and specific MAO-B inhibitor selegiline (L-deprenyl) whose studies initiated the revival of interests in MAO-A and -B inhibitors as psychotropic drugs. Here, he concentrates on selegiline metabolites elaborating their potential for neuroprotection (desmethylselegiline, selegiline-*N*-oxide) as well as their possible harmful action (amphetamine, metamphetamine). As one of the discoverers of selegiline, Magyar reviews the pharmacokinetic properties of selegiline in experimental studies and thus critically views some unknown properties of this irreversible MAO-B inhibitor without the cheese effect.

Molecular biological and genetic aspects of selegiline's action are covered by Naoi *et al.* In this chapter, they not only point to differentiation of selegiline and rasagiline molecular properties, but rather focus on the so far overlooked important aspects, namely, the role of MAO-A in neuronal death and protection mechanisms by MAO-B inhibitors. This aspect is of major importance as understanding of the interaction uncovers new lines of strategies to fight neurodegeneration.

Silvia Mandel's working group concentrates on the molecular pharmacology and molecular biology of rasagiline and selegiline and follow-up inhibitor compound. In addition, the role of the presumptive neuroprotective and neurorestorative metabolite of rasagiline aminoindan is highlighted and that aminoindan could contribute to the neuroprotective activity of the parent compound rasagiline. Evidence for Parkinson's disease-modifying action of rasagiline is presented as being the first neuroprotective drug for this disorder.

This line of arguments for a protective role of rasagiline is further enlarged by Chahine and Stern by focusing on clinical trials, their outcome, and future developments.

Unzeta and Sanz presents new data on another propargylamine-derived MAO inhibitor, PF 9601 N, a compound with a high potency and selectivity to inhibit MAO-B. *In vivo* and *in vitro* models demonstrate neuroprotective potential so that there is a new promising candidate for the treatment of Parkinson's disease.

Some 20 to 30 years ago, we were confronted with what was called "dirty drugs" meaning that the multiple pharmacological action of many medications not only improved the patient's condition but rather and frequently led to unwanted side effects and adverse reactions. Although potent in their clinical beneficial effects specific/monocausal drugs were developed with the implementation to create medication with both a significant beneficial effect and less side effects/adverse reactions. This concept, however, has to be reconsidered on the basis that neurodegenerative disorders at least are multitransmitter disorders, they are mostly sporadic and therefore multigenetic, and they are very much dependent on multiple environmental influences. Therefore, Van der Schyf and Geldenhuys focus the important chapter on "multimodal," "multifunctional", designed multiple ligands (DMLs) drug developments. Delineating arguments and ways for this concept means that, by using modern knowledge about receptor subtypes, their preferred localization in (sub)regions of the human brain, etc., multiple pharmacological actions will indeed lead to more specific treatment of such devastating neurodegenerative disorders like Parkinson's and Alzheimer's diseases and a lower potential for side effects/adverse reactions. Thus the "dirty drugs" have lost their "dirty part," that is, side effects/adverse reactions and have converted to admire the positive part of "multifunctional drugs," that is, the beneficial effect.

Weinreb and her cooperators have used this concept in developing new multitarget drugs for Parkinson's disease, for example, M30, M30 S, and HLA-20, and Alzheimer's disease, for example, ladostigil. These drugs are the first brain-selective MAO-A and -B-inhibitors, with little inhibition of systemic MAO, that do not give a cheese reaction in contrast to the older drug such as phenelzine and tranylcypromine. These drugs have true antidepressant, anti-Parkinson, and anti-Alzheimer activity in animal model studies. The M30 series of compounds are brain permeable and have the unique property that they have neurorestorative activity *in vivo.* They increase endogenous BDNF, GDNF, VEGF, HIF, and erythropoetin. There is new hope that these drugs will help our patients at least to minimize their burden if it would not be possible to protect their neurons from degenerative processes.

Altogether, this book offers unique aspects and a wide range of novel research and developments into the field of MAO-A and MAO-B inhibitors in order to create new and more specific medication for mental as well as neurodegenerative disorders.

STRUCTURAL PROPERTIES OF HUMAN MONOAMINE OXIDASES A AND B

Claudia Binda[1], Andrea Mattevi[1] and Dale E. Edmondson[2]

[1]Department of Genetics and Microbiology, University of Pavia, Pavia, Italy
[2]Department of Biochemistry, Emory University, Atlanta, Georgia, USA

Abstract

The structural elucidations of human monoamine oxidases A and B (MAO-A and -B) have provided novel insights into their similarities and differences. Although the enzymes exhibit ~70% sequence identities, highly conserved chain folds, and are structurally identical in their flavin adenine dinucleotide (FAD)-binding sites, they differ considerably in the structures of their active sites opposite the flavin cofactor. MAO-A has a monopartite cavity of ~550 $Å^3$, and MAO-B exhibits a bipartite cavity structure with an entrance cavity of 290 $Å^3$ and a substrate cavity of ~400 $Å^3$. Ile199 functions as a conformational "gate" separating the two cavities. Both enzymes are anchored to the outer mitochondrial membrane via C-terminal helical tails. Loop structures are found at the entrances to their active sites at the membrane surface. Although the crystal structure of human MAO-A is monomeric while MAO-B is dimeric, both enzymes are dimeric in their membrane-bound forms. Dimerization may be important for the favorable orientation of the resultant protein dipole moment toward the anionic membrane surface.

DOI: 10.1016/B978-0-12-386467-3.00001-7

0074-7742/11 $35.00

I. Introduction

A major advancement in the field of mitochondrial MAOs was the development of high-level expression system for the genes encoding human MAO-A (Li *et al.*, 2002) and human MAO-B (Newton-Vinson *et al.*, 2000) in the methylotrophic organism *Pichia pastoris*. Previous work had shown that it is possible to obtain the expression of human MAO-A in *Saccharomyces cerevisiae* (Weyler *et al.*, 1990); however, attempts to achieve high-level expression of MAO-B in that organism resulted in only low levels of expressed enzyme (Urban *et al.*, 1991). Both enzymes have also been successfully expressed in baculovirus-infected insect cells (Rebrin *et al.*, 2001) and that system has resulted in the commercial availability of membrane preparations for the analysis of inhibitor-binding affinities. The yeast systems allow for purification of reagent quantities of either enzyme that were required for crystallization trials and the successful crystallization of human MAO-B (Binda *et al.*, 2002a) and, subsequently, human MAO-A (De Colibus *et al.*, 2005; Son *et al.*, 2008). The availability of structural data on both enzymes provided the basis for computational docking studies that have now become a common component of developmental studies of MAO inhibitors as neuroprotectants or as cardioprotectants.

II. Crystallization of Purified MAO-B and MAO-A

The original success of MAO-B crystallization utilized screens with various detergents. LDAO (lauryldimethylamine oxide) was found to be the first detergent system that resulted in diffracting crystals. On analysis of synchrotron data collected, it was found that the detergent was bound to the enzyme in its active site traversing both cavities of a bipartite cavity structure (Fig. 1A). This observation led to the realization that successful crystal growth requires a ligand in the active site of MAO-B. Subsequent successful crystallizations of human MAO-B incorporated the detergent Zwittergent 3–12 (*N*-dodecyl-*N*,*N*-dimethyl-3-amino-1-propanesulfonate) which did not bind in the MAO-B active site and allowed structural determinations of complexes with both covalent and reversible MAO-B inhibitors. Subsequent work illustrated the advantage of using polypropylene crystallization bridges for crystal growth. MAO-B crystallizations in polystyrene bridges resulted in the extraction of a potent MAO-B-specific inhibitor (1,4-diphenyl-2-butene) from the plastic which interfered with structural determinations of reversible inhibitors (Hubálek *et al.*, 2003). All crystal forms of human MAO-B show the purified enzyme

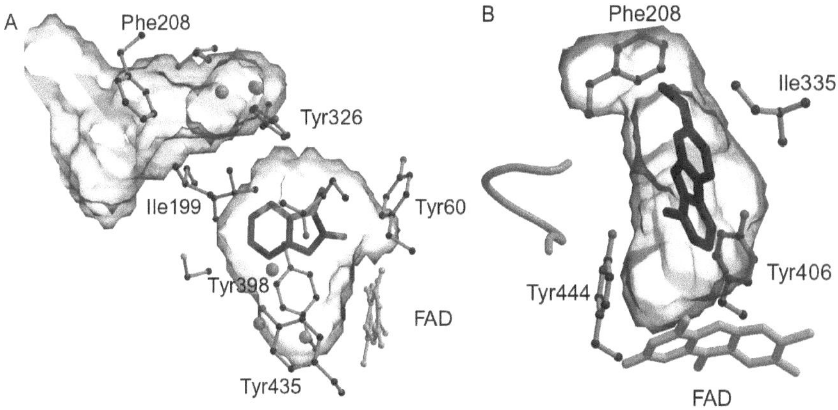

FIG. 1. (A) Active site structure for the bipartite cavity of human MAO-B (PDB Code 1OJA). Isatin is the bound ligand in the substrate cavity and the conformation of the Ile199 gating residue is in the closed conformation. (B) Structure of the monopartite active site cavity of human MAO-A (PDB Code 2Z5X). Harmine is shown bound in the active site.

to be dimeric which has been subsequently verified for its membrane-bound form using pulsed electron paramagnetic resonance-double electron–electron resonance (EPR-DEER) spectroscopic distance measurements (Upadhyay *et al.*, 2008).

Encouraged by the success with MAO-B, extensive crystallization trials with human MAO-A were initiated. Given the ~70% sequence identities of the two enzymes, it was expected that MAO-A should crystallize using the same techniques employed for MAO-B. This prediction proved to be erroneous as MAO-A crystallized in different detergent systems yielding data to a resolution of 3.2 Å (De Colibus *et al.*, 2005). Higher diffracting crystals of human MAO-A were found by Tsukihara's group (Son *et al.*, 2008) (2.2 Å) using the detergent *N*-dodecylphosphocholine. One possible contribution to the higher resolution MAO-A structure is that Tsukihara's group used expressed enzyme with an N-terminal His_6 tag. Current crystal data on human MAO-A from both laboratories show it to crystallize as a monomer rather than the dimeric structure found with human MAO-B (Binda *et al.*, 2002a,b) and with rat MAO-A (Ma *et al.*, 2004). Distance measurements by pulsed EPR-DEER experiments on membrane-bound forms of MAO-A (both human and rat) show it is dimeric in the membrane. It therefore appears that purification of human MAO-A in detergent results in its dissociation to a mixture of monomer and dimer forms with the monomeric form more readily crystallized for the human enzyme, whereas the dimeric form crystallizes more readily for rat MAO-A.

III. Structure of Human MAO-B

A total of 34 structures of human MAO-B differing in mutations and in nature of inhibitor binding have been deposited in the Protein Data Bank with resolutions ranging from 3.0 to 1.6 Å. The structure of MAO-B follows a *p*-hydroxybenzoate hydroxylase (PHBH) fold with a C-terminal helix that is predicted to be an anchoring transmembrane helix (Fig. 2A). Disorder does not allow visualization of the terminal 20 residues of this helix. The enzyme is dimeric in all crystal forms with a contact surface area between monomers of 2095 Å^2 ($\sim$15% of the monomer accessible surface). Substrate entry into the catalytic site involves negotiation of a protein loop (residues 99–112) covering the opening to the "entrance cavity" and traversing the opening between the entrance cavity and the substrate cavity involving a "gating" Ile199 side chain (Fig. 1A). These cavities comprise the bipartite cavity structure of human MAO-B and are probably similar for MAO-B preparations from other sources such as rat or bovine. In this context, Ile199 is replaced by a Phe in the bovine enzyme which accounts for differences observed in its binding affinities with reversible inhibitors expected to traverse both substrate and entrance cavities (Hubálek *et al.*, 2005; Krueger *et al.*, 1995).

To date, only one inhibitor, 2-benzofuranyl-imidazoline (Bonivento *et al.*, 2010), has been shown to specifically bind to the entrance cavity with no evidence for any binding into the substrate cavity (Fig. 3). Closure of the Ile199 gate and a

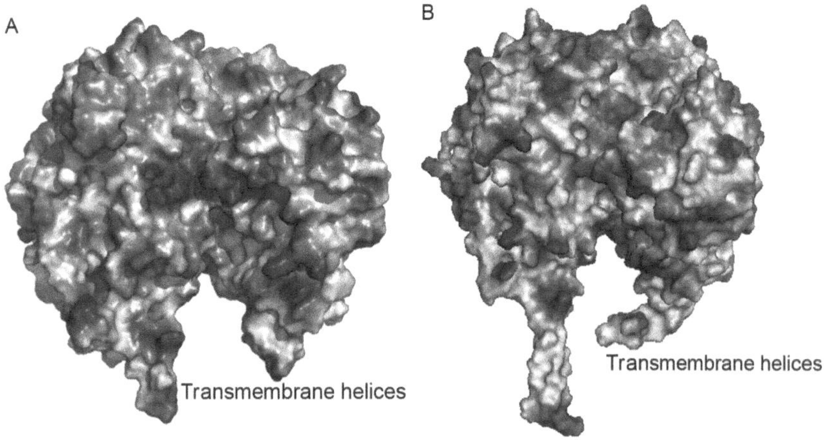

FIG. 2. (A) Electrostatic surface representation of human MAO-B (PDB Code 2V5Z) where neutral residues are white, basic positively charged residues in blue, and anionic acidic residues in red. The transmembrane helices are denoted for the dimeric structure. (B) Electrostatic surface representation of rat MAO-A structure (PDB Code 1O5W). The same coloring scheme as in A is presented. The transmembrane helices are denoted for the dimeric structure. (For interpretation of the references to color in this figure legend, the reader is referred to the Web version of this chapter.)

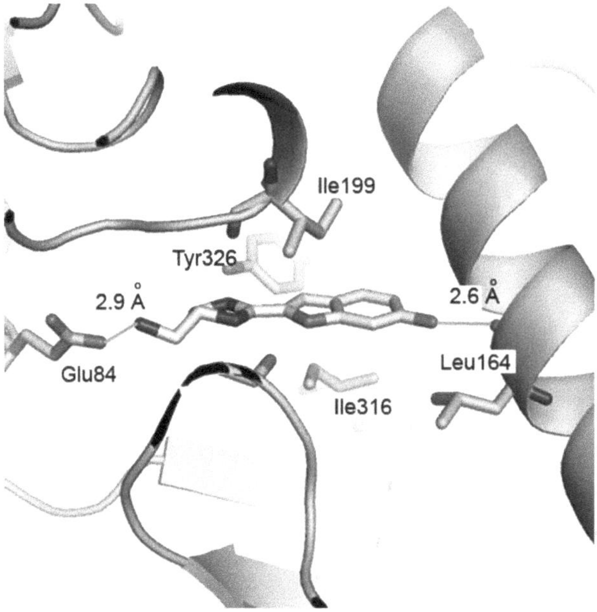

FIG. 3. Structure of 2-benzofuranyl-imidazoline bound to the entrance cavity of human MAO-B (PDB Code 2XFN). The carbon skeleton of the bound ligand is in yellow, oxygens in red, and nitrogens in blue. (For interpretation of the references to color in this figure legend, the reader is referred to the Web version of this chapter.)

conformational alteration in the side chain of Gln206 lead to an increase in binding affinity of 2-benzofuranyl-imidazoline from ~8 μM to 15 nM (Bonivento *et al.*, 2010). This increase in affinity is suggested to originate from synergistic H-bonding and hydrophobic interactions as a result of these alterations in side chain conformation about the entrance cavity. These structural studies provide insights into protein–ligand interactions that would result in improved specificity and affinity.

IV. Structure of Human MAO-A and Comparison with Rat MAO-A

The highest resolution structure of human MAO-A is the 2.2 Å structure published by Tsukihara's group (Son *et al.*, 2008). In agreement with earlier structural studies at 3.0 Å (De Colibus *et al.*, 2005), human MAO-A crystallizes in its monomeric form. In contrast with MAO-B, MAO-A has a monopartite active site cavity (Fig. 1B), which contains a protein loop (residues 108–118) at its entrance. Mutagenesis studies demonstrate this loop to be functionally important in catalysis; however, the detailed molecular dynamics and substrate interactions have yet to be determined. To date, the structures of MAO-A in complex with clorgyline (an irreversible acetylenic inhibitor) and harmane (a reversible inhibitor) have been determined. A goal then for future structural studies is to provide a more detailed description of the molecular interactions of MAO-A with a more diverse number of MAO-A-specific inhibitors at higher resolution. Such information is important to provide detailed molecular insights into MAO-A interactions with both reversible and irreversible MAO-A inhibitors and in the design of highly specific reversible inhibitors.

V. Insights into Membrane Binding of MAO-A and MAO-B

The structural work on purified enzyme preparations provides only suggestions for the membrane interactions of MAO-A and MAO-B. The α-helices observed at the C-termini of both enzymes and truncation studies suggest these domains constitute the transmembrane helixes for insertion into the outer mitochondrial membrane (Fig. 2A and B). Electron density data show the C-terminal helix of MAO-A to be more rigid than that of MAO-B and may be relevant to previous data on MAO chimeras that show the MAO-A C-terminus cannot be swapped with that of MAO-B (Gottowik *et al.*, 1995; Shih and Chen, 2002). The electrostatic surface maps of the enzymes (Fig. 2A and B) show positively charged domains that should form favorable electrostatic interactions with the negatively charged membrane surface. In addition, a comparison of protein dipole moment directions shows the dipole moment of isolated monomers of MAO-B to dramatically change on dimerization (Fig. 4A and B). The orientation of the dimer dipole has its positive end directed toward the anionic membrane surface, thus providing additional favorable interactions. These alterations in dipole moment direction for the isolated chains are less pronounced on dimerization in MAO-A (Fig. 4C and D) which may be relevant to the differential oligomeric state of the two enzymes on extraction from their native membrane environments.

VI. Structural Basis for Inhibitor-Binding Specificities of MAO-A and MAO-B

The available crystallographic data on the two human enzymes show the catalytic sites to exhibit similarities and differences. The structural environments about the covalent flavin sites are quite similar. Both FAD cofactors are bound with the adenine rings in extended conformations relative to the isoalloxazine rings as found in many flavin adenine dinucleotide (FAD) enzyme structures. Ionic interactions between the pyrophosphates and Arg residues exist as do extensive H-bonding between the ribityl and ribose side chains and the enzyme. In both enzymes, the

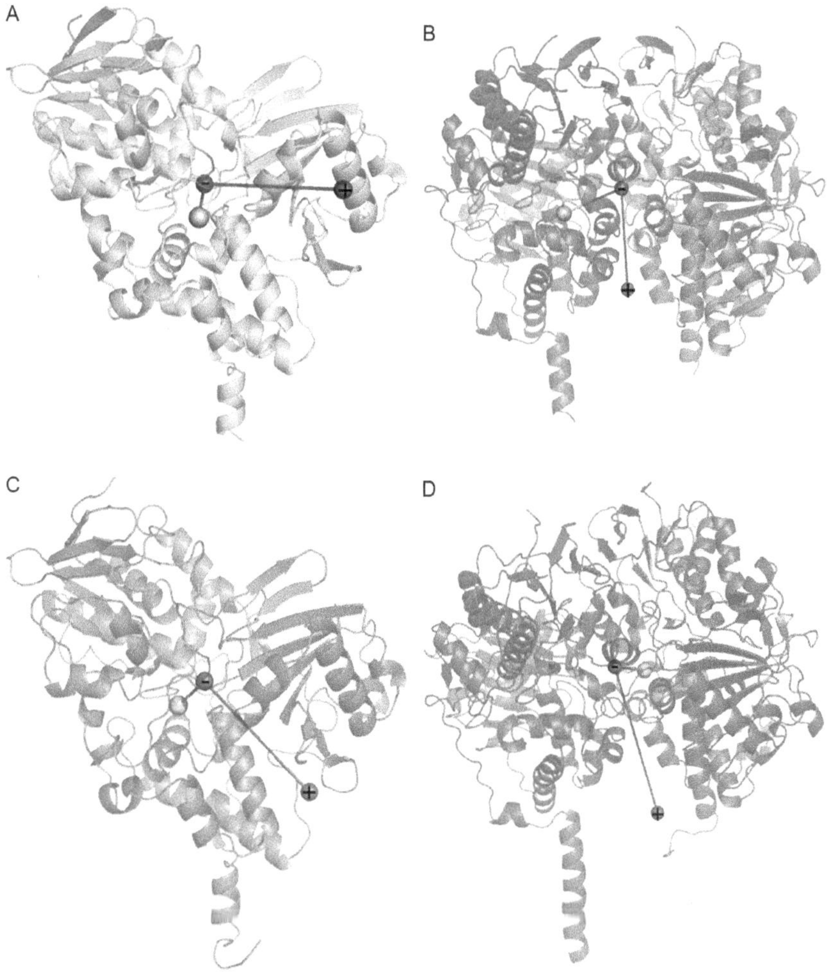

FIG. 4. Representation of the calculated dipole moment for the single chain structure of human MAO-B (A) and (B) for its dimeric form (PDB Code 2V5Z). (C) Single chain of rat MAO-A and (D) the dimeric form of rat MAO-A (PDB Code 1O5W). The resultant dipole in all structures is represented by a red line connecting two red spheres which are labeled as positive and negative ends of the dipole moment. These calculations were performed by the server at the Weizmann Institute, Rehovot, Israel (http://bioinfo.weizmann.ac.il/dipol/). (For interpretation of the references to color in this figure legend, the reader is referred to the Web version of this chapter.)

covalently bound isoalloxazine ring is in a "bent" conformation about the N(5)–N(10) axis ~30° from planarity. The site for covalent attachment is a thioether linkage with Cys398 in MAO-B and Cys406 in MAO-A. In both enzymes, strain is introduced via *cis* peptide linkages between these Cys residues and adjacent Tyr residues.

Also common in both enzymes is the presence of an "aromatic cage" situated on the re-face of the isoalloxazine ring where the substrate is expected to react with the flavin (Binda *et al.*, 2002a,b). In MAO-B, this cage structure comprises Tyr398 and Tyr435 and in MAO-A, Tyr406 and Tyr444. Mutagenesis studies show this cage structure to be important in catalysis (Li *et al.*, 2006) by polarization of the lone pair electrons on the substrate amine prior to oxidation to the imine by the enzymes. Structural work on other flavin-dependent amine oxidases also demonstrate aromatic residues forming a cage structure on the face of the flavin reacting with the substrate and therefore thought to be a general structural feature contributing to the catalytic function of flavin-dependent amine oxidizing enzymes important in their reductive half-reactions Edmondson *et al.*, 2009.

Differences exist, however, when comparing the structural features of MAO-A and MAO-B in regions of their active sites away from the flavin. As discussed above, MAO-B has a bipartite cavity structure, whereas MAO-A has a monopartite cavity. In both cases, the residues lining the active sites lead to a hydrophobic environment with H-bonding residues placed in differing positions of the active sites. The Ile199 gate in MAO-B separates the two cavities as does the Tyr326 side chain. Inhibitors that are bound in conformations that encompass both cavities exhibit tighter binding and a higher degree of MAO-B specificity. When the two MAO-B cavities are fused through the opening of the gating residues, a total volume of $\sim$700 $Å^3$ is formed. Unpublished mutagenesis studies in our laboratory show that conversion of the Ile and Tyr gating side chains to Ala residues results in no major structural alteration in the geometry of the active site and the double-mutated enzyme exhibits inhibitor-binding properties more in common to that of MAO-A.

The current structural data on MAO-A are not as extensive as that of MAO-B. A monopartite cavity is found with both human and rat MAO-A (Son *et al.*, 2008) with respective volumes of 550 $Å^3$ (human MAO-A) and 490 $Å^3$ (rat MAO-A). Thus, a common feature for MAO-A selectivity for substrates such as serotonin and inhibitor binding is the presence of a monopartite substrate cavity that facilitates binding of ligands that contain hydrophobic aromatic rings and do not exist in an extended conformation. These conditions appear to be met by inhibitors such as clorgyline, harmane, methylene blue, pirindole, and, more recently, CX157 (Fowler *et al.*, 2009).

Although both MAO-A and MAO-B have hydrophobic active sites, structural data on both enzymes show the presence of several nonmobile, conserved water molecules within each active site. Direct H-bonding is observed between individual water molecules with bound ligands, and the positions and numbers of immobile waters appear to change in MAO-B depending on the nature of the bound ligand. It can therefore be concluded that these water molecules play important but still poorly understood roles in MAO inhibitor affinities and specificities.

VII. Conclusions and Future Prospects

Progress in the past decade on structural properties of human MAO-B and human MAO-A has resulted in unique insights into the functional properties of these membrane-bound enzymes. The existing structural data serve as an important resource for molecular modeling and docking studies to explain experimentally observed differences in inhibitor affinities to each enzyme. Our understanding of inhibitor binding to MAO-B is at a more advanced level, as the structures of more than 20 different reversible and irreversible inhibitors have been solved to resolutions ranging from 2.3 to 1.6 Å with all coordinates deposited in the Protein Data Bank. Our knowledge of inhibitor binding to MAO-A is more fragmentary with structures of harmine and clorgyline complexes with the enzyme published at resolutions ranging from 3.2 to 2.2 Å. More information on the structures of tight binding reversible inhibitors specific to MAO-A is required to achieve a more in-depth molecular understanding of enzyme–inhibitor interactions. This latter goal is of importance with recent discoveries relating increased MAO-A levels in cardiac mitochondria that occur on aging (Bianchi *et al.*, 2005; Maurel *et al.*, 2003) and suggested to have a causative effect on cardiovascular disease in man. Future advances in understanding specific inhibition of extraneuronal MAO-A will be an important contribution to the development of cardioprotectants for an aging population.

Acknowledgments

The published and unpublished work cited in this review was supported by grants from the National Institutes of Health (GM-29433) and Fondazione Cariplo. The authors acknowledge the valuable contributions of current and former coworkers who are listed in publication citations in the reference section.

References

Bianchi, P., Kunduzova, O., Masini, E., Cambon, C., Bani, D., Raimondi, L., Seguelas, M.H., Nistri, S., Colucci, W., Leducq, N., and Parini, A. (2005). Oxidative stress by monoamine oxidase mediates receptor-independent cardiomyocyte apoptosis by serotonin and postischemic myocardial injury. *Circulation* **112**, 3297–3305.

Binda, C., Newton-Vinson, P., Hubálek, F., Edmondson, D.E., and Mattevi, A. (2002a). Structure of human monoamine oxidase B, a drug target for the treatment of neurological disorders. *Nat. Struct. Biol.* **9**, 22–26.

Binda, C., Mattevi, A., and Edmondson, D.E. (2002b). Structure-function relationships in flavoenzyme-dependent amine oxidations: a comparison of polyamine oxidase and monoamine oxidase. *J. Biol. Chem.* **277**, 23973–23976.

Bonivento, D., Milczek, E.M., McDonald, G.R., Binda, C., Holt, A., Edmondson, D.E., and Mattevi, A. (2010). Potentiation of ligand binding through cooperative effects in monoamine oxidase B. *J. Biol. Chem.* **285**, 36849–36856.

De Colibus, L., Li, M., Binda, C., Lustig, A., Edmondson, D.E., and Mattevi, A. (2005). Three-dimensional structure of human monoamine oxidase A (MAO A): relation to the structures of rat MAO A and human MAO B. *Proc. Natl. Acad. Sci. USA* **102**, 12684–12689.

Edmondson, D.E., Binda, C., Wang, J., Upadhyay, A.K., and Mattevi, A. (2009). Molecular and mechanistic properties of the membrane-bound mitochondrial monoamine oxidases. *Biochemistry* **48**, 4220–4230.

Fowler, J.S., Logan, J., Azzaro, A.J., Fielding, R.M., Zhu, W., Posthura, A.K., Burch, D., Brand, D., Free, J., Agharnejad, M., Wang, G.J., Telang, F., *et al.* (2009). Reversible inhibitors of monoamine oxidase-A (RIMAs): robust, reversible inhibition of human brain MAO-A by CX157. *Neuropsychopharmacology* **35**, 623–631.

Gottowik, K.J., Malherbe, P., Lang, G., Da Prada, M., and Cesura, A.M. (1995). Structure/function relationships of mitochondrial monoamine oxidase A and B chimeric forms. *Eur. J. Biochem.* **230**, 934–942.

Hubálek, F., Binda, C., Li, M., Mattevi, A., and Edmondson, D.E. (2003). Polystyrene microbridges used in sitting-drop crystallization release 1,4-diphenyl-2-butene, a novel inhibitor of human MAO B. *Acta Crystallogr. D Biol. Crystallogr.* **10**, 1874–1876.

Hubálek, F., Binda, C., Khalil, A., Li, M., Mattevi, A., Castagnoli, N., and Edmondson, D.E. (2005). Demonstration of isoleucine 199 as a structural determinant for the selective inhibition of human monoamine oxidase B by specific reversible inhibitors. *J. Biol. Chem.* **280**, 15761–15766.

Krueger, M.P., Mazouz, F., Ramsay, R.R., Milcent, R., and Singer, T.P. (1995). Dramatic differences in the susceptibility of monoamine oxidase B to a group of powerful inhibitors. *Biochem. Biophys. Res. Commun.* **206**, 556–562.

Li, M., Hubálek, F., Newton-Vinson, P., and Edmondson, D.E. (2002). High-level expression of human liver monoamine oxidase A in *Pichia pastoris*: comparison with the enzyme expressed in *Saccharomyces cerevisiae*. *Protein Expr. Purif.* **24**, 152–162.

Li, M., Binda, C., Mattevi, A., and Edmondson, D.E. (2006). Functional role of the aromatic cage in human monoamine oxidase B: structures and catalytic properties of Tyr435 mutant proteins. *Biochemistry* **45**, 4775–4784.

Ma, J., Yoshimura, M., Yamashita, E., Nakagawa, A., Ito, A., and Tsukihara, T. (2004). Structure of rat monoamine oxidase A and its specific recognitions for substrates and inhibitors. *J. Mol. Biol.* **338**, 103–114.

Maurel, A., Hernandez, C., Kunduzova, O., Bompart, G., Cambon, C., Parini, A., and Frances, B. (2003). Age-dependent increase in hydrogen peroxide production by cardiac monoamine oxidase A in rats. *Am. J. Physiol. Heart Circ. Physiol.* **284**, H1460–H1467.

Newton-Vinson, P., Hubálek, F., and Edmondson, D.E. (2000). High-level expression of human liver monoamine oxidase B in *Pichia pastoris*. *Protein Expr. Purif.* **20**, 334–345.

Rebrin, I., Geha, R.M., Chen, K., and Shih, J.C. (2001). Effects of carboxy-terminal truncations on the activity and solubility of human monoamine oxidase B. *J. Biol. Chem.* **276**, 29499–29506.

Shih, J.C., and Chen, K. (2002). In vivo and in vitro functions of two types of MAO. *Adv. Behav. Biol (Catecholamine Research)* **53**, 147–155.

Son, S.Y., Ma, J., Kondou, Y., Yoshimura, M., Yamashita, E., and Tsukihara, T. (2008). Structure of human monoamine oxidase A at 2.2-Angstrom resolution: the control of opening the entry for substrates/inhibitors. *Proc. Natl. Acad. Sci. USA* **105**, 5739–5744.

Upadhyay, A., Borbat, P., Wang, J., Freed, J., and Edmondson, D.E. (2008). Determination of the oligomeric states of human and rat monoamine oxidases in the outer mitochondrial membrane and octyl β-D-glucopyranoside micelles using pulsed dipolar electron spin resonance spectroscopy. *Biochemistry* **47**, 1554–1566.

Urban, P., Andersen, J.K., Hsu, H.P.P., and Pompon, D. (1991). Comparative membrane locations and activities of human monoamine oxidases expressed in yeast. *FEBS Lett.* **286**, 142–146.

Weyler, W., Titlow, C.C., and Salach, J.I. (1990). Catalytically active monoamine oxidase type a from human liver expressed in *Saccharomyces cerevisiae* contains covalent FAD. *Biochem. Biophys. Res. Commun.* **173**, 1205–1211.

BEHAVIORAL OUTCOMES OF MONOAMINE OXIDASE DEFICIENCY: PRECLINICAL AND CLINICAL EVIDENCE

Marco Bortolato[1] and Jean C. Shih[1,2]

[1]Department of Pharmacology and Pharmaceutical Sciences, School of Pharmacy, University of Southern California, Los Angeles, California, USA
[2]Department of Cell and Neurobiology, Keck School of Medicine, University of Southern California, Los Angeles, California, USA

Abstract

Monoamine oxidase (MAO) isoenzymes A and B are mitochondrial-bound proteins, catalyzing the oxidative deamination of monoamine neurotransmitters as well as xenobiotic amines. Although they derive from a common ancestral progenitor gene, are located at X-chromosome and display 70% structural identity, their substrate preference, regional distribution, and physiological role are divergent. In fact, while MAO-A has high affinity for serotonin and norepinephrine, MAO-B primarily serves the catabolism of 2-phenylethylamine (PEA) and contributes to the degradation of other trace amines and dopamine. Convergent lines of preclinical and clinical evidence indicate that variations in MAO enzymatic activity—due to

INTERNATIONAL REVIEW OF NEUROBIOLOGY, VOL. 100
DOI: 10.1016/B978-0-12-386467-3.00002-9

either genetic or environmental factors—can exert a profound influence on behavioral regulation and play a role in the pathophysiology of a large spectrum of mental and neurodegenerative disorders, ranging from antisocial personality disorder to Parkinson's disease. Over the past few years, numerous advances have been made in our understanding of the phenotypical variations associated with genetic polymorphisms and mutations of the genes encoding for both isoenzymes. In particular, novel findings on the phenotypes of MAO-deficient mice are highlighting novel potential implications of both isoenzymes in a broad spectrum of mental disorders, ranging from autism and anxiety to impulse-control disorders and ADHD. These studies will lay the foundation for future research on the neurobiological and neurochemical bases of these pathological conditions, as well as the role of gene × environment interactions in the vulnerability to several mental disorders.

I. General Characteristics of Monoamine Oxidase

A. Catalytic Reaction

Monoamine oxidases [MAOs; amine: oxygen oxidoreductase (deaminating) (flavin-containing); EC 1.4.3.4] are a family of mitochondrial-bound flavoproteins catalyzing the oxidative deamination of monoamine neurotransmitters, neuromodulators, and hormones to the corresponding aldehydes:

$$RCH_2NH_2 + O_2 + H_3O^+ \rightarrow RCHO + NH_4{}^+ + H_2O_2$$

This reaction requires flavin adenine dinucleotide (FAD) as a covalently bound redox cofactor and consists of three main steps (for a detailed analysis of the current knowledge on the catalytic mechanisms of MAO, see Edmondson *et al.*, 2009):

1. Following the formation of a FAD-substrate adduct, the cofactor is reduced to its hydroquinone form ($FADH_2$), while the amine is converted into the corresponding imine.

$$Enz - FAD + RCH_2NH_2 \rightarrow Enz - FADH_2 + RCH = NH$$

2. Once dissociated from the enzyme, the imine is spontaneously hydrolyzed, with production of aldehyde and ammonium:

$$RCH = NH + H_3O^+ \rightarrow RCHO + NH_4{}^+$$

3. FADH2 is reoxidized to FAD, with formation of hydrogen peroxide from molecular oxygen. This reaction is the rate-limiting step of the whole enzymatic process:

$$Enz - FADH_2 + O_2 \rightarrow Enz - FAD + H_2O_2$$

As shown in Table I, the endogenous substrates of MAO include key brain neurotransmitters, such as serotonin (5-hydroxytryptamine, 5-HT), dopamine (DA), norepinephrine (NE), and epinephrine (E), as well as a number of trace amines, such as tyramine, tryptamine, 2-phenylethylamine (PEA), octopamine, and 3-iodothyronamine (T1AM). Notably, the oxidative deamination of short-chain primary amines (including PEA, tyramine, and T1AM) is not exclusively mediated by MAO but also contributed by the copper/topaquinone-containing semicarbazide-sensitive amine oxidase (SSAO; encoded by the gene *AOC3*; Obata, 2002; Saba *et al.*, 2010). The role of MAO in the homeostasis of these compounds is essential to modulate the neuroendocrine regulation of the central nervous system and many peripheral organs.

The aldehydes produced by MAO are toxic species (for a review on the pathogenic potential of aldehydes, see O'Brien *et al.*, 2005) which need to be converted in less harmful metabolites. Thus, this enzyme is functionally coupled with a $NAD(P)^+$-dependent aldehyde dehydrogenase (ALDH), which oxidizes the aldehyde to the corresponding carboxylic acid; alternatively (depending on the location and the intracellular conditions), aldehydes can be reduced to alcohols or glycols by aldehyde reductase (ALR) or alcohol dehydrogenase (ADH) (Table I). The main metabolic pathway of 5-HT consists in the conversion of this monoamine into 5-hydroxyindolacetic acid (5-HIAA) by joint action of MAO and ALDH. Like other MAO metabolites, 5-HIAA is rapidly eliminated by diffusion into the bloodstream and excreted through the kidneys by glomerular filtration and active tubular excretion (Udenfriend *et al.*, 1956; Despopoulos and Weissbach, 1957). Given the predominance of the MAO-ALDH pathway in 5-HT metabolism, urinary levels of 5-HIAA are used as an index for measurement of plasma 5-HT content (with diagnostic value as a biomarker for carcinoid syndrome, a paraneoplastic disorder caused by gastrointestinal apudomas secreting 5-HT). Small amounts of 5-HT (1–5%) are converted into 5-hydroxyindolethanol (5-HIET, also termed 5-hydroxytryptophol) by either ALR or ADH (Feldstein and Williamson, 1968; Beck *et al.*, 1984; Consalvi *et al.*, 1986; Svensson *et al.*, 1999) (Table I). Interestingly, the amount of 5-HIET can be enhanced by compounds that compete with endogenous 5-HT metabolite for ALDH, such as ethanol (Helander *et al.*, 1993).

The metabolism of catecholamines (DA, NE, E) is served by both MAO (in conjunction with either ALDH or ALR) and catecholamine-*O*-methyl-transferase (COMT). The combined action of the two enzymes converts DA into either homovanillic acid (HVA; MAO/ALDH + COMT pathway) or, less frequently, into 3-methoxy-4-hydroxyphenylethanol (MHPE; MAO/ALR + COMT pathway). The latter can be processed by ADH into HVA (Fig. 1).

NE and E undergo similar degradation pathways (Fig. 2). Specifically, MAO converts both monoamines into 3,4-dihydroxyphenylglycol aldehyde (DOPGAL), which is further processed by ALR into 3,4-dihydroxylphenylethylene glycol (DOPEG). A much smaller aliquote of DOPGAL is oxidized to

Table I

Synoptic View of the Main Substrates and Products of MAO-Mediated Metabolism (Coupled with Aldehyde Dehydrogenase (ALDH) or Aldehyde Reductase (ALR)).

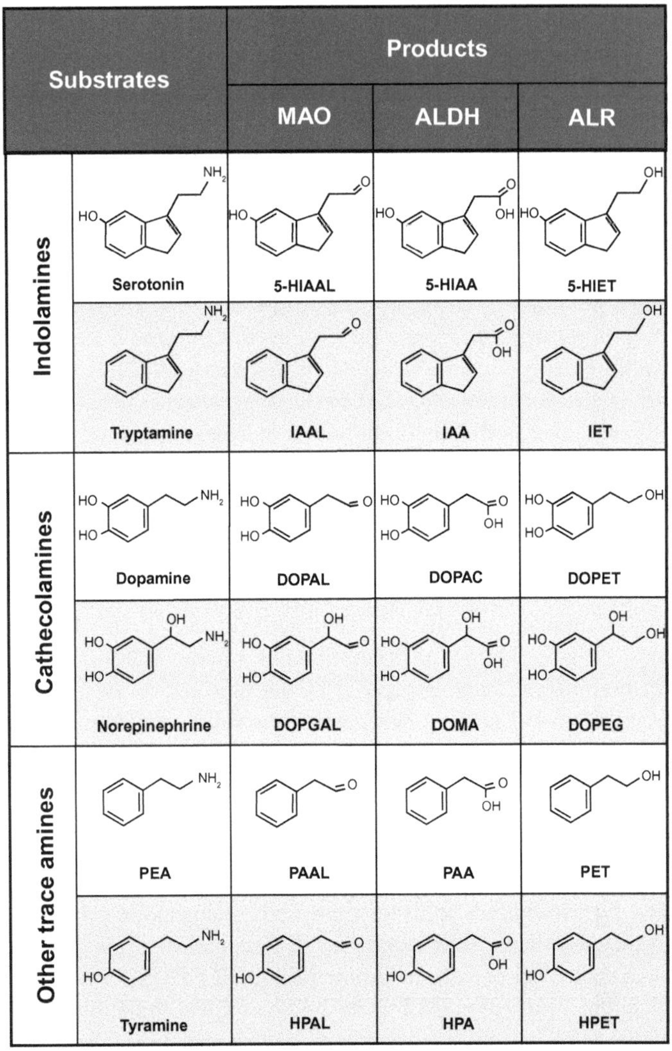

	Substrates	Products: MAO	Products: ALDH	Products: ALR
Indolamines	Serotonin	5-HIAAL	5-HIAA	5-HIET
Indolamines	Tryptamine	IAAL	IAA	IET
Cathecolamines	Dopamine	DOPAL	DOPAC	DOPET
Cathecolamines	Norepinephrine	DOPGAL	DOMA	DOPEG
Other trace amines	PEA	PAAL	PAA	PET
Other trace amines	Tyramine	HPAL	HPA	HPET

Epinephrine is not listed, as its metabolites are the same as those indicated for norepinephrine. Abbreviations: 5-HIAAL, 5-hydroxyindolaldehyde; 5-HIAA, 5-hydroxyindolacetic acid; 5-HIET, 5-hydroxyindolethanol; IAAL, indole-3-acetaldehyde; IAA, indole-3-acetic acid, IET, indole-3-ethanol (tryptophol); DOPAL, 3,4-dihydroxyphenylacetaldehyde; DOPAC, 3,4-dihydroxyphenylacetic acid; DOPET, 3,4-dihydroxyphenylethanol; DOPGAL, 3,4-dihydroxyphenylglycolaldehyde; DOMA, 3,4-dihydroxymandelic acid; DOPEG, 3,4-dihydroxylphenylethyleneglycol; PEA, 2-phenylethylamine; PAAL, 2-phenylacetaldehyde; PAA, 2-phenylacetic acid; PET, 2-phenylethanol; HPAL, 4-hydroxy-phenylaldehyde; HPA, 4-hydroxyphenylacetic acid; HPET, 4-hydroxyphenylethanol.

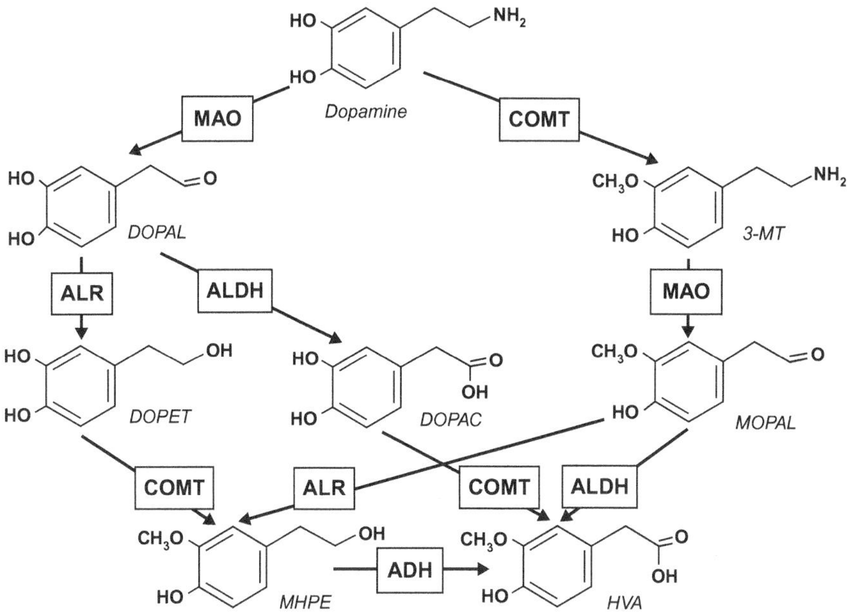

FIG. 1. Metabolic pathways of dopamine. DOPAL, 3,4-dihydroxyphenylacetaldehyde; 3-MT, 3-methoxytyramine; DOPET, 3,4-dihydroxyphenylethanol; DOPAC, 3,4-dihydroxyphenylacetic acid; MOPAL, 3-methoxy-4-hydroxyphenylacetaldehyde; MHPE, 3-methoxy-4-hydroxyphenylethanol; HVA, homovanillic acid.

3,4-dihydroxymandelic acid (DOMA). COMT converts DOPEG into 3-methoxy-4-hydroxyphenylethylene glycol (MHPG) and DOMA into vanillyl mandelic acid (VMA). Alternatively, NE and E can be methylated by COMT to normetanephrine and metanephrine, respectively. These metabolites can be conjugated with sulfate groups by sulfatransferase 1A3 (SULT1A3) or processed by either MAO/ALR or MAO/ALDH into MHPG and VMA.

MAO function is highly critical for the regulation the intracellular redox state in neurons and other cells; indeed, one of the byproducts of MAO-mediated reaction, hydrogen peroxide, is a potent oxidizer which can trigger the formation of superoxide radicals and other reactive oxygen species, which can in turn induce mitochondrial and cytoplasmic damage. Under physiological conditions, the overall redox potential is kept in equilibrium by antioxidant enzymes, such as superoxide dismutase, catalase, and glutathione peroxidase; nevertheless, high concentrations of ammonia (the other by-product of the reaction) have been shown to decrease the activity of these enzymes and lead to the formation of superoxide radicals (Kosenko *et al.*, 1997). The excess of oxidizing species in the central nervous system leads to permanent damages through death of neurons and glia. These mechanisms lay the theoretical foundations for the implication of MAO in the pathophysiology of certain neurodegenerative disorders, such as

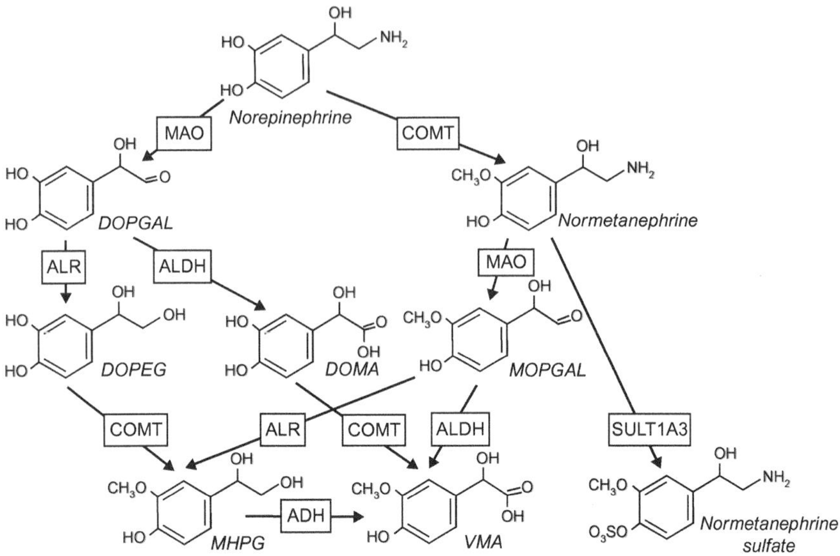

FIG. 2. Metabolic pathways of norepinephrine. DOPGAL, 3,4-dihydroxyphenylglycol aldehyde; DOPEG, 3,4-dihydroxylphenylethylene glycol; DOMA, 3,4-dihydroxymandelic acid; MOPGAL, 3-methoxy-4-hydroxyphenylglycol aldehyde; MHPG, 3-methoxy-4-hydroxyphenylethylene glycol; VMA, vanillyl mandelic acid.

Parkinson's disease (PD) and dementias (Danielczyk *et al.*, 1988). In line with this concept, an increase in the activity of the isoenzyme MAO-B in platelets has been found in Alzheimer's disease patients, leading to the proposal that this parameter may be an early biomarker for diagnosis of this condition (Grünblatt *et al.*, 2005).

Finally, MAO serves a primary role in the degradation of primary, secondary, and some tertiary xenobiotic amines, which is particularly important to preventing their cardio- and neurotoxicity. A well-characterized example of these detrimental effects is the "cheese reaction," a vasoconstrictive crisis (often lethal) caused by the absorption of sympathomimetic amines in fermented food (such as cheese, wine, etc.) following administration of irreversible MAO inhibitors (Anderson *et al.*, 1993).

B. MAO ISOENZYMES

In higher vertebrates, the MAO family comprises two isoenzymes, termed A and B, which, despite a substantial structural overlap, are remarkably different for substrate preference, inhibitor selectivity, anatomical distribution, and functional role in behavioral regulation.

The existence of multiple MAO isoenzymes was initially postulated to account for a number of experimental data indicating divergent neurochemical effects of different inhibitors. Indeed, MAO-A is selectively blocked by low doses of clorgyline (Johnston, 1968), while MAO-B is inhibited by nanomolar concentrations of (*R*)-deprenyl (Knoll and Magyar, 1972). The two isoenzymes can be separated electrophoretically (Shih and Eiduson, 1969; Youdim *et al.*, 1969).

Subsequent studies showed that MAO-A had very high affinity for 5-HT (120 folds higher than MAO-B) and, to a lower degree, NE; by contrast, MAO-B preferred PEA and benzylamine as its substrates. The degradation of DA, tryptamine, and tyramine is mediated by both MAOs, but the relative contribution of each isoenzyme appears to vary greatly in relation to the species and the tissue under consideration. For example, DA metabolism is prevalently served by MAO-A in rodents and by MAO-B in humans and other primates (Garrick and Murphy, 1980; Fornai *et al.*, 1999). The dichotomy between MAO-A and -B in terms of substrate preference is not absolute; in fact, in the absence of one isoenzyme, the other can deaminate a certain amount of its nonpreferred monoamine substrates (Chen *et al.*, 2004). This mechanism of partial compensation mechanism is fully revealed by the neurochemical outcomes of MAO deficiency in murine models (see below) and indicates the physiological significance of the presence of two isoenzymes *in vivo*. The unequivocal demonstration of the existence of two isoenzymes came in 1988, with the cloning of human *MAO-A* and *MAO-B* genes (Bach *et al.*, 1988). Subsequent studies elucidated the structural configuration of both genes, revealing that they are located on the locus Xp11.23, in a tail-to-tail arrangement, with the 3′-coding sequences separated by about 50 kb (Ozelius *et al.*, 1988; Lan *et al.*, 1989a,b; Levy *et al.*, 1989; Chen *et al.*, 1992). *MAO-A* and *MAO-B* encode for two proteins of 527 and 520 amino acids, with molecular weights of 59.7 and 58.8 kDa, respectively. Interestingly, the two genes share ~70% sequence identity and an identical intron–exon organization, with 15 exons and 14 introns; these findings provided one of the first lines of evidence to define the structure and evolution of the isoenzymes, by suggesting that both genes derive from duplication of a common ancestor gene (Grimsby *et al.*, 1991). In agreement with this interpretation, numerous phylogenetic studies have revealed the existence of only one MAO in early eukaryotes (Schilling and Lerch, 1995), invertebrates (Boutet *et al.*, 2004), and teleost fish (Chen *et al.*, 1994; Setini *et al.*, 2005; Anichtchik *et al.*, 2006). Conversely, the presence of two MAO isoenzymes can be traced back to anuran amphibians (Kobayashi *et al.*, 1981); in frogs, MAO-A is predominant in the tadpole stage, while MAO-B expression increases through metamorphosis (Nicotra and Senatori, 1988), pointing at the possibility that the duplication of the gene may have been selected as an advantageous trait to maintain redox homeostasis in response to the development of lung-based respiration and the consequent hyperoxic shift. Interestingly, MAO-B has a much higher K_m (lower affinity) for O_2 (~ 250 μM) than MAO-A

(~ 6 μM) (Edmondson *et al.*, 2004). Given the role of monoamines in the regulation of cardiovascular function, the development of substrate-selective isoenzymes may have also been instrumental to withstand the new challenges posed by terrestrial life to blood circulation.

A more detailed insight in the structural characteristics of MAOs was afforded by several mutagenesis and chimerization studies (Bach *et al.*, 1988; Gottowik *et al.*, 1993; Wu *et al.*, 1993; Chen *et al.*, 1996; Geha *et al.*, 2000), as well as by the crystallization of the two isoenzymes (Binda *et al.*, 2002, 2003; Ma *et al.*, 2004; De Colibus *et al.*, 2005).

Nascent MAO polypeptidic chains undergo a number of posttranslational modifications; the best-characterized processes are the removal of the initiator methionine in MAO-B (but not in MAO-A) and the acetylation of the N-terminus in both molecules (methionine for MAO-A and serine for MAO-B) (Newton-Vinson *et al.*, 2000; Li *et al.*, 2002). Another critical modification is the covalent attachment of FAD to cysteinyl residues 406 in MAO-A and 397 in MAO-B; both amino acids are encoded by exon 12 of the respective gene, and their mutation ablates enzymatic activity (Wu *et al.*, 1993). The coenzyme is attached by a thioether bond with the 8α-methylene group of its isoalloxazine ring (Kearney *et al.*, 1971), and maintained in a position opposite to the entrance of the monoamine-binding cavity within the active site (Edmondson *et al.*, 2007). The substrate preference and inhibitor specificity appear to be conferred by a number of internal residues, such as Ile 335 in MAO-A and Tyr 326 in MAO-B (Geha *et al.*, 2001).

Both proteins are anchored to the mitochondrial outer membrane through a transmembrane helix located within the carboxyl-terminal domain. In their membrane-bound conformations, human MAO-A and MAO-B are both dimeric (Ma *et al.*, 2004; De Colibus *et al.*, 2005; Binda *et al.*, 2007; Upadhyay *et al.*, 2008; Edmondson *et al.*, 2009).

C. Anatomical Localization of MAO-A and -B

Although both isoenzymes are expressed in most tissues, only MAO-A is characteristically abundant in fibroblasts and placenta; in contrast, MAO-B is the only isoenzyme expressed in platelets and lymphocytes (Bond and Cundall, 1977; Donnelly and Murphy, 1977). MAO-A and -B are present in most brain regions; however, certain areas display only one isoenzyme. MAO-A is found mainly in DAergic and NEergic neurons; conversely, MAO-B is the only isoenzyme expressed in the cell bodies of 5-HTergic neurons (as well as in histaminergic neurons and astrocytes) (Konradi *et al.*, 1989); the significance of this localization remains partially unclear, since 5-HT appears to be mainly metabolized by MAO-A *in vivo*. To account for this apparent mismatch, we have hypothesized that MAO-A protein may be translated in the cell body and segregated to the axon terminals (Bortolato *et al.*, 2010); this hypothesis is supported by the discovery that MAO-B is absent from the mitochondria of the axon terminals (Arai *et al.*, 2002), as well as by the documentation of MAO-A mRNA in the 5-HTergic cells (Luque *et al.*, 1995, 1996; Jahng *et al.*, 1997; Filipenko *et al.*, 2002; Wylie *et al.*, 2010). The proposed compartmentalization may facilitate the specific degradation of 5-HT in the synaptic terminal; further, the expression of MAO-B in the somata of 5-HTergic neurons may serve protective functions for 5-HT.

II. Phenotypical Outcomes of MAO-A Deficit

A. Clinical Findings

The serendipitous discovery of the mood-enhancing effects elicited by MAO pharmacological blockade (Fox and Gibas, 1953) was a historical breakthrough in the pharmacotherapy of mental disorders and gave impetus to the first investigations on the role of MAO in behavioral regulation. It was subsequently discovered that the antidepressant properties of MAO inhibitors were mainly due to the inactivation of MAO-A, which resulted in increased synaptic 5-HT concentrations (Sharp *et al.*, 1997) and modifications of the firing rate of 5-HTergic neurons (Blier and de Montigny, 1985).

The significant side effects of MAO inhibitors, however, led to a progressive decline in the employment of these agents in the therapy of depression, in favor of other categories of antidepressants. The discussion of the antidepressant properties of MAO-A inhibitors and their therapeutic usage is beyond the scope of the present chapter; the interested reader is referred to Amrein *et al.* (1993) and Kennedy (1997).

Interest in the clinical implications of MAO was rekindled by a number of reports on the implications of its deficiency in atypical Norrie disease (ND) patients. This recessive X-linked disease is caused by loss-of-function mutations of *NDP* (Norrie disease pseudoglioma) gene, which encodes for norrin, a protein involved in the development and vascularization of the retina and inner ear. In affected males, total norrin deficiency results in congenital blindness, cataracts, and progressive deterioration of the iris; additionally, several patients experience progressive hearing loss and other abnormalities of the cardiovascular, respiratory, and digestive systems.

As a result of the close proximity of *NDP* (located in Xp11.4) and the two *MAO* genes, a relatively sizable contingent of ND patients are reported to harbor deletion of these genes. In particular, the deletion of *MAO-A* and *MAO-B* in ND patients is conducive to severe mental retardation, growth failure, alterations of sleep pattern, and autistic-like symptoms (Lan *et al.*, 1989a,b; Sims *et al.*, 1989a.b; Murphy *et al.*, 1990; Collins *et al.*, 1992).

Further insight into the phenotypical outcomes of selective MAO-A deficiency was gained with the discovery of a behavioral syndrome in eight males of a large Dutch kindred (Brunner *et al.*, 1993a,b), characterized by borderline mental retardation and maladaptive regulation of impulsive aggression. The genetic defect was a point mutation in exon 8 of the *MAO-A* gene, resulting in the substitution of a glutamine codon (CAG) with a stop codon (TAG) at position 296 of the amino acid sequence. The main nosographic feature of the disorder was a high proclivity to engage in violent and antisocial behavior (including arson, attempted rape and murder, exhibitionism and voyeurism), often in response to minor stressors. The affected individuals also exhibited stereotyped hand movements and sleep disturbances. These alterations were paralleled by a set of abnormalities in the urinary concentrations of monoamine metabolites, including decreased content of 5-HIAA, HVA, and VMA and increased levels of 5-HT (fivefolds higher than the normal range) and normetanephrine (from COMT metabolism of NE). To the best of our knowledge, no other case of Brunner syndrome has been described in the medical literature to date, even despite specific attempts to identify the disorder in cohorts of aggressive individuals (Mejia *et al.*, 2001).

The sequencing of *MAO-A* gene allowed the characterization of its variants (for a review, see Shih and Thompson, 1999) and their different influence in behavioral regulation; among the numerous allelic variants identified to date, four polymorphisms have been particularly studied as potential biomarkers/risk factors for psychiatric disorders:

1. *MAO-A* (*CA*)n, a dinucleotide repeat polymorphism in intron 2 (Black *et al.*, 1991);
2. a 23 bp variable-number tandem repeat (VNTR) near exon 1 (Hinds *et al.*, 1992);
3. *Fnu*4HI and *Eco*RV, two restriction fragment length polymorphisms (Lim *et al.*, 1994);
4. *MAO-A*-uVNTR, a 30 bp VNTR polymorphism located 1.2 kb upstream of MAO-A transcription initiation site (Sabol *et al.*, 1998).

Variants of the first three polymorphisms (localized in *MAO-A* gene) have been associated to higher susceptibility to several mental conditions; in particular, a robust association was found between bipolar disorder and MAO-A (CA)n and 23 bp-VNTR polymorphisms (Lim *et al.*, 1994, 1995; Kawada *et al.*, 1995; Rubinsztein *et al.*, 1996; Preisig *et al.*, 2000). This association, albeit not confirmed by few studies (Craddock *et al.*, 1995; Muramatsu *et al.*, 1997), was also supported by a meta-analysis study (Furlong *et al.*, 1999).

The *MAO-A*-uVNTR promoter polymorphism has been extensively studied, in consideration of its well-characterized functional nature. Six *MAO-A*-uVNTR variants have been characterized based on a different number of repeats (2, 3, 3.5,

4, 5, and 6) (Huang *et al.*, 2004); in particular, the 3-repeat (3R) and 4-repeat (4R) alleles are the most common in the population (Sabol *et al.*, 1998; Deckert *et al.*, 1999; Jonsson *et al.*, 2000); of these, the 4R variant has been associated to higher transcriptional efficiency and enzymatic activity (Sabol *et al.*, 1998; Deckert *et al.*, 1999; Denney *et al.*, 1999). In line with this finding, a number of studies have shown that 4R carriers have higher levels of 5-HIAA in the cerebrospinal fluid (Williams *et al.*, 2003), as well as a higher prevalence of panic disorder and major depression (in females), with poor response to chronic fluoxetine treatment (Deckert *et al.*, 1999; Yu *et al.*, 2005).

The 3R variant, which has been found to result in lower MAO-A catalytic activity in fibroblasts, has been linked to a higher risk for behavioral traits related to Brunner syndrome symptoms, namely impulsive aggressiveness and antisocial personality (Samochowiec *et al.*, 1999; Contini *et al.*, 2006; Oreland *et al.*, 2007; Buckholtz and Meyer-Lindenberg, 2008; Williams *et al.*, 2009), as well as impaired stress response (Brummett *et al.*, 2008), lower cognitive functioning (Cohen *et al.*, 2003), and maladaptive emotional processing of affect (Lee and Ham, 2008). Interestingly, this variant has also been shown to influence the clinical course and severity of mental disorders; for example, 3R autistic children exhibit higher severity of the pathological manifestations (Cohen *et al.*, 2003), lower levels of anxiety and attentional deficits (Roohi *et al.*, 2009), and larger cortical volumes (Davis *et al.*, 2008).

The behavioral changes associated to *MAO-A*-uVNTR polymorphic variants have been documented to be related to a number of morphological and functional differences between the brains of 3R and 4R carriers. In particular, several studies have shown that male individuals with the 3R haplotype exhibit morphological alterations of the orbitofrontal cortex (Meyer-Lindenberg *et al.*, 2006; Cerasa *et al.*, 2008, 2010), as well as functional abnormalities in several cortical and limbic regions, including prefrontal cortex, amygdala, and hippocampus (Meyer-Lindenberg *et al.*, 2006; Passamonti *et al.*, 2006).

Recent findings have challenged the association between *MAO-A*-uVNTR variants and MAO-A brain activity. For example, postmortem studies showed that, while the average MAO-A catalytic activity in brain samples from 4R carriers was higher than 3R, this difference was not significant (Balciuniene *et al.*, 2002). Similarly, investigations conducted on populations of 3R and 4R male carriers with positron emission tomography [PET] for [^{11}C]clorgyline revealed no significant difference in brain MAO-A activity between the two groups (Fowler *et al.*, 2007; Alia-Klein *et al.*, 2008). However, irrespective of the genetic components, MAO-A activity in cortical and subcortical brain regions was shown to be inversely correlated with the degree of self-reported aggression in men (Alia-Klein *et al.*, 2008).

These data strongly suggest that polymorphic variants, rather than dictating MAO-A activity, may only confer a predisposition to a higher or lower baseline level of this index. Indeed, a large number of environmental factors have been shown to modify MAO-A expression and activity, including stress (Maura and Vaccari, 1975), diet changes (Jahng *et al.*, 1998), tobacco smoking (Fowler *et al.*, 1996), physical exercise (Morishima *et al.*, 2006), social environment (Filipenko *et al.*, 2002), and aging itself (Saura *et al.*, 1994). Thus, the interaction between a genetic predisposition and specific environmental determinants could induce variations of MAO-A activity and increase the vulnerability to develop aggressive conduct and antisocial personality.

In line with this possibility, Caspi and coworkers reported that male 3R-carriers with a history of abuse during childhood had a significantly higher prevalence of aggressive behavior in adulthood than 3R-carriers with no history of early maltreatment or 4R-carriers with history of abuse (Caspi *et al.*, 2002). This important finding has been confirmed by subsequent studies (Foley *et al.*, 2004; Huang *et al.*, 2004; Kim-Cohen *et al.*, 2006; Edwards *et al.*, 2010). Further, recent evidence shows that the effects of early stress on impulsivity are reduced by high levels of perceived parental care in individuals with the 3R-, but not the 4R, allelic variant (Kinnally *et al.*, 2009). These results highlight the importance of gene × environment interactions in the pathophysiology of the conditions associated with low MAO-A activity.

B. Preclinical Findings

The analysis of the psychopathological implications of MAO-A deficiency and their neurobiological underpinnings is greatly limited by the rarity of Brunner syndrome and its elusive nosographic description (Hebebrand and Klug, 1995; Schuback *et al.*, 1999). A useful experimental tool to partially obviate this limitation has been afforded by MAO-A knockout (KO) mice. The first line of these mutants was generated in the C3H/HeJ strain by the insertion of an interferon-β minigene into exon 2 of *Maoa* gene (Cases *et al.*, 1995). More recently, another murine line has been developed in 129S6 background, carrying a spontaneous nonsense point mutation of the exon 8 (in a position close to that documented in Brunner syndrome patients) (Scott *et al.*, 2008). Although strain differences may play a modulatory role in phenotypical manifestations, our results have shown that the neurochemical and behavioral phenotypes of both lines bear striking resemblances (Scott *et al.*, 2008).

The selective loss of MAO-A enzymatic function leads to high levels of brain 5-HT and NE, as well as a broad spectrum of phenotypical aberrances, highly reminiscent of the symptoms described in Brunner syndrome. The most evident behavioral alteration in MAO-A KO mice consists in their elevated aggressiveness, toward both foreign mice and cage mates (Cases *et al.*, 1995; Scott *et al.*, 2008). Further, MAO-A KO mice display marked reduction in exploratory activity (Godar *et al.*, 2011), low levels of depression-like behavior in the forced swim test

(Cases *et al.*, 1995) and high level of mnemonic retention of aversive events (Kim *et al.*, 1997; Dubrovina *et al.*, 2006).

Recent studies in our laboratory have begun to elucidate the potential psychopathological bases of the emotional alterations in MAO-A KO mice. In particular, we have observed that these mutants exhibit maladaptive defensive reactivity to different contextual cues. In particular, MAO-A KO mice display high levels of neophobia and antagonistic behavior in presence of novel neutral objects (particularly if introduced within a familiar context); conversely, they increase their level of exploration (with no aggressive or defensive postures) in the presence of cues associated with high danger, such as objects impregnated with predator urine or an anesthetized rat.

In line with this background, MAO-A deficient mice display exaggerated freezing reactions to relatively minor stressors (such as a mild footshock) (Kim *et al.*, 1997), but reduced endocrine responses to major environmental stress, such as physical restraint, water deprivation, cold temperature, and chronic variable stress (Popova *et al.*, 2006).

In conflict-based models of anxiety (which are based on the contrast of exploratory drive and neophobia), MAO-A KO mice did not display major behavioral alterations associated with anxiety-like behavior, but did show a reduction in risk-assessment postures (Popova *et al.*, 2001; Godar *et al.*, 2011). This phenomenon may be partially explained by the simultaneous decrease in avoidance/fear-related behaviors as well as approach/exploration in most contextual settings (Godar *et al.*, 2011).

Taken together, these studies suggest that most behavioral alterations featured in MAO-A KO mice may depend on their inability to attune their responses to environmental inputs. In particular, their maladaptive responses are similar to the deficits in facial affect processing in schizophrenia and autism patients (Phillips *et al.*, 1999; Bolte and Poustka, 2003; Dawson *et al.*, 2004; Surguladze *et al.*, 2006; Gur *et al.*, 2007; Hall *et al.*, 2008).

Although the assessment of the morphological characteristics in the brain of MAO-A KO mice is still incomplete, the most remarkable feature identified to date is the dysmorphogenesis of the barrel fields in layer IV of the somatosensory cortex (Cases *et al.*, 1995). Barrel fields are the cortical representations of the mystacial vibrissae in the rodent snout, and their formation depends on the thalamocortical projections arising from ventrobasal thalamic nuclei (Erzurumlu and Jhaveri, 1990).

The impairment of barrel fields has been associated with alterations of perceptual processing, exploratory activity, and sensory integration (Hurwitz *et al.*, 1990; Sanders *et al.*, 2001; Dowman and Ben-Avraham, 2008; Straube *et al.*, 2009), suggesting similar deficits in MAO-A KO mice. Nevertheless, we showed that these animals were able to recognize familiar objects both in the presence or absence of environmental light in a fashion comparable with their wild-type (WT) littermates (Godar *et al.*, 2011). This finding challenges the possibility that the behavioral alterations observed in MAO-A KO mice may be strictly reflective of their deficits in vibrissal function.

The analysis of sensory modalities in both MAO-A KO lines has revealed impairments in acoustic reactivity (Cases *et al.*, 1995), in line with defects in the auditory pathways (Thompson, 2008; Thompson and Thompson, 2009). Although MAO-A deficiency may also result in subtle developmental alterations of retinal projections (Upton *et al.*, 1999), these deficits do not seem to affect the visual acuity of the mice, as assessed with the visual cliff paradigm (Godar *et al.*, 2011). Similarly, MAO-A KO mice do not exhibit any overt change in olfactory discrimination (Godar *et al.*, 2011).

Previous studies have shown that long-term treatment with MAO-A inhibitors in adult rodents induces a decrease in defensive behavior against predators (Griebel *et al.*, 1998), and an enhancement in exploratory activity (Steckler *et al.*, 2001). These alterations are distinctly different from those observed in MAO-A KO mice, highlighting the likely contribution of early developmental mechanisms in the alterations associated with MAO-A deficiency. Accordingly, early treatment with clorgyline and other MAO inhibitors was found to induce behavioral alterations and impairments in thalamocortical development similar to those observed in MAO-A KO mice (Whitaker-Azmitia *et al.*, 1994; Boylan *et al.*, 2000; Mejia *et al.*, 2002). Moreover, the neurobehavioral alterations of MAO-A KO mice begin at very early stages, with intense head bobbing, prolonged righting, and trembling and delayed maturation of motor skills (Cases *et al.*, 1995; Cazalets *et al.*, 2000). In addition, MAO-A KO mice feature abnormalities of respiratory activity (Bou-Flores *et al.*, 2000). Between postnatal days 11 and 16, MAO-A KO mice display hyperlocomotion, jumping, abnormal postures, and hyperreactivity to stimuli (Cases *et al.*, 1995).

Several studies have shown that the sensorimotor cortex deficits in these animals are due to neurodevelopmental alterations based on the excessive 5-HT levels and 5-HT1B receptor hyperactivation in the first days of postnatal life (Cases *et al.*, 1995; Vitalis *et al.*, 1998; Salichon *et al.*, 2001). In contrast, the morphological alterations of the respiratory centers in the medulla and the cervical phrenic motoneurons in MAO-A KO are reversed by 5-HT2A receptor inhibition (Bou-Flores *et al.*, 2000).

III. Phenotypical Outcomes of MAO-B Deficit

A. Clinical Findings

Although the functional role of MAO-B in brain and behavioral regulation is more elusive than MAO-A, numerous lines of evidence point to its role in emotional regulation. The main MAO-B substrate, PEA, is widely regarded as an endogenous amphetamine, in view of its similar chemical structure and effects *in vivo,* which include increased alertness, euphoria, insomnia, and tremor (Baud *et al.*, 1985; Zucchi *et al.*, 2006). In line with this concept, numerous studies have highlighted a key role of this trace amine in the pathophysiology of schizophrenia and other neuropsychiatric disorders (Beckmann *et al.*, 1983; Szymanski *et al.*, 1987; O'Reilly *et al.*, 1991; Berry, 2007).

Alterations of MAO-B activity and expression have been associated with a broad constellation of neuropsychiatric manifestations, including psychotic disorders, depression, alcoholism, impulsivity, and neurodegenerative diseases (Mann and Chiu, 1978; Adolfsson *et al.*, 1980; Sandler *et al.*, 1993). In addition to its well-characterized neuroprotective role in PD therapy, the prototypical MAO-B inhibitor selegiline has been shown to exert mood-enhancing and anxiolytic effects in depression (Mendlewicz and Youdim, 1983; Quitkin *et al.*, 1984; Robinson *et al.*, 2007) and other pathological conditions (Tariot *et al.*, 1987; Tolbert and Fuller, 1996).

Most clinical studies on MAO-B function have focused on the activity of this enzyme in platelets, as these bodies can be easily collected and display exclusively this isoenzyme. Other advantages of this index lie in its high heritability (Oxenstierna *et al.*, 1986; Pedersen *et al.*, 1993), as well as in its potential association to brain MAO-B catalytic activity (af. Klintenberg *et al.*, 2004). Rich evidence shows a robust correlation between low MAO-B platelet activity and a spectrum of psychological traits related to behavioral disinhibition, such as sensation-seeking and novelty-seeking personality, extraversion, poor impulse control, and proclivity to engage in risky behaviors (Buchsbaum *et al.*, 1976; Fowler *et al.*, 1980; von Knorring *et al.*, 1984; Reist *et al.*, 1990; for a review, see Oreland and Hallman, 1995). While the discovery that smoking can reduce MAO-B activity (Simpson *et al.*, 1999; Hauptmann and Shih, 2001) partially tempered this notion, in view of the high prevalence of this habit among sensation-seekers, further studies confirmed that the association between low MAO-B activity and novelty-seeking remain even after controlling for this environmental factor (Ruchkin *et al.*, 2005).

To the best of our knowledge, there have been no reports of clinical conditions characterized by selective MAO-B deficiency. However, in few cases of atypical ND with MAO-B deletion, the latter deficit was reported to result in increased urinary excretion of PEA, but no overt behavioral abnormalities or cognitive deficits (Berger *et al.*, 1992; Lenders *et al.*, 1996).

Several MAO-B polymorphic variants have been significantly associated with a number of psychological traits, such as negative emotionality (Dlugos *et al.*, 2009), and neuropsychiatric conditions, such as attention-deficit hyperactivity disorder (Li *et al.*, 2008; Ribasés *et al.*, 2009). This association is particularly intriguing, as other authors have documented associations between low platelet MAO-B activity with attention-deficit hyperactivity (Shekim *et al.*, 1986; Coccini *et al.*, 2009; Nedic *et al.*, 2010). Notably, sensation-seeking personality and poor impulse control are key features of ADHD psychopathology.

One of the best-characterized functional MAO-B polymorphisms consists in a single-base variation (A or G) in the intron 13 (Kurth *et al.*, 1993); the A allele has been associated with lower MAO-B catalytic activity in platelets (Garpenstrand *et al.*, 2000) and higher activity in the brain (Balciuniene *et al.*, 2002). However, other studies have failed to detect any connection between this polymorphism and the enzymatic activity (Girmen *et al.*, 1992).

An increase in MAO-B activity and/or expression has been associated with PD; accordingly, few studies have revealed associations between MAO-B polymorphisms and specific clusters of this disease (Kang *et al.*, 2006; Bialecka *et al.*, 2007). Nevertheless, the results on the potential connection between the polymorphism of intron 13 and PD remain controversial (Kurth *et al.*, 1993; Ho *et al.*, 1995; Costa *et al.*, 1997).

B. Preclinical Findings

A complementary line of evidence on the phenotypical implications of MAO-B deficits has been provided by the generation of MAO-B KO mice (Grimsby *et al.*, 1997), achieved by targeted insertion of a foreign neomycin resistance cassette in exon 6 of *Maob* gene. In this line of mice, the lack of MAO-B catalytic activity resulted in significantly higher levels of PEA in brain. However, levels of major monoamine neurotransmitters (5-HT, NE, and DA) were comparable with those in WT mice.

The initial studies aimed at the behavioral characterization of MAO-B KO mice did not highlight many overt behavioral and cognitive changes. For example, unlike MAO-A-deficient animals, MAO-B KO mice display no significant alteration of aggressive behavior. The most remarkable alterations were a reduced level of immobility in the forced swim test (Grimsby *et al.*, 1997). This phenomenon was originally interpreted as an enhancement in stress responsiveness; however, recent studies have shown that MAO-B KO mice display low levels of restraint-induced hyperthermia, a typical parameter of stress reactivity (Bortolato *et al.*, 2009). Taken together, these data suggest that the behavioral performance in MAO-B KO mice was likely reflective of their increased ability to counteract the stress induced by hazardous situations.

We recently documented that MAO-B KO mice display responses reminiscent of behavioral disinhibition, increased novelty-seeking, and reduced anxiety in a number of complementary behavioral paradigms aimed at capturing different aspects of emotional reactivity. For example, MAO-B KO mice displayed significant reduction in anxiety-like behaviors in an elevated plus maze, as well as in the defensive withdrawal, marble burying, hole-board. In addition, MAO-B KO mice displayed a high inclination to explore unfamiliar objects and displayed low novelty-induced grooming, suggesting an enhancement of novelty-seeking behavior (or reduction of neophobia) in these animals. These results are in agreement with the numerous findings on the correlation between low MAO-B platelet activity and novelty-seeking personality, and point to a causal link between the two phenomena.

Interestingly, the abnormal characteristics of MAO-B KO mice were best observed in the presence of several environmental adjustments, such as the reduction of environmental light, a strong anxiogenic factor in mice. This consideration suggests that the ablation of MAO-B may result in variations of personality, rather than actual pathological outcomes.

However, it is extremely likely that low MAO-B activity may be a key vulnerability factor for several conditions associated with sensation-seeking conduct. Future studies will have to evaluate what factors (both genetic and environmental) can interact with low MAO-B activity to induce pathological outcomes.

PEA has been implicated in the regulation of emotional responses, including exploratory activity, arousal, and behavioral reinforcement (Sabelli and Javaid, 1995). MAO-B KO mice display high levels of this monoamine particularly in striatum. Given the implication of PEA in the regulation of DA functions (Kuroki *et al.*, 1990; Sotnikova *et al.*, 2004) and the relevance of DA in behavioral disinhibition (Megens *et al.*, 1992; Black *et al.*, 2002; van Gaalen *et al.*, 2006) and anxiolysis (Shabanov *et al.*, 2005; Picazo *et al.*, 2009), it is possible that DA may play an important role in the behavioral features of MAO-B KO mice. In line with this possibility, MAO-B KO mice have been shown to feature hypersensitivity of D1 receptor (Chen *et al.*, 1999), which have been implicated in the motivational aspects of novelty-seeking behavior (Peters *et al.*, 2007; Olsen and Winder, 2009).

IV. Phenotypical Outcomes of Combined MAO-A and MAO-B Deficit

A. Clinical Findings

As outlined above, the phenotypical outcomes of joint MAO-A and MAO-B deficiency could be initially surmised by comparisons between atypical ND patients with deletion of both genes and their counterparts with mutations restricted to *NDP* gene. These studies suggested that total congenital MAO deficit resulted in a spectrum of severe developmental deficits, with profound mental retardation and autistic-like behavior (Sims *et al.*, 1989a,b; Murphy *et al.*, 1990;

Collins *et al.*, 1992). The ultimate description of the consequences of total MAO deficiency, however, was recently provided by Whibley and colleagues, who reported the case of two male siblings carrying a 240 kb deletion in Xp11.3, which encompassed exons 2–15 of *MAO-A* and all exons of *MAO-B*, but no mutations of NDP or other adjacent genes (Whibley *et al.*, 2010).

The two affected children were born to healthy, nonconsanguineous Caucasian parents and presented with a spectrum of severe abnormalities, including major developmental delay (with height and weight on very low percentiles), mental retardation, and stereotypical movements (hand-flapping and lip-smacking) akin to those featured in Rett syndrome and other pervasive developmental disorders (Whibley *et al.*, 2010). Both brothers had exhibited several episodes of profound hypotonia since perinatal stages, which was typically resistant to phenobarbital, sodium valproate, and lamotrigine. One patient, who displayed minor dysmorphic features (inner canthal folds and an extra incisor) and EEG alterations, died at 5 years of age. His autopsy revealed small foci of perivascular calcification with loss of Purkinje cells in the cerebellum and neurons in the cortex (Whibley *et al.*, 2010).

B. Preclinical Findings

Following the fortuitous discovery of a mouse with a spontaneous mutation in *Maoa* gene within the MAO-B KO colony (Chen *et al.*, 2004), a colony of MAO-A/B KO mice was established, and their phenotypes were characterized. The lack of catalytic activity for either MAO isoenzymes was confirmed by a high increase in brain levels of all monoamines (5-HT: 850%; NE: 220%; DA: 170%; PEA: 1570%) in comparison to WT littermates. Interestingly, the magnitude of these enhancements is significantly greater than those observed in either MAO-A or MAO-B KO mice, suggesting that, in the absence of one isoenzyme, the other can partially overtake its catalytic role. This concept underscores that the alterations induced by joint MAO-A and MAO-B deficiency do not simply result from the summation of the aberrant phenotypes related to each mutation. In fact, it is likely that the abnormalities exhibited by MAO-A/B KO mice may be largely mediated by the exposure to extremely high concentrations of monoamines (in particular 5-HT) in early developmental stages. Preliminary phenotypical analyses of MAO-A/B KO mice revealed that these animals display major developmental alterations (not observed in either MAO-A or MAO-B KO mice), with weight and size at birth significantly lower than WT littermates. These animals also exhibit a complex array of behavioral abnormalities, including aberrant emotional response to novelty, high levels of anxiety-like behaviors in select tasks and low latency to attack in the resident-intruder paradigm (Chen *et al.*, 2004). Recent studies conducted in our laboratory suggest that MAO-A/B KO mice may display alterations in emotional reactivity and informational processing similar (but more severe) to those observed in MAO-A KO mice (unpublished observations). In contrast, these mutants show no patent analogies with the behavioral changes of MAO-B KO mice, likely due to the phenotypical predominance of the defects associated with MAO-A deficiency.

V. Conclusions

Since the cloning of *MAO-A* and *MAO-B* genes in 1988, the employment of complementary approaches has paved the way for the elucidation of the structural and functional characteristics of these two genes and their products. Over the past few years, the advances in the analysis of the phenotypical variations associated with the polymorphisms of both genes has revealed a number of promising leads for the understanding of their role in mental illness and behavioral regulation. Novel findings in the phenotypes of MAO-deficient mice are highlighting novel potential implications of both isoenzymes in a broad spectrum of mental disorders, ranging from autism and anxiety to impulse-control disorders and ADHD. These studies will lay the foundation for future research on the neurobiological and neurochemical bases of these pathological conditions, as well as the role of gene × environment interactions in the vulnerability to several mental disorders.

Acknowledgments

The present study was supported by National Institute of Health grants R01MH39085 (to JCS) and R21HD070611 (to MB), as well as the Boyd and Elsie Welin Professorship (to JS), and the USC Zumberge Research Individual Grant (to MB). We are grateful to Kevin Chen for his unique contributions in MAO research, and particularly in generating various lines of MAO-deficient mice.

References

Adolfsson, R., Gottfries, C.G., Oreland, L., Wiberg, A., and Winblad, B. (1980). Increased activity of brain and platelet monoamine oxidase in dementia of Alzheimer type. *Life Sci.* **27**, 1029–1034.

Alia-Klein, N., Goldstein, R.Z., Kriplani, A., Logan, J., Tomasi, D., Williams, B., Telang, F., Shumay, E., Biegon, A., Craig, I.W., Henn, F., Wang, G.J., *et al.* (2008). Brain monoamine oxidase A activity predicts trait aggression. *J. Neurosci.* **28**, 5099–5104.

Amrein, R., Hetzel, W., Stabl, M., and Schmid-Burgk, W. (1993). RIMA—a new concept in the treatment of depression with moclobemide. *Int. Clin. Psychopharmacol.* **7**(3–4), 123–132.

Anderson, M.C., Hasan, F., McCrodden, J.M., and Tipton, K.F. (1993). Monoamine oxidase inhibitors and the cheese effect. *Neurochem. Res.* **18**, 1145–1149.

Anichtchik, O., Sallinen, V., Peitsaro, N., and Panula, P. (2006). Distinct structure and activity of monoamine oxidase in the brain of zebrafish (Danio rerio). *J. Comp. Neurol.* **498**, 593–610.

Arai, R., Karasawa, N., Kurokawa, K., Kanai, H., Horiike, K., and Ito, A. (2002). Differential subcellular location of mitochondria in rat serotonergic neurons depends on the presence and the absence of monoamine oxidase type B. *Neuroscience* **114**, 825–835.

Bach, A.W., Lan, N.C., Johnson, D.L., Abell, C.W., Bembenek, M.E., Kwan, S.W., Seeburg, P.H., and Shih, J.C. (1988). cDNA cloning of human liver monoamine oxidase A and B: molecular basis of differences in enzymatic properties. *Proc. Natl. Acad. Sci. USA* **85**, 4934–4938.

Balciuniene, J., Emilsson, L., Oreland, L., Petterson, U., and Jazin, E.E. (2002). Investigation of the functional effect of monoamine oxidase polymorphisms in human brain. *Hum. Genet.* **110**, 1–7.

Baud, P., Arbilla, S., Cantrill, R.C., Scatton, B., and Langer, S.Z. (1985). Trace amines inhibit the electrically evoked release of [3H]acetylcholine from slices of rat striatum in the presence of pargyline: similarities between beta-phenylethylamine and amphetamine. *J. Pharmacol. Exp. Ther.* **235**, 220–229.

Beck, O., Borg, S., Edman, G., Fyro, B., Oxenstierna, G., and Sedvall, G. (1984). 5-hydroxytryptophol in human cerebrospinal fluid: conjugation, concentration gradient, relationship to 5-hydroxyindoleacetic acid, and influence of hereditary factors. *J. Neurochem.* **43**, 58–61.

Beckmann, H., Waldmeier, P., Lauber, J., and Gattaz, W.F. (1983). Phenylethylamine and monoamine metabolites in CSF of schizophrenics: effects of neuroleptic treatment. *J. Neural Transm.* **57**, 103–110.

Berger, W., Meindl, A., van de Pol, T.J., Cremers, F.P., Ropers, H.H., Döerner, C., Monaco, A., Bergen, A.A., Lebo, R., Warburg, M., *et al.* (1992). Isolation of a candidate gene for Norrie disease by positional cloning. *Nat. Genet.* **1**(3), 199–203.

Berry, M.D. (2007). The potential of trace amines and their receptors for treating neurological and psychiatric diseases. *Rev. Recent Clin. Trials* **2**, 3–19.

Bialecka, M., Klodowska-Duda, G., Honczarenko, K., Gawrońska-Szklarz, B., Opala, G., Safranow, K., and Droździk, M. (2007). Polymorphisms of catechol-0-methyltransferase (COMT), monoamine oxidase B (MAOB), N-acetyltransferase 2 (NAT2) and cytochrome P450 2D6 (CYP2D6) gene in patients with early onset of Parkinson's disease. *Parkinsonism Relat. Disord.* **13**, 224–229.

Binda, C., Newton-Vinson, P., Hubálek, F., Edmondson, D.E., and Mattevi, A. (2002). Structure of human monoamine oxidase B, a drug target for the treatment of neurological disorders. *Nat. Struct. Biol.* **9**, 22–26.

Binda, C., Li, M., Hubalek, F., Restelli, N., Edmondson, D.E., and Mattevi, A. (2003). Insights into the mode of inhibition of human mitochondrial monoamine oxidase B from high-resolution crystal structures. *Proc. Natl. Acad. Sci. USA* **100**, 9750–9755.

Binda, C., Wang, J., Pisani, L., Caccia, C., Carotti, A., Salvati, P., Edmondson, D.E., and Mattevi, A. (2007). Structures of human monoamine oxidase B complexes with selective noncovalent inhibitors: safinamide and coumarin analogs. *J. Med. Chem.* **50**, 5848–5852.

Black, G.C., Chen, Z.Y., Craig, I.W., and Powell, J.F. (1991). Dinucleotide repeat polymorphism at the MAOA locus. *Nucleic Acids Res.* **19**, 689.

Black, K.J., Hershey, T., Koller, J.M., Videen, T.O., Mintun, M.A., Price, J.L., and Perlmutter, J.S. (2002). A possible substrate for dopamine-related changes in mood and behavior: prefrontal and limbic effects of a D3-preferring dopamine agonist. *Proc. Natl. Acad. Sci. USA* **99**(26), 17113–17118.

Blier, P., and de Montigny, C. (1985). Serotoninergic but not noradrenergic neurons in rat central nervous system adapt to long-term treatment with monoamine oxidase inhibitors. *Neuroscience* **16**, 949–955.

Bolte, S., and Poustka, F. (2003). The recognition of facial affect in autistic and schizophrenic subjects and their first-degree relatives. *Psychol. Med.* **33**(5), 907–915.

Bond, P.A., and Cundall, R.L. (1977). Properties of monoamine oxidase (MAO) in human blood platelets, plasma. *Clin. Chim. Acta* **80**, 317–326.

Bortolato, M., Godar, S.C., Davarian, S., Chen, K., and Shih, J.C. (2009). Behavioral disinhibition and reduced anxiety-like behaviors in monoamine oxidase B-deficient mice. *Neuropsychopharmacology* **34**(13), 2746–2757.

Bortolato, M., Chen, K., and Shih, J.C. (2010). The degradation of serotonin: role of MAO. In: C.P. Müller and B.L. Jacobs (Eds.), *Handbook of the Behavioral Neurobiology of Serotonin.* Elsevier, Amsterdam, pp. 203–218.

Bou-Flores, C., Lajard, A.M., Monteau, R., De Maeyer, E., Seif, I., Lanoir, J., and Hilaire, G. (2000). Abnormal phrenic motoneuron activity and morphology in neonatal monoamine oxidase A-deficient transgenic mice: possible role of a serotonin excess. *J. Neurosci.* **20**, 4646–4656.

Boutet, I., Tanguy, A., and Moraga, D. (2004). Molecular identification and expression of two non-P450 enzymes, monoamine oxidase A and flavin-containing monooxygenase 2, involved in phase I of xenobiotic biotransformation in the Pacific oyster, Crassostrea gigas. *Biochim. Biophys. Acta* **1679**, 29–36.

Boylan, C.B., Bennett-Clarke, C.A., Crissman, R.S., Mooney, R.D., and Rhoades, R.W. (2000). Clorgyline treatment elevates cortical serotonin and temporarily disrupts the vibrissae-related pattern in rat somatosensory cortex. *J. Comp. Neurol.* **427**, 139–149.

Brummett, B.H., Boyle, S.H., Siegler, I.C., Kuhn, C.M., *et al.* (2008). HPA axis function in male caregivers: effect of the monoamine oxidase-A gene promoter (MAOA-uVNTR). *Biol. Psychol.* **79**(2), 250–255.

Brunner, H.G., Nelen, M., Breakefield, X.O., Ropers, H.H., and van Oost, B.A. (1993a). Abnormal behavior associated with a point mutation in the structural gene for monoamine oxidase A. *Science* **262**, 578–580.

Brunner, H.G., Nelen, M.R., van Zandvoort, P., Abeling, N.G., van Gennip, A.H., Wolters, E.C., Kuiper, M.A., Ropers, H.H., and van Oost, B.A. (1993b). X-linked borderline mental retardation with prominent behavioral disturbance: phenotype, genetic localization, and evidence for disturbed monoamine metabolism. *Am. J. Hum. Genet.* **52**, 1032–1039.

Buchsbaum, M.S., Coursey, R.D., and Murphy, D.L. (1976). The biochemical high-risk paradigm: behavioral and familial correlates of low platelet monoamine oxidase activity. *Science* **194**, 339–341.

Buckholtz, J.W., and Meyer-Lindenberg, A. (2008). MAO A and the neurogenetic architecture of human aggression. *Trends Neurosci.* **31**, 120–129.

Cases, O., Seif, I., Grimsby, J., Gaspar, P., Chen, K., Pournin, S., Muller, U., Aguet, M., Babinet, C., and Shih, J.C. (1995). Aggressive behavior and altered amounts of brain serotonin and norepinephrine in mice lacking MAOA. *Science* **268**, 1763–1766.

Caspi, A., McClay, J., Moffitt, T.E., Mill, J., Martin, J., Craig, I.W., Taylor, A., and Poulton, R. (2002). Role of genotype in the cycle of violence in maltreated children. *Science* **297**, 851–854.

Cazalets, J.R., Gardette, M., and Hilaire, G. (2000). Locomotor network maturation is transiently delayed in the MAOA-deficient mouse. *J. Neurophysiol.* **83**, 2468–2470.

Cerasa, A., Gioia, M.C., Labate, A., Lanza, P., Magariello, A., Muglia, M., and Quattrone, A. (2008). MAO A VNTR polymorphism and variation in human morphology: a VBM study. *Neuroreport* **19**, 1107–1110.

Cerasa, A., Cherubini, A., Quattrone, A., Gioia, M.C., Magariello, A., Muglia, M., Manna, I., Assogna, F., Caltagirone, C., and Spalletta, G. (2010). Morphological correlates of MAO A VNTR polymorphism: new evidence from cortical thickness measurement. *Behav. Brain Res.* **211**, 118–124.

Chen, K., Wu, H.F., Grimsby, J., and Shih, J.C. (1994). Cloning of a novel monoamine oxidase cDNA from trout liver. *Mol. Pharmacol.* **46**, 1226–1233.

Chen, K., Wu, H.F., and Shih, J.C. (1996). Influence of C terminus on monoamine oxidase A and B catalytic activity. *J. Neurochem.* **66**, 797–803.

Chen, L., He, M., Sibille, E., Thompson, A., Sarnyai, Z., Baker, H., Shippenberg, T., and Toth, M. (1999). Adaptive changes in postsynaptic dopamine receptors despite unaltered dopamine dynamics in mice lacking monoamine oxidase B. *J. Neurochem.* **73**, 647–655.

Chen, K., Holschneider, D.P., Wu, W., Rebrin, I., and Shih, J.C. (2004). A spontaneous point mutation produces monoamine oxidase A/B knock-out mice with greatly elevated monoamines and anxiety-like behavior. *J. Biol. Chem.* **279**, 39645–39652.

Chen, Z.Y., Powell, J.F., Hsu, Y.P., Breakefield, X.O., and Craig, I.W. (1992). Organization of the human monoamine oxidase genes and long-range physical mapping around them. *Genomics* **14**(1), 75–82.

Coccini, T., Crevani, A., Rossi, G., Assandri, F., Balottin, U., Nardo, R.D., and Manzo, L. (2009). Reduced platelet monoamine oxidase type B activity and lymphocyte muscarinic receptor binding in unmedicated children with attention deficit hyperactivity disorder. *Biomarkers* **14**, 513–522.

Cohen, I.L., Liu, X., Schutz, C., White, B.N., *et al.* (2003). Association of autism severity with a monoamine oxidase A functional polymorphism. *Clin. Genet.* **64**(3), 190–197.

Collins, F.A., Murphy, D.L., Reiss, A.L., Sims, K.B., Lewis, J.G., Freund, L., Karoum, F., Zhu, D., Maumenee, I.H., and Antonarakis, S.E. (1992). Clinical, biochemical, and neuropsychiatric evaluation of a patient with a contiguous gene syndrome due to a microdeletion Xp11.3 including the Norrie disease locus and monoamine oxidase (MAOA and MAOB) genes. *Am. J. Med. Genet.* **42**, 127–134.

Consalvi, V., Mardh, G., and Vallee, B.L. (1986). Human alcohol dehydrogenases and serotonin metabolism. *Biochem. Biophys. Res. Commun.* **139**, 1009–1016.

Contini, V., Marques, F.Z., Garcia, C.E., Hutz, M.H., and Bau, C.H. (2006). MAOA-uVNTR polymorphism in a Brazilian sample: further support for the association with impulsive behaviors and alcohol dependence. *Am. J. Med. Genet. B Neuropsychiatr. Genet.* **141B**(3), 305–308.

Costa, P., Checkoway, H., Levy, D., Smith-Weller, T., Franklin, G.M., Swanson, P.D., and Costa, L.G. (1997). Association of a polymorphism in intron 13 of the monoamine oxidase B gene with Parkinson disease. *Am. J. Med. Genet.* **74**, 154–156.

Craddock, N., Daniels, J., Roberts, E., Rees, M., McGuffin, P., and Owen, M.J. (1995). No evidence for allelic association between bipolar disorder and monoamine oxidase A gene polymorphisms. *Am. J. Med. Genet.* **60**, 322–324.

Danielczyk, W., Streifler, M., Konradi, C., Riederer, P., and Moll, G. (1988). Platelet MAO-B activity and the psychopathology of Parkinson's disease, senile dementia and multi-infarct dementia. *Acta Psychiatr. Scand.* **78**, 730–736.

Davis, L.K., Hazlett, H.C., Librant, A.L., Nopoulos, P., Sheffield, V.C., Piven, J., and Wassink, T.H. (2008). Cortical enlargement in autism is associated with a functional VNTR in the monoamine oxidase A gene. *Am. J. Med. Genet. B Neuropsychiatr. Genet.* **147B**, 1145–1151.

Dawson, G., Webb, S.J., Carver, L., Panagiotides, H., *et al.* (2004). Young children with autism show atypical brain responses to fearful versus neutral facial expressions of emotion. *Dev. Sci.* **7**(3), 340–359.

De Colibus, L., Li, M., Binda, C., Lustig, A., Edmondson, D.E., and Mattevi, A. (2005). Three-dimensional structure of human monoamine oxidase A (MAO A): relation to the structures of rat MAO A and human MAO B. *Proc. Natl. Acad. Sci. USA* **102**, 12684–12689.

Deckert, J., Catalano, M., Syagailo, Y.V., Bosi, M., Okladnova, O., Di Bella, D., Nöthen, M.M., Maffei, P., Franke, P., Fritze, J., Maier, W., Propping, P., *et al.* (1999). Excess of high activity monoamine oxidase A gene promoter alleles in female patients with panic disorder. *Hum. Mol. Genet.* **8**, 621–624.

Denney, R.M., Koch, H., and Craig, I.W. (1999). Association between monoamine oxidase A activity in human male skin fibroblasts and genotype of the MAO-A promoter-associated variable number tandem repeat. *Hum. Genet.* **105**, 542–551.

Despopoulos, A., and Weissbach, H. (1957). Renal metabolism of 5-hydroxyindolacetic acid. *Am. J. Physiol.* **189**, 548–550.

Dlugos, A.M., Palmer, A.A., and de Wit, H. (2009). Negative emotionality: monoamine oxidase B gene variants modulate personality traits in healthy humans. *J. Neural Transm.* **116**, 1323–1334.

Donnelly, C.H., and Murphy, D.L. (1977). Substrate- and inhibitor-related characteristics of human platelet. *Biochem. Pharmacol.* **26**, 853–858.

Dowman, R., and Ben-Avraham, D. (2008). An artificial neural network model of orienting attention toward threatening somatosensory stimuli. *Psychophysiology* **45**(2), 229–239.

Dubrovina, N.I., Popova, N.K., Gilinskii, M.A., Tomilenko, R.A., and Seif, I. (2006). Acquisition and extinction of a conditioned passive avoidance reflex in mice with genetic knockout of monoamine oxidase A. *Neurosci. Behav. Physiol.* **36**, 335–339.

Edmondson, D.E., Mattevi, A., Binda, C., Li, M., and Hubálek, F. (2004). Structure and mechanism of monoamine oxidase. *Curr. Med. Chem.* **11**, 1983–1993.

Edmondson, D.E., Binda, C., and Mattevi, A. (2007). Structural insights into the mechanism of amine oxidation by monoamine oxidases A and B. *Arch. Biochem. Biophys.* **464**, 269–276.

Edmondson, D.E., Binda, C., Wang, J., Upadhyay, A.K., and Mattevi, A. (2009). Molecular and mechanistic properties of the membrane-bound mitochondrial monoamine oxidases. *Biochemistry* **48**, 4220–4230.

Edwards, A.C., Dodge, K.A., Latendresse, S.J., Lansford, J.E., Bates, J.E., Pettit, G.S., Budde, J.P., Goate, A.M., and Dick, D.M. (2010). MAOA-uVNTR and early physical discipline interact to influence delinquent behavior. *J. Child Psychol. Psychiatry* **51**(6), 679–687.

Erzurumlu, R.S., and Jhaveri, S. (1990). Thalamic axons confer a blueprint of the sensory periphery onto the developing rat somatosensory cortex. *Brain Res. Dev. Brain Res.* **56**, 229–234.

Feldstein, A., and Williamson, O. (1968). 5-Hydroxytryptamine metabolism in rat brain and liver homogenates. *Br. J. Pharmacol.* **34**, 38–42.

Filipenko, M.L., Beilina, A.G., Alekseyenko, O.V., Dolgov, V.V., and Kudryavtseva, N.N. (2002). Repeated experience of social defeats increases serotonin transporter and monoamine oxidase A mRNA levels in raphe nuclei of male mice. *Neurosci. Lett.* **321**, 25–28.

Foley, D.L., Eaves, L.J., Wormley, B., Silberg, J.L., *et al.* (2004). Childhood adversity, monoamine oxidase a genotype, and risk for conduct disorder. *Arch. Gen. Psychiatry* **61**(7), 738–744.

Fornai, F., Chen, K., Giorgi, F.S., Gesi, M., Alessandri, M.G., and Shih, J.C. (1999). Striatal dopamine metabolism in monoamine oxidase B-deficient mice: a brain dialysis study. *J. Neurochem.* **73**, 2434–2440.

Fowler, C.J., von Knorring, L., and Oreland, L. (1980). Platelet monoamine oxidase activity in sensation seekers. *Psychiatry Res.* **3**(3), 273–279.

Fowler, J.S., Volkow, N.D., Wang, G.-J., Pappas, N., Logan, J., Shea, C., Alexoff, D., MacGregor, R.R., Schlyer, D.J., Zezulkova, I., and Wolf, A.P. (1996). Brain monoamine oxidase A inhibition in cigarette smokers. *Proc. Natl. Acad. Sci. USA* **93**, 14065–14069.

Fowler, J.S., Alia-Klein, N., Kriplani, A., Logan, J., Williams, B., Zhu, W., Craig, I.W., Telang, F., Goldstein, R., Volkow, N.D., Vaska, P., and Wang, G.J. (2007). Evidence that brain MAO A activity does not correspond to MAO A genotype in healthy male subjects. *Biol. Psychiatry* **62**, 355–358.

Fox, H.H., and Gibas, J.T. (1953). Synthetic tuberculostats. V. Alkylidene derivatives of isonicotinylhydrazine. *J. Org. Chem.* **18**, 983–989.

Furlong, R.A., Ho, L., Rubinsztein, J.S., Walsh, C., Paykel, E.S., and Rubinsztein, D.C. (1999). Analysis of the monoamine oxidase A (MAOA) gene in bipolar affective disorder by association studies, meta-analyses, and sequencing of the promoter. *Am. J. Med. Genet.* **88**, 398–406.

Garpenstrand, H., Ekblom, J., Forslund, K., Rylander, G., and Oreland, L. (2000). Platelet monoamine oxidase activity is related to MAOB intron 13 genotype. *J. Neural Transm.* **107**, 523–530.

Garrick, N.A., and Murphy, D.L. (1980). Species differences in the deamination of dopamine and other substrates for monoamine oxidase in brain. *Psychopharmacology* **72**, 27–33.

Geha, R.M., Chen, K., and Shih, J.C. (2000). Phe(208) and Ile(199) in human monoamine oxidase A and B do not determine substrate and inhibitor specificities as in rat. *J. Neurochem.* **75**, 1304–1309.

Geha, R.M., Rebrin, I., Chen, K., and Shih, J.C. (2001). Substrate and inhibitor specificities for human monoamine oxidase A and B are influenced by a single amino acid. *J. Biol. Chem.* **276**, 9877–9882.

Girmen, A.S., Baenziger, J., Hotamisligil, G.S., Konradi, C., Shalish, C., Sullivan, J.L., and Breakefield, X.O. (1992). Relationship between platelet monoamine oxidase B activity and alleles at the MAOB locus. *J. Neurochem.* **59**, 2063–2066.

Godar, S.C., Bortolato, M., Frau, R., Dousti, M., Chen, K., and Shih, J.C. (2011). Maladaptive defensive behaviours in monoamine oxidase A-deficient mice. *Int. J. Neuropsychopharmacol.* in press.

Gottowik, J., Cesura, A.M., Malherbe, P., Lang, G., and Da Prada, M. (1993). Characterisation of wild-type and mutant forms of human monoamine oxidase A and B expressed in a mammalian cell line. *FEBS Lett.* **317**, 152–156.

Griebel, G., Curet, O., Perrault, G., and Sanger, D.J. (1998). Behavioral effects of phenelzine in an experimental model for screening anxiolytic and anti-panic drugs: correlation with changes in monoamine-oxidase activity and monoamine levels. *Neuropharmacology* **37**(7), 927–935.

Grimsby, J., Chen, K., Wang, L.J., Lan, N.C., and Shih, J.C. (1991). Human monoamine oxidase A and B genes exhibit identical exon-intron organization. *Proc. Natl. Acad. Sci. USA* **88**, 3637–3641.

Grimsby, J., Toth, M., Chen, K., Kumazawa, T., Klaidman, L., Adams, J.D., Karoum, F., Gal, J., and Shih, J.C. (1997). Increased stress response and b-phenylethylamine in MAOB-deficient mice. *Nat. Genet.* **17**, 206–210.

Grünblatt, E., Schlösser, R., Fischer, P., Fischer, M.O., Li, J., Koutsilieri, E., Wichart, I., Sterba, N., Rujescu, D., Möller, H.J., Adamcyk, W., Dittrich, B., *et al.* (2005). Oxidative stress related markers in the "VITA" and the centenarian projects. *Neurobiol. Aging* **26**, 429–438.

Gur, R.E., Loughead, J., Kohler, C.G., Elliott, M.A., *et al.* (2007). Limbic activation associated with misidentification of fearful faces and flat affect in schizophrenia. *Arch. Gen. Psychiatry* **64**(12), 1356–1366.

Hall, J., Whalley, H.C., McKirdy, J.W., Romaniuk, L., *et al.* (2008). Overactivation of fear systems to neutral faces in schizophrenia. *Biol. Psychiatry* **64**(1), 70–73.

Hauptmann, N., and Shih, J.C. (2001). 2-Naphthylamine, a compound found in cigarette smoke, decreases both monoamine oxidase A and B catalytic activity. *Life Sci.* **68**, 1231–1241.

Hebebrand, J., and Klug, B. (1995). Specification of the phenotype required for men with monoamine oxidase type A deficiency. *Hum. Genet.* **96**(3), 372–376.

Helander, A., Beck, O., Jacobsson, G., Lowenmo, C., and Wikstrom, T. (1993). Time course of ethanol-induced changes in serotonin metabolism. *Life Sci.* **53**, 847–855.

Hinds, H.L., Hendriks, R.W., Craig, I.W., and Chen, Z.Y. (1992). Characterization of a highly polymorphic region near the first exon of the human MAOA gene containing a GT dinucleotide and a novel VNTR motif. *Genomics* **13**(3), 896–897.

Ho, S.L., Kapadi, A.L., Ramsden, D.B., and Williams, A.C. (1995). An allelic association study of monoamine oxidase B in Parkinson's disease. *Ann. Neurol.* **37**, 403–405.

Huang, Y.-y., Cate, S.P., Battistuzzi, C., Oquendo, M.A., Brent, D., and Mann, J.J. (2004). An association between a functional polymorphism in the monoamine oxidase A gene promoter, impulsive traits and early abuse experiences. *Neuropsychopharmacology* **29**, 1498–1505.

Hurwitz, B.E., Dietrich, W.D., McCabe, P.M., Watson, B.D., *et al.* (1990). Sensory-motor deficit and recovery from thrombotic infarction of the vibrissal barrel-field cortex. *Brain Res.* **512**(2), 210–220.

Jahng, J.W., Houpt, T.A., Wessel, T.C., Chen, K., Shih, J.C., and Joh, T.H. (1997). Localization of monoamine oxidase A and B mRNA in the rat brain by in situ hybridization. *Synapse* **25**, 30–36.

Jahng, J.W., Houpt, T.A., Joh, T.H., and Son, J.H. (1998). Differential expression of monoamine oxidase A, serotonin transporter, tyrosine hydroxylase and norepinephrine transporter mRNA by anorexia mutation and food deprivation. *Dev. Brain Res.* **107**, 241–246.

Johnston, J.P. (1968). Some observations upon a new inhibitor of monoamine oxidase in brain tissue. *Biochem. Pharmacol.* **17**, 1285–1297.

Jonsson, E.G., Norton, N., Gustavsson, J.P., Oreland, L., Owen, M.J., and Sedvall, G.C. (2000). A promoter polymorphism in the monoamine oxidase A gene and its relationships to monoamine metabolite concentrations in CSF of healthy volunteers. *J. Psychiatr. Res.* **34**, 239–244.

Kang, S.J., Scott, W.K., Li, Y.J., Hauser, M.A., van der Walt, J.M., Fujiwara, K., Mayhew, G.M., West, S.G., Vance, J.M., and Martin, E.R. (2006). Family-based case-control study of MAOA and MAOB polymorphisms in Parkinson disease. *Mov. Disord.* **21**, 2175–2180.

Kawada, Y., Hattori, M., Dai, X.Y., and Nanko, S. (1995). Possible association between monoamine oxidase A gene and bipolar affective disorder. *Am. J. Hum. Genet.* **56**, 335–336.

Kearney, E.B., Salach, J.I., Walker, W.H., Seng, R.L., Kenney, W., Zeszotek, E., and Singer, T.P. (1971). The covalently-bound flavin of hepatic monoamine oxidase. 1. Isolation and sequence of a flavin peptide and evidence for binding at the 8alpha position. *Eur. J. Biochem.* **24**, 321–327.

Kennedy, S.H. (1997). Continuation and maintenance treatments in major depression: the neglected role of monoamine oxidase inhibitors. *J Psychiatry Neurosci.* **22**, 127–131.

Kinnally, E.L., Huang, Y.Y., Haverly, R., Burke, A.K., Galfalvy, H., Brent, D.P., Oquendo, M.A., and Mann, J.J. (2009). Parental care moderates the influence of MAOA-uVNTR genotype and childhood stressors on trait impulsivity and aggression in adult women. *Psychiatr. Genet.* **19**(3), 126–133.

Kim, J.J., Shih, J.C., Chen, K., Chen, L., Bao, S., Maren, S., Anagnostaras, S.G., Fanselow, M.S., De Maeyer, E., Seif, I., and Thompson, R.F. (1997). Selective enhancement of emotional, but not motor, learning in monoamine oxidase A-deficient mice. *Proc. Natl. Acad. Sci. USA* **94**(11), 5929–5933.

Kim-Cohen, J., Caspi, A., Taylor, A., Williams, B., Newcombe, R., Craig, I.W., and Moffitt, T.E. (2006). MAOA, maltreatment, and gene-environment interaction predicting children's mental health: new evidence and a meta-analysis. *Mol. Psychiatry.* **11**(10), 903–913.

Klinteberg, B., von Knorring, L., and Oreland, L. (2004). On the psychobiology of impulsivity. In: R.M. Stelmack (Ed.), *On the psychobiology of personality: essays in honour of Marvin Zuckerman.* Elsevier, Amsterdam, pp. 455–478.

Knoll, J., and Magyar, K. (1972). Some puzzling pharmacological effects of monoamine oxidase inhibitors. *Adv. Biochem. Psychopharmacol.* **5**, 393–408.

Kobayashi, S., Takahara, K., and Kamijo, K. (1981). Monoamine oxidase in frog liver and brain. *Comp. Biochem. Physiol. C* **69**, 179–183.

Konradi, C., Kornhuber, J., Froelich, L., Fritze, J., Heinsen, H., Beckmann, H., Schulz, E., and Riederer, P. (1989). Demonstration of monoamine oxidase-A and -B in the human brainstem by a histochemical technique. *Neuroscience* **33**, 383–400.

Kosenko, E., Kaminsky, Y., Kaminsky, A., Valencia, M., Lee, L., Hermenegildo, C., and Felipo, V. (1997). Superoxide production and antioxidant enzymes in ammonia intoxication in rats. *Free Radic. Res.* **27**, 637–644.

Kurth, J.H., Kurth, M.C., Poduslo, S.E., and Schwankhaus, J.D. (1993). Association of a monoamine oxidase B allele with Parkinson's disease. *Ann. Neurol.* **33**, 368–372.

Kuroki, T., Tsutsumi, T., Hirano, M., Matsumoto, T., Tatebayashi, Y., Nishiyama, K., Uchimura, H., Shiraishi, A., Nakahara, T., and Nakamura, K. (1990). Behavioral sensitization to beta-phenylethylamine (PEA): enduring modifications of specific dopaminergic neuron systems in the rat. *Psychopharmacology (Berl).* **102**(1), 5–10.

Lan, N.C., Chen, C.H., and Shih, J.C. (1989a). Expression of functional human monoamine oxidase A and B cDNAs in mammalian cells. *J. Neurochem.* **52**, 1652–1654.

Lan, N.C., Heinzmann, C., Gal, A., Klisak, I., Orth, U., Lai, E., Grimsby, J., Sparkes, R.S., Mohandas, T., and Shih, J.C. (1989b). Human monoamine oxidase A and B genes map to Xp 11.23 and are deleted in a patient with Norrie disease. *Genomics* **4**, 552–559.

Lee, B.T., and Ham, B.J. (2008). Monoamine oxidase A-uVNTR genotype affects limbic brain activity in response to affective facial stimuli. *Neuroreport* **19**(5), 515–519.

disorder and monoamine oxidase A gene polymorphisms: results of a multicenter study. *Am. J. Psychiatry* **157**, 948–955.

Quitkin, F.M., Liebowitz, M.R., Stewart, J.W., McGrath, P.J., Harrison, W., Rabkin, J.G., Markowitz, J., and Davies, S.O. (1984). L-Deprenyl in atypical depressives. *Arch. Gen. Psychiatry* **41**, 777–781.

Reist, C., Haier, R.J., DeMet, E., and Chicz-DeMet, A. (1990). Platelet MAO activity in personality disorders and normal controls. *Psychiatry Res.* **33**, 221–227.

Ribasés, M., Ramos-Quiroga, J.A., Hervás, A., Bosch, R., Bielsa, A., Gastaminza, X., Artigas, J., Rodriguez-Ben, S., Estivill, X., Casas, M., Cormand, B., and Bayés, M. (2009). Exploration of 19 serotoninergic candidate genes in adults and children with attention-deficit/hyperactivity disorder identifies association for 5HT2A, DDC and MAOB. *Mol. Psychiatry* **14**, 71–85.

Robinson, D.S., Gilmor, M.L., Yang, Y., Moonsammy, G., Azzaro, A.J., Oren, D.A., and Campbell, B.J. (2007). Treatment effects of selegiline transdermal system on symptoms of major depressive disorder: a meta-analysis of short-term, placebo-controlled, efficacy trials. *Psychopharmacol. Bull.* **40**, 15–28.

Roohi, J., DeVincent, C.J., Hatchwell, E., and Gadow, K.D. (2009). Association of a monoamine oxidase-a gene promoter polymorphism with ADHD and anxiety in boys with autism spectrum disorder. *J. Autism Dev. Disord.* **39**(1), 67–74.

Rubinsztein, D.C., Leggo, J., Goodburn, S., Walsh, C., Jain, S., and Paykel, E.S. (1996). Genetic association between monoamine oxidase A microsatellite and RFLP alleles and bipolar affective disorder: analysis and meta-analysis. *Hum. Mol. Genet.* **5**, 779–782.

Ruchkin, V.V., Koposov, R.A., af Klinteberg, B., Oreland, L., and Grigorenko, E.L. (2005). Platelet MAO-B, personality, and psychopathology. *J. Abnorm. Psychol.* **114**, 477–482.

Saba, A., Chiellini, G., Frascarelli, S., Marchini, M., Ghelardoni, S., Raffaelli, A., Tonacchera, M., Vitti, P., Scanlan, T.S., and Zucchi, R. (2010). Tissue distribution and cardiac metabolism of 3-iodothyronamine. *Endocrinology* **151**(10), 5063–5073.

Sabelli, H.C., and Javaid, J.I. (1995). Phenylethylamine modulation of affect: therapeutic and diagnostic implications. *J. Neuropsychiatry Clin. Neurosci.* **7**(1), 6–14.

Sabol, S.Z., Hu, S., and Hamer, D. (1998). A functional polymorphism in the monoamine oxidase A gene promoter. *Hum. Genet.* **103**, 273–279.

Salichon, N., Gaspar, P., Upton, A.L., Picaud, S., Hanoun, N., Hamon, M., De Maeyer, E., Murphy, D. L., Mossner, R., Lesch, K.P., Hen, R., and Seif, I. (2001). Excessive activation of serotonin (5-HT) 1B receptors disrupts the formation of sensory maps in monoamine oxidase a and 5-ht transporter knock-out mice. *J. Neurosci.* **21**, 884–896.

Samochowiec, J., Lesch, K.P., Rottmann, M., Smolka, M., Syagailo, Y.V., Okladnova, O., Rommelspacher, H., Winterer, G., Schmidt, L.G., and Sander, T. (1999). Association of a regulatory polymorphism in the promoter region of the monoamine oxidase A gene with antisocial alcoholism. *Psychiatry Res.* **86**(1), 67–72.

Sanders, M.J., Dietrich, W.D., and Green, E.J. (2001). Behavioral, electrophysiological, and histopathological consequences of mild fluid-percussion injury in the rat. *Brain Res.* **904**(1), 141–144.

Sandler, M., Glover, V., Clow, A., and Jarman, J. (1993). Monoamine oxidase-B, monoamine oxidase-B inhibitors, and Parkinson's disease. A role for superoxide dismutase? *Adv. Neurol.* **60**, 238–241.

Saura, J., Richards, J.G., and Mahy, N. (1994). Age-related changes on MAO in Bl/C57 mouse tissues: a quantitative radioautographic study. *J. Neural Transm.* **41**(Suppl.), 89–94.

Schilling, B., and Lerch, K. (1995). Cloning, sequencing and heterologous expression of the monoamine oxidase gene from Aspergillus niger. *Mol. Gen. Genet.* **247**(4), 430–438.

Schuback, D.E., Mulligan, E.L., Sims, K.B., Tivol, E.A., Greenberg, B.D., Chang, S.F., Yang, S.L., Mau, Y.C., Shen, C.Y., Ho, M.S., Yang, N.H., Butler, M.G., *et al.* (1999). Screen for MAOA mutations in target human groups. *Am. J. Med. Genet.* **88**, 25–28.

Scott, A.L., Bortolato, M., Chen, K., and Shih, J.C. (2008). Novel monoamine oxidase A knock out mice with human-like spontaneous mutation. *Neuroreport* **19**, 739–743.

Setini, A., Pierucci, F., Senatori, O., and Nicotra, A. (2005). Molecular characterization of monoamine oxidase in zebrafish (Danio rerio). *Comp. Biochem. Physiol. B Biochem. Mol. Biol.* **140**, 153–161.

Shabanov, P.D., Lebedev, A.A., Meshcherov, Sh.K., and Strel'tsov, V.F. (2005).The effects of neurochemical lesioning of dopaminergic terminals in early ontogenesis on behavior in adult rats. *Neurosci. Behav. Physiol.* **35**(5), 535–544.

Sharp, T., Gartside, S.E., and Umbers, V. (1997). Effects of co-administration of a monoamine oxidase inhibitor and a 5-HT1A receptor antagonist on 5-hydroxytryptamine cell firing and release. *Eur. J. Pharmacol.* **320**, 15–19.

Shekim, W.O., Bylund, D.B., Alexson, J., Glaser, R.D., Jones, S.B., Hodges, K., and Perdue, S. (1986). Platelet MAO and measures of attention and impulsivity in boys with attention deficit disorder and hyperactivity. *Psychiatry Res.* **18**, 179–188.

Shih, J.C., and Eiduson, S. (1969). Multiple forms of monoamine oxidase in the developing brain. *Nature* **224**, 1309–1310.

Shih, J.C., and Thompson, R.F. (1999). Monoamine oxidase in neuropsychiatry and behavior. *Am. J. Hum. Genet.* **65**, 593–598.

Simpson, G.M., Shih, J.C., Chen, K., Flowers, C., Kumazawa, T., and Spring, B. (1999). Schizophrenia, monoamine oxidase activity, and cigarette smoking. *Neuropsychopharmacology* **20**, 392–394.

Sims, K.B., de la Chapelle, A., Norio, R., Sankila, E.M., Hsu, Y.P., Rinehart, W.B., Corey, T.J., Ozelius, L., Powell, J.F., Bruns, G., *et al.* (1989a). Monoamine oxidase deficiency in males with an X chromosome deletion. *Neuron* **2**, 1069–1076.

Sims, K.B., Ozelius, L., Corey, T., Rinehart, W.B., Liberfarb, R., Haines, J., Chen, W.J., Norio, R., Sankila, E., de la Chapelle, A., *et al.* (1989b). Norrie disease gene is distinct from the monoamine oxidase genes. *Am. J. Hum. Genet.* **45**, 424–434.

Sotnikova, T.D., Budygin, E.A., Jones, S.R., Dykstra, L.A., Caron, M.G., and Gainetdinov, R.R. (2004). Dopamine transporter-dependent and -independent actions of trace amine beta-phenylethylamine. *J. Neurochem.* **91**(2), 362–373.

Steckler, T., Rammes, G., Sauvage, M., van Gaalen, M.M., *et al.* (2001). Effects of the monoamine oxidase A inhibitor moclobemide on hippocampal plasticity in GR-impaired transgenic mice. *J. Psychiatr. Res.* **35**(1), 29–42.

Straube, T., Schmidt, S., Weiss, T., Mentzel, H.J., *et al.* (2009). Dynamic activation of the anterior cingulate cortex during anticipatory anxiety. *Neuroimage* **44**(3), 975–981.

Surguladze, S., Russell, T., Kucharska-Pietura, K., Travis, M.J., *et al.* (2006). A reversal of the normal pattern of parahippocampal response to neutral and fearful faces is associated with reality distortion in schizophrenia. *Biol. Psychiatry* **60**(5), 423–431.

Svensson, S., Some, M., Lundsjo, A., Helander, A., Cronholm, T., and Hoog, J.O. (1999). Activities of human alcohol dehydrogenases in the metabolic pathways of ethanol and serotonin. *Eur. J. Biochem.* **262**, 324–329.

Szymanski, H.V., Naylor, E.W., and Karoum, F. (1987). Plasma phenylethylamine and phenylalanine in chronic schizophrenic patients. *Biol. Psychiatry* **22**, 194–198.

Tariot, P.N., Cohen, R.M., Sunderland, T., Newhouse, P.A., Yount, D., Mellow, A.M., Weingartner, H., Mueller, E.A., and Murphy, D.L. (1987). L-deprenyl in Alzheimer's disease. Preliminary evidence for behavioral change with monoamine oxidase B inhibition. *Arch. Gen. Psychiatry* **44**, 427–433.

Thompson, A.M. (2008). Serotonin immunoreactivity in auditory brainstem neurons of the postnatal monoamine oxidase-A knockout mouse. *Brain Res.* **1228**, 58–67.

Thompson, A.M., and Thompson, G.C. (2009). Serotonin-immunoreactive neurons in the postnatal MAO-A KO mouse lateral superior olive project to the inferior colliculus. *Neurosci. Lett.* **460**, 47–51.

Tolbert, S.R., and Fuller, M.A. (1996). Selegiline in treatment of behavioral and cognitive symptoms of Alzheimer disease. *Ann. Pharmacother.* **30**, 1122–1129.

Udenfriend, S., Titus, E., Weissbach, H., and Peterson, R.E. (1956). Biogenesis and metabolismof 5-hydroxyindole compounds. *J. Biol. Chem.* **219**, 335–344.

Upadhyay, A.K., Wang, J., and Edmondson, D.E. (2008). Comparison of the structural properties of the active site cavities of human and rat monoamine oxidase A and B in their soluble and membrane-bound forms. *Biochemistry* **47**, 526–536.

Upton, A.L., Salichon, N., Lebrand, C., Ravary, A., Blakely, R., Seif, I., and Gaspar, P. (1999). Excess of serotonin (5-HT) alters the segregation of ispilateral and contralateral retinal projections in monoamine oxidase A knock -out mice: possible role of 5-HT uptake in retinal ganglion cells during development. *J. Neurosci.* **19**, 7007–7024.

van Gaalen, M.M., Brueggeman, R.J., Bronius, P.F., Schoffelmeer, A.N., and Vanderschuren, L.J. (2006). Behavioral disinhibition requires dopamine receptor activation. *Psychopharmacology (Berl).* **187**(1), 73–85.

Vitalis, T., Cases, O., Callebert, J., Launay, J.M., Price, D.J., Seif, I., and Gaspar, P. (1998). Effects of monoamine oxidase A inhibition on barrel formation in the mouse somatosensory cortex: determination of a sensitive developmental period. *J. Comp. Neurol.* **393**, 169–184.

von Knorring, L., Oreland, L., and Winblad, B. (1984). Personality traits related to monoamine oxidase activity in platelets. *Psychiatry Res.* **12**, 11–26.

Whibley, A., Urquhart, J., Dore, J., Willatt, L., Parkin, G., Gaunt, L., Black, G., Donnai, D., and Raymond, F.L. (2010). Deletion of MAOA and MAOB in a male patient causes severe developmental delay, intermittent hypotonia and stereotypical hand movements. *Eur. J. Hum. Genet.* **18**, 1095–1099.

Whitaker-Azmitia, P.M., Zhang, X., and Clarke, C. (1994). Effects of gestational exposure to monoamine oxidase inhibitors in rats: preliminary behavioral and neurochemical studies. *Neuropsychopharmacology* **11**(2), 125–132.

Williams, L.M., Gatt, J.M., Kuan, S.A., Dobson-Stone, C., *et al.* (2009). A polymorphism of the MAOA gene is associated with emotional brain markers and personality traits on an antisocial index. *Neuropsychopharmacology* **34**(7), 1797–1809.

Williams, R.B., Marchuk, D.A., Gadde, K.M., Barefoot, J.C., Grichnik, K., Helms, M.J., Kuhn, C.M., Lewis, J.G., Schanberg, S.M., Stafford-Smith, M., Suarez, E.C., Clary, G.L., Svenson, I.K., and Siegler, I.C. (2003). Serotonin-related gene polymorphisms and central nervous system serotonin function. *Neuropsychopharmacology* **28**(3), 533–541.

Wu, H.F., Chen, K., and Shih, J.C. (1993). Site-directed mutagenesis of monoamine oxidase A and B: role of cysteines. *Mol. Pharmacol.* **43**, 888–893.

Wylie, C.J., Hendricks, T.J., Zhang, B., Wang, L., Lu, P., Leahy, P., Fox, S., Maeno, H., and Deneris, E.S. (2010). Distinct transcriptomes define rostral and caudal serotonin neurons. *J. Neurosci.* **30**(2), 670–684.

Youdim, M.B., Collins, G.G., and Sandler, M. (1969). Multiple forms of rat brain monoamine oxidase. *Nature* **223**, 626–628.

Yu, Y.W., Tsai, S.J., Hong, C.J., Chen, T.J., Chen, M.C., and Yang, C.W. (2005). Association study of a monoamine oxidase a gene promoter polymorphism with major depressive disorder and antidepressant response. *Neuropsychopharmacology* **30**(9), 1719–1723.

Zucchi, R., Chiellini, G., Scanlan, T.S., and Grandy, D.K. (2006). Trace amine-associated receptors and their ligands. *Br. J. Pharmacol.* **149**, 967–978.

KINETIC BEHAVIOR AND REVERSIBLE INHIBITION OF MONOAMINE OXIDASES—ENZYMES THAT MANY WANT DEAD

Keith F. Tipton, Gavin P. Davey and Andrew G. McDonald

Department of Biochemistry, Trinity College, Dublin, Ireland

Abstract

Monoamine oxidase (MAO) inhibitors have proven to be valuable tools in pharmacology and therapeutics. This account concerns the behavior of the different types of reversible inhibitor and how an understanding of the kinetic mechanisms of MAO may help in their design.

I. Introduction

For many research workers, interest in the kinetic behavior of monoamine oxidase (MAO; EC 1.4.3.4), if they have any at all, is restricted to the desire to stop the enzyme working. This is reasonable, since the discovery that inhibitors MAO were antidepressants (see Tipton, 1989b for review) resulted in the synthesis of

INTERNATIONAL REVIEW OF NEUROBIOLOGY, VOL. 100
DOI: 10.1016/B978-0-12-386467-3.00003-0

large numbers of compounds for potential clinical use. This was given added impetus when it was recognized that there were two forms of the enzyme with different substrate and inhibitor specificities (see Youdim *et al.*, 2006 for review). Inhibitors of MAO-A have been shown to be effective antidepressants (see Tipton, 1989b), whereas inhibitors of MAO-B appear to be of value in the treatment of Parkinson's disease (see Youdim *et al.*, 2006).

The two isoenzymes of MAO, which are present in most mammalian tissues, were originally distinguished by their sensitivities to inhibition by the acetylenic inhibitors clorgyline and deprenyl and by their substrate specificities. MAO-A is inhibited by low concentrations of clorgyline and catalyzes the oxidative deamination of 5-hydroxytryptamine (5-HT), whereas MAO-B is inhibited by low concentrations of L-deprenyl and is active toward 2-phenylethylamine (PEA) and the xenobiotic benzylamine. These substrate specificities are not absolute. For example, 5-HT is a substrate for MAO-B as well as MAO-A in rat brain and liver. However, in the rat brain the K_m and V_{max} values for MAO-B are sixfold higher and ninefold lower, respectively, than those for MAO-A. Hence, relatively low concentrations of 5-HT will be oxidized essentially by MAO-A alone, whereas at very high concentrations the B form can contribute about 10% of the total activity. Similarly, PEA is a substrate for MAO-A but with a considerably higher K_m and a V_{max} value that is five- to sixfold lower than the corresponding values for MAO-B. Thus, although the A form can contribute to a percentage of the total activity at high concentrations of PEA, only the activity of the B form is important at low concentrations of this substrate. Similarly, in some species, it is possible to detect low levels of benzylamine oxidation by MAO-A (Miller and Edmondson, 1999; Parkinson *et al.*, 1980). These factors are important in interpreting the results of attempts to determine the relative proportions of the two enzymes in tissues from their activities toward such substrates.

The selectivities of the MAOs for a range of substrates are summarized in Table I. These refer to the enzymes from human and rat brain. However, species differences may exist between the specificities of the enzymes, so it cannot be assumed that the behavior will necessarily be identical in all tissues and species. Significant differences between the behavior of human MAO-B and that enzyme from rat in its deamination of some substrates (Garrick and Murphy, 1982) and their inhibition by tricyclic antidepressants (Reid *et al.*, 1988) have been reported. Human and rat MAO-A have also been shown to differ in their sensitivities to inhibition by (*S*)-4-alkylthioamphetamine derivatives (Fierro *et al.*, 2007). There are also marked species differences in the selectivities of MAO-A and -B for a series of derivatives of the anticonvulsant drug milacemide (O'Brien *et al.*, 1995) and the oxidation of the protoxin MPTP and some of its analogues (Sullivan and Tipton, 1992; Krueger *et al.*, 1995; Inoue *et al.*, 1999).

Table I
SUBSTRATE SPECIFICITIES OF THE HUMAN BRAIN MONOAMINE OXIDASES.

Substrate	MAO-A		MAO-B	
Human (Cortex)	K_m (μM)	V_{max} (pmol $min^{-1}mg$ $protein^{-1}$)	K_m (μM)	V_{max} (pmol min^{-1} mg $protein^{-1}$)
Adrenaline	125 ± 42	379 ± 54	266 ± 9	465 ± 61
Dopamine	212 ± 33	680 ± 123	229 ± 33	702 ± 158
5-Hydroxytryptamine	137 ± 24	228 ± 31	1093 ± 20	6.6 ± 1.3
Noradrenaline	284 ± 17	561 ± 42	238 ± 30	321 ± 13
2-Phenylethylamine	140 ± 22	20 ± 8	4 ± 3	309 ± 24
Tryptamine	35 ± 6	58 ± 5	35 ± 8	108 ± 2
Tyramine	127 ± 18	182 ± 28	107 ± 21	343 ± 48
Rat				
Adrenaline	610 ± 59	558 ± 93	840 ± 76	531 ± 66
Dopamine	150 ± 9	889 ± 74	290 ± 14	396 ± 18
5-Hydroxytryptamine	340 ± 28	997 ± 59	1533 ± 39	125 ± 32
Noradrenaline	582 ± 168	1253 ± 214	1954 ± 46	383 ± 42
2-Phenylethylamine	327 ± 14	53 ± 8	5 ± 2	89 ± 12
Tyramine	116 ± 13	258 ± 12	328 ± 16	687 ± 16

Values determined at 37 °C and pH 7.2 in air-saturated 0.1 M phosphate buffer.

Knowledge of the kinetic behavior of the enzyme can be a useful aid to the design and development of effective inhibitors. The kinetics of enzyme inhibition have been covered in exhausting detail elsewhere (Tipton, 1996, 2001; McDonald and Tipton, 2002) and this account will consider the relevance of the kinetics of MAO to the behavior of reversible inhibitors, since detailed considerations of the kinetic analysis of irreversible inhibitors can be found elsewhere (Tipton, 1989a, 2001; Tipton *et al.*, 1993).

II. Some Basics of Enzyme Inhibition

Enzyme inhibitors may act by combining with the enzyme either reversibly or irreversibly. The different types of inhibitor are summarized in Table II. It is important to distinguish between these types for any interpretation of their behavior, either *in vivo* or *in vitro*. This can most simply be done by determining whether dilution or dialysis results in reversal of the inhibition.

Table II
BASIC TYPES OF ENZYME INHIBITOR.

Inhibitor type	Mechanism	Examples
Reversible	$E + I \underset{k_{-1}}{\overset{k_{+1}}{\rightleftharpoons}} EI$	Substrate/Product analogues. Also tight-binding inhibitors
Irreversible Nonspecific	$E + I \overset{k}{\rightarrow} E - I$	Group-specific reagents
Irreversible Specific	$E + I \underset{k_{-1}}{\overset{k_{+1}}{\rightleftharpoons}} EI \overset{k}{\rightarrow} E - I$	Active-site-directed, "suicide" (k_{cat}, mechanism-based) inhibitors

A. REVERSIBLE INHIBITORS

For reversible inhibition, the full inhibition is usually obtained extremely rapidly since there is no chemical reaction involved; simply a noncovalent interaction, where the dissociation constant for the reaction (k_{-1}/k_{+1}), is defined as the inhibitor constant, K_i.

Inhibition is reversed by dialysis of the enzyme–inhibitor mixture or by dilution to lower the concentration of the inhibitor. *In vivo*, the rate of recovery from the effects of a reversible inhibitor will be governed by the rate that it is removed from the tissues by metabolism and elimination.

There are several basic types of reversible inhibitor. For a simple single-substrate reaction the possible modes of inhibitor binding are shown in Scheme 1.

They are normally distinguished by their effects on the Michaelis–Menten relationship:

$$v = \frac{V_{max}\boldsymbol{s}}{K_m + \boldsymbol{s}} \tag{1}$$

or in its double-reciprocal form:

$$\frac{1}{v} = \frac{K_m}{V_{max}\boldsymbol{s}} + \frac{1}{V_{max}} \tag{2}$$

where V_{max} is the maximum or limiting velocity, $\boldsymbol{s}$ is the substrate concentration, and K_m is the Michaelis constant, corresponding to the substrate concentration that gives half maximum velocity. The double-reciprocal form is commonly used because it yields linear plots of $1/v$ against $1/\boldsymbol{s}$, often known as the Lineweaver–Burk plots. It is useful for displaying the differences between the kinetic behavior of the different inhibitor types, but it is an extremely inaccurate way of determining kinetic parameters, for which a direct, nonlinear regression fit to the Michaelis–Menten equation is to be preferred.

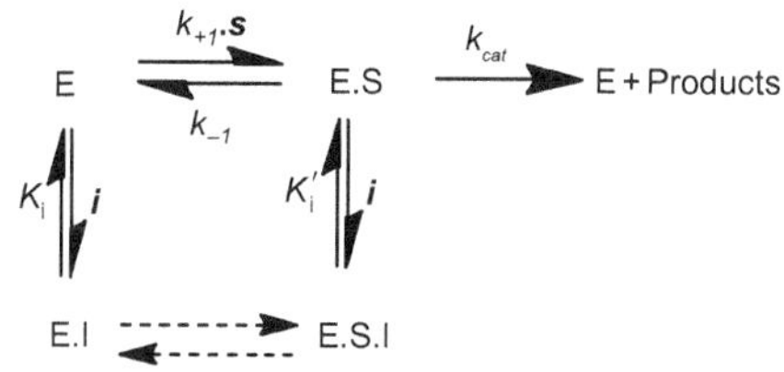

SCHEME 1. Simple modes of reversible inhibition.

1. *Competitive inhibitors*, represented by the inhibitor constant K_i in Scheme 1, bind to the same site on the enzyme as the substrate; substrate analogues and products are often competitive inhibitors The enzyme can either bind substrate or inhibitor but not both. At very high substrate concentrations, all inhibitors will be displaced from the enzyme; therefore, V_{max} is unchanged but in the presence of inhibitors more substrate will be needed to get to V_{max}. Thus, you need more substrate to get to $V_{max}/2$. So for competitive inhibitors, V_{max} is unchanged but K_m is increased. Mathematically:

$$v = \frac{V_{max}\boldsymbol{s}}{\boldsymbol{s} + K_m(1 + \boldsymbol{i}/K_i)} \tag{3}$$

where $\boldsymbol{i}$ is the inhibitor concentration.

Thus, at any fixed inhibitor concentration, the degree of inhibition will decrease as the substrate concentration is increased, tending to zero as $\boldsymbol{s}$ becomes very large, as illustrated in Fig. 1.

2. *Uncompetitive inhibitors*, represented by the inhibitor constant K_i' in Scheme 1. The inhibitor does not bind to the free enzyme but only to an enzyme–substrate complex. Both the K_m and the maximum velocity values are decreased by the same amount, as shown in Eq. (4). Thus, inhibition will increase as the concentration of the ES complex increases.

$$v = \frac{V_{max}\boldsymbol{s}/(1 + \boldsymbol{i}/K_i')}{\boldsymbol{s} + K_m/(1 + \boldsymbol{i}/K_i')} = \frac{V_{max}\boldsymbol{s}}{\boldsymbol{s}\left(1 + \boldsymbol{i}/K_i'\right) + K_m} \tag{4}$$

In double-reciprocal form uncompetitive and competitive inhibition are easily distinguished, as shown in Fig. 2.

As $\boldsymbol{s}$ tends to ∞, $\boldsymbol{i}_{50} = K_i'$. Inhibition will increase as the concentration of substrate is increased. As shown in Fig. 1, this is the exact opposite of the competitive case.

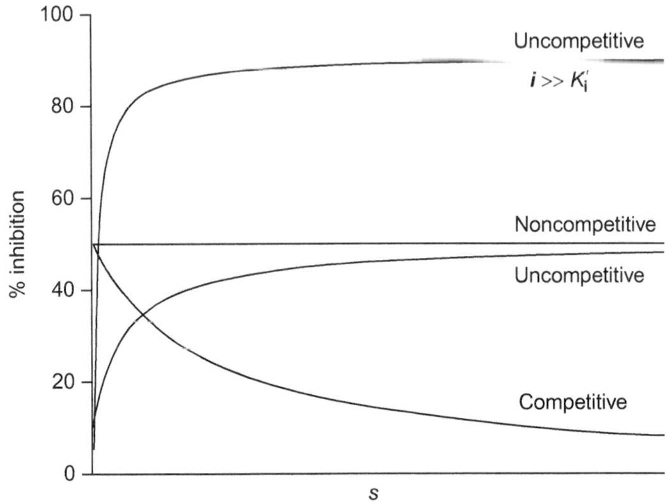

FIG. 1. The effects of substrate concentration of the inhibition observed with the different types of reversible inhibitor. Except where indicated, the inhibitor concentration was equal to its K_i value.

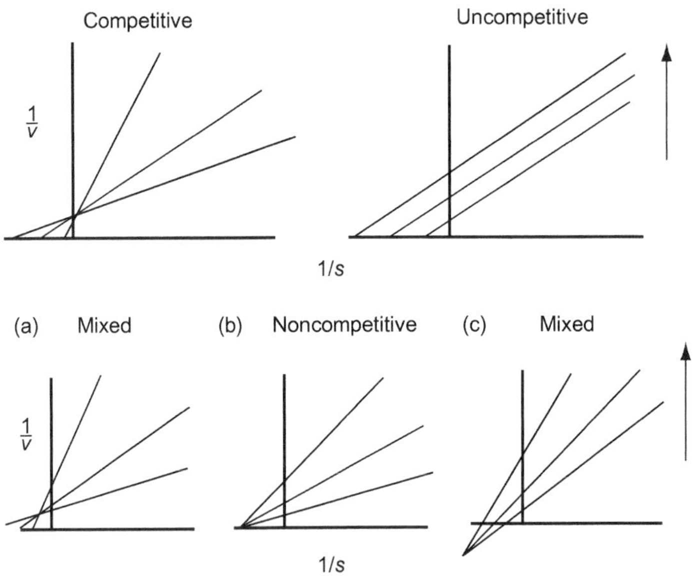

FIG. 2. Double-reciprocal plot behavior of reversible inhibitors. The arrows indicate increasing inhibitor concentrations. In the mixed and noncompetitive cases, the point where these lines intersect will depend on the relative values of K_i and K_i'. If $K_i' > K_i$ case (a) will apply; If $K_i' < K_i$ case (c) will be seen. Case (b) represents noncompetitive inhibition, where $K_i' = K_i$.

3. *Mixed or noncompetitive inhibitors* Mixed inhibitors bind both to free enzyme and to the ES as represented by all the inhibitor constants K_i plus K_i' in Scheme 1.

$$v = \frac{V_{max}\boldsymbol{s}}{\boldsymbol{s}(1 + \boldsymbol{i}K_i') + K_m(1 + \boldsymbol{i}/K_i)} \tag{5}$$

The kinetic behavior looks like a mixture of competitive and uncompetitive inhibition. Thus, the double-reciprocal plot can look like those shown in Fig. 2.

If $K_i = K_i'$ Eq. (5) simplifies to

$$v = \frac{V_{max}\boldsymbol{s}/(1 + \boldsymbol{i}K_i)}{\boldsymbol{s} + K_m} \tag{6}$$

where V_{max} is decreased in the presence of the inhibitor but K_m is unaffected. This is often known as noncompetitive inhibition or true noncompetitive inhibition. In this case, the degree of inhibition at any fixed inhibitor concentration will be unaffected by changes in the substrate concentration (see Fig. 1).

The mixed and noncompetitive inhibition systems will give more complex relationships unless the step represented by the broken arrow in Scheme 1 does not occur or all binding steps occur under conditions of thermodynamic equilibrium. When steady-state conditions may obtain, the rate equation can be represented as:

$$v = \frac{(C_1\boldsymbol{s} + C_2\boldsymbol{s}^2 + C_3\boldsymbol{si})\boldsymbol{e}}{C_4 + C_5\boldsymbol{s} + C_6\boldsymbol{i} + C_7\boldsymbol{i}^2 + C_8\boldsymbol{si} + C_9\boldsymbol{s}^2 + C_{10}\boldsymbol{si}^2 + C_{11}\boldsymbol{s}^2\boldsymbol{i}} \tag{7}$$

where $\boldsymbol{e}$ is the enzyme concentration and C_1–C_{11} are combinations of rate constants. For a fixed inhibitor concentration, this may be written as:

$$v = \frac{a\boldsymbol{s}^2 + b\boldsymbol{s}}{x\boldsymbol{s}^2 + y\boldsymbol{s} + z} \tag{8}$$

Such systems would not follow the Michaelis–Menten equation and, therefore, would not be expected to give linear double-reciprocal plots in the presence of inhibitor. The type of curve obtained will depend on the values of the individual rate constants and a variety of curve shapes are possible (see Tipton, 1996). In some cases, deviations from simple behavior may not be easy to see. For example, at very high inhibitor concentrations variation of the substrate concentrations at low levels may make the terms in $\boldsymbol{s}^2$ less important and may give apparent Michaelis–Menten behavior (Frieden, 1964; Pettersson, 1972). Any system in which there is more than one route by which substrate can bind to form a productive complex

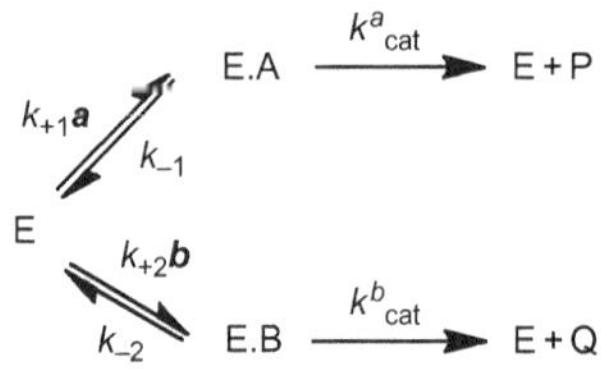

(9)

In this case, each substrate will act as a competitive inhibitor of the other. For example, if the initial rates of the formation of P (v_a) are measured at a series of different concentrations of A in the presence of different concentrations of B, the kinetic equation is of the same form as for simple competitive inhibition (Eq. (3)):

$$v_a = \frac{k^a_{\text{cat}}\boldsymbol{ea}}{1 + K^a_{\text{m}}\left(1 + \frac{\boldsymbol{b}}{K^b_{\text{m}}}\right)} = \frac{k^a_{\text{cat}}\boldsymbol{ea}/K^a_{\text{m}}}{1 + \frac{\boldsymbol{a}}{K^a_{\text{m}}} + \frac{\boldsymbol{b}}{K^b_{\text{m}}}} \tag{10}$$

where $\boldsymbol{a}$ and $\boldsymbol{b}$ are the concentrations of A and B, respectively, $\boldsymbol{e}$ is enzyme concentration, and ${K_{\text{m}}}^a$ and ${K_{\text{m}}}^b$ are the Michaelis constants for A and B ($\{k_{-1} + {k_{\text{cat}}}^a\}/k_{+1}$ and $(\{k_{-2} + {k_{\text{cat}}}^b\}/k_{+2}$, respectively). Similarly, the initial rates of Q formation (v_b) in the presence of A will be:

$$v_b = \frac{k^b_{\text{cat}}\boldsymbol{eb}}{1 + K^b_{\text{m}}\left(1 + \frac{\boldsymbol{a}}{K^a_{\text{m}}}\right)} = \frac{k^b_{\text{cat}}\boldsymbol{eb}/K^b_{\text{m}}}{1 + \frac{\boldsymbol{a}}{K^a_{\text{m}}} + \frac{\boldsymbol{b}}{K^b_{\text{m}}}} \tag{11}$$

From Eqs. (10) and (11), the rates at which the enzyme will use each substrate in a mixture of the two will be given by the relative substrate concentrations and the ratio of their $k_{\text{cat}}/K_{\text{m}}$ (or, since $V = k_{\text{cat}}\boldsymbol{e}$, V/K_{m}) values:

$$\frac{v_a}{v_b} = \frac{\left(k^a_{\text{cat}}/K^a_{\text{m}}\right)\boldsymbol{a}}{\left(k^b_{\text{cat}}/K^b_{\text{m}}\right)\boldsymbol{b}} = \frac{\left(V_a/K^a_{\text{m}}\right)\boldsymbol{a}}{\left(V_b/K^b_{\text{m}}\right)\boldsymbol{b}} \tag{12}$$

The relationships also provide a way of determining whether the same enzyme is involved in the metabolism of two different substrates, which has been applied to MAO (Houslay and Tipton, 1975a; Houslay *et al.*, 1974). If each substrate is assayed at its K_{m} concentration ($\boldsymbol{a} = {K_{\text{m}}}^a$ and $\boldsymbol{b} = {K_{\text{m}}}^b$), $\boldsymbol{a}/{K_{\text{m}}}^a$ and $\boldsymbol{b}/{K_{\text{m}}}^b$ will be unity. The rates for each substrate assayed separately will each be one half their limiting velocities, and the sum of the two rates would be

$$v_a + v_b = \frac{V_a}{2} + \frac{V_b}{2} = \frac{V_a + V_b}{2} \tag{13}$$

where V_a and V_b are the maximum velocities for A and B measured separately. If rate of formation of both products ($v_{(a+b)}$) from mixture of the two substrates, each at its K_{m} concentration, is determined, the above relationships will be true if they were metabolized by separate enzymes. However, if the same enzyme was responsible for the metabolism of both the substrates, substitution into Eqs. (10) and (11) show that the rate would be

$$v_{(a+b)} = \frac{V_a}{3} + \frac{V_b}{3} = \frac{V_a + V_b}{3} \quad (14)$$

and so the rate with the mixture will be two-thirds of the sum of the separate rates under the same conditions.

B. Enzyme Competition

Since the substrate specificities of MAO-A and MAO-B overlap, it is possible to have a situation where the two isoenzymes will be competing for the same substrate. If the same reaction is catalyzed by two enzymes that have different K_m values, as indicated below

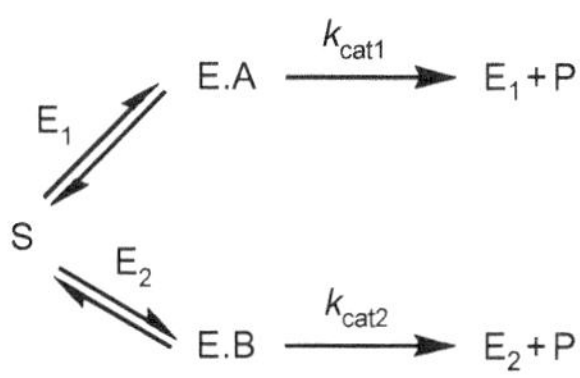

(15)

the rate of the overall reaction (v_{total}) will be given by the sum of the velocities of the reactions catalyzed by E_1 and E_2 (v_1 and v_2, respectively)

$$v_{total} = v_1 + v_2 = \frac{V_1 \boldsymbol{s}}{\boldsymbol{s} + K_{m1}} + \frac{V_2 \boldsymbol{s}}{\boldsymbol{s} + K_{m2}} \quad (16)$$

where K_{m1} and K_{m2} are the K_m values for E_1 and E_2, respectively, $V_1 = k_{cat}\boldsymbol{e}_1$ and $V_2 = k_{cat}\boldsymbol{e}_2$. This equation can be rearranged to

$$v_{total} = \frac{(V_1 + V_2)\boldsymbol{s} + V_1 K_{m2} + V_2 K_{m1}}{\boldsymbol{s} + K_{m1} + K_{m2} + K_{m1}K_{m2}/\boldsymbol{s}} \quad (17)$$

This equation represents the sum of two Michaelis–Menten rectangular hyperbolas. In double-reciprocal form, it will be written as

$$\frac{1}{v_{total}} = \frac{1 + \frac{K_{m1}+K_{m2}}{\boldsymbol{s}} + \frac{K_{m1}+K_{m2}}{\boldsymbol{s}^2}}{V_1 + V_2 + \frac{V_1 K_{m2} + V_2 K_{m1}}{\boldsymbol{s}}} \quad (18)$$

which predicts a smooth, downwardly curving (hyperbolic) double-reciprocal plot. The degree of curvature will depend on the relative values of the two Michaelis constants. If $K_{m1} = K_{m2}$, the dependence of $1/v_{total}$ upon $1/\boldsymbol{s}$ will give a straight line that intersects the $1/v_{total}$ axis at a value corresponding to $1/(V_1 + V_2)$ and

$$v = \frac{V_{\max} \boldsymbol{ax}}{\boldsymbol{ax} + K_{\mathrm{m}}^{Ax}} \tag{22}$$

and the K_{m} for the amine substrate will be independent of the oxygen concentration unless the latter is reduced to very low levels.

The release of product(s), leaving the free enzyme in the reduced form, before binding of oxygen would be consistent with the double-displacement, or ping-pong, mechanism, as shown in Fig. 4. In this case, Eq. (20) would reduce to

$$V = \frac{V_{\max}}{1 + \frac{K_{\mathrm{m}}^{Ax}}{\boldsymbol{ax}} + \frac{K_{\mathrm{m}}^{B}}{\boldsymbol{b}}} \tag{23}$$

and any inhibitor that bound solely to the free enzyme would be expected to be competitive with the amine substrate and any compound that bound solely to the free reduced form of the enzyme would be expected to be an uncompetitive inhibitor. A compound that bound to both these forms would be a mixed inhibitor. MAO preparations from several sources have been reported to follow such a mechanism (Fischer *et al.*, 1968; Oi *et al.*, 1971; Houslay and Tipton, 1973, 1975b) . Transient and steady-state kinetic studies are also consistent the double-displacement mechanism being followed by beef liver MAO-B with PEA as substrate (Husain *et al.*, 1982; Pearce and Roth, 1985). It appears that MAO-A may also follow a basic ping-pong mechanism but that substrate binding to the reduced enzyme enhances the reoxidation step (Ramsay 1991; Tan and Ramsay 1993). This would clearly have consequences for the design of MAO-A inhibitors, since an analogue of the substrate could behave similarly, binding both to the free

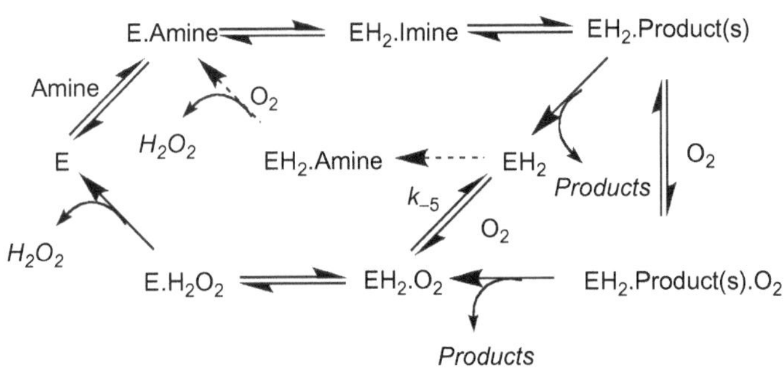

FIG. 4. Some alternative pathways for amine oxidation by MAO. The top half of the scheme is common for all mechanism. The pathway through the free reduced enzyme (EH_2) represents the double-displacement, or ping-pong mechanism. The alternative path through an E.Product(s).Oxygen ternary complex is also shown. It is possible that hydrogen peroxide may be released before the imine-derived products in this mechanism. The broken arrows indicate the substrate activation pathway for MAO-A. Other variations of these basic mechanisms are discussed in the references cited in the text.

enzyme and the reduced enzyme as a mixed inhibitor, but also inhibiting the substrate-dependent rate enhancement.

In contrast, it appears that the mechanism followed by beef liver MAO-B with benzylamine as the substrate follows a mechanism that involves the binding of oxygen before the release of the product(s) (Husain *et al.* 1982; Pearce and Roth, 1985) resulting in a kinetically significant ternary (E.imine.Oxygen) ternary complex (see Fig. 4). Such a mechanism would follow Eq. (20).

Thus, it appears that there may be competing alternative pathways for substrate oxidation and that the reaction mechanism followed may depend upon the MAO isoenzyme, the assay conditions, the substrate used and, possibly, the enzyme source or preparation. Since the kinetic mechanism followed will be determined by the values of individual rate constants, the possibility exists that the different reaction pathways might be considered as competing processes. In this case, the question might not be "which mechanism is followed by the enzyme under stated conditions?" but "what fraction of the total oxidation occurs through each pathway?" That could lead to intriguing kinetic complexities.

As discussed above, these complexities have implications for inhibitor design. A further complexity is the presence of a high-affinity binding site for imidazoline I_2 ligands on a proportion of MAO-B molecules from some sources (Bonivento *et al.*, 2010; McDonald *et al.*, 2010). These appear to mediate allosteric inhibition of the enzyme, but the significance of this is, as yet, unclear.

V. Tight-Binding Inhibitors

Some inhibitors have such high affinities for the enzyme that the concentrations required for inhibition are comparable to those of the enzyme. Harmaline (Nelson *et al.*, 1979), Ro 41-1049, and lazabemide (see Saura *et al.*, 1992, 1996) have been reported to behave as tight-binding inhibitors of MAO, In such cases, the binding of the inhibitor to the enzyme will significantly reduce the free inhibitor concentration and so the assumption that the total inhibitor concentration is equal to the free inhibitor concentration, which has been implicit in all the above treatments, is no longer valid, and it is necessary to derive relationships in terms of the concentrations of free enzyme ($\boldsymbol{e}_{\mathrm{f}}$) and free inhibitor ($\boldsymbol{i}_{\mathrm{f}}$) and the enzyme–inhibitor complex concentration ($\boldsymbol{ei}$), where

$$\boldsymbol{i}_{\mathbf{t}} = \boldsymbol{i}_{\mathbf{f}} + \boldsymbol{ei} \tag{24}$$

Inhibition will not follow a simple hyperbolic curve, but graphical and computational procedures have been developed for the analysis of such inhibitors (Dixon, 1972; Henderson, 1973; Williams and Morrison, 1979; Baici, 1987; Szedlacsek and Duggleby, 1995).

The low concentrations of inhibitor and enzyme used may result in the rate of inhibition being relatively slow. Further, tight binding inhibitors will, by definition, dissociate from the enzyme rather slowly following removal of the excess inhibitor. Thus, it may be possible to determine the rate constants for binding and dissociation directly from these time courses.

VI. Recovery from Inhibition

A factor that may be important in the responses to MAO inhibitors is that the rate of recovery from the effects of an irreversible MAO inhibitor can be quite slow, since it will depend on the rate of synthesis of new enzyme. Measurements of the rates of recovery of MAO activity after administration of a single dose of a selective inhibitor have shown the process to follow first-order kinetics and given similar values for the recovery half-times of MAO-A and -B, of about 2.5–3.5 and 9–13 days for rat liver (Erwin and Deitrich, 1971; Della Corte and Tipton, 1980; Youdim and Tipton, 2002) and brain (Neff and Goridis, 1972; Felner and Waldmeier, 1979; Youdim and Tipton, 2002), respectively. The turnover rates of MAO-A and -B have been reported to be different in the rat intestine, with values of 2.2 and 7.7 days, respectively, being determined for the half-lives (Callingham *et al.*, 1985). A very much slower rate of recovery (half-time about 40 days) has been reported for the MAO-B activity in human brain after irreversible inhibition by L-deprenyl (Fowler *et al.*, 1994) and a value consistent with the 30-day half-life of the enzyme in baboon brain had been reported (Arnett *et al.*, 1987). Such slow rates of recovery from the effects of irreversible inhibitors might cast doubt on the interpretation of wash-out experiments used to test inhibitor efficacy. However, it has been reported that it is necessary to inhibit 80% of MAO before alterations in neurotransmitter levels and behavioral effects are seen (Green *et al.*, 1977; Youdim and Tipton, 2002; Riederer and Lachenmayer, 2003). In that case, the time for recovery of 20% of the activity after complete inhibition would be relevant. This would be about 10–13 days for baboon and human brain MAO-B and 1–4 days for rat liver and brain MAO, respectively. It should also be borne in mind that the neurorotective effects of irreversible MAO-B inhibitors, such as L-deprenyl and rasagyline, may be independent of their inhibitory actions on the enzyme (see Youdim *et al.*, 2006).

The rate of recovery from the effects of a reversible inhibitor will be dependent on the rate at which it is eliminated from the tissues, since removal of the free drug will result in it dissociating from the enzyme. Thus, the effects of reversible inhibitors would be expected to be shorter in duration. For example, the rates of recovery of MAO activity in the rat after a single doses of the reversible inhibitor brofaromine correspond to half-lives of about 12 and 8 h in liver and brain, respectively (Waldmeier *et al.*, 1983).

VII. Conclusions

Our knowledge of the behavior of reversible enzyme inhibitors and the structures of MAO-A and -B (see De Colibus *et al.*, 2005) should allow the rational design of effective reversible inhibitors for drug use. Standard modeling and docking procedures should make this a relatively straightforward task. However, the complexities of the kinetic mechanisms followed by the enzymes mean that the trial-and-error approach still has an important role in establishing that designed inhibitors behave as predicted.

References

Anderson, M.C., Hasan, F., McCrodden, J.M., and Tipton, K.F. (1993). Monoamine oxidase inhibitors and the cheese effect. *Neurochem. Res.* **18**, 1145–1149.

Arnett, C.D., Fowler, J.S., MacGreggor, R.R., Schyler, D.J., Wolf, A.P., Langström, B., and Haldin, C.J. (1987). Turnover of brain monoamine oxidase measured *in vivo* by positron emission tomography using L-[^{14}C] deprenyl. *J. Neurochem.* **49**, 522–527.

Baici, A. (1987). Graphical and statistical analysis of hyperbolic tight-binding inhibition. *Biochem. J.* **244**, 793–796.

Binda, C., Wang, J., Li, M., Hubalek, F., Mattevi, A., and Edmondson, D.E. (2008). Structural and mechanistic studies of arylalkylhydrazine inhibition of human monoamine oxidases A and B. *Biochemistry* **47**, 5616–5625.

Blackwell, B. (1963). Hypertensive crisis due to monoamine oxidase inhibitors. *Lancet* **II**, 849–851.

Bonivento, D., Milczek, E.M., McDonald, G.R., Binda, C., Holt, A., Edmondson, D.E., and Mattevi, A. (2010). Potentiation of ligand binding through cooperative effects in monoamine oxidase B. *J. Biol. Chem.* **285**, 36849–36856.

Boocock, M.R., and Coggins, J.R. (1983). Kinetics of inhibition of 5-enolpyruvylshikimate- 3-phosphate synthase by glyphosate. *FEBS Lett.* **154**, 127–133.

Callingham, B.A., Mazel, P., and Porter, J.C. (1985). Some properties of amine oxidase activities in the rat intestine. *Br. J. Pharmacol.* **86**, 553.

Da Prada, M., Zürcher, G., Würthrich, I., and Haefely, W.E. (1988). On tyramine, food beverages and the reversible MAO inhibitor moclobemide. *J. Neural Transm.* **26**(Suppl.), 33–56.

Davis, D.S., Tasuhara, H., Boobis, A.R., and George, C.F. (1984). The effects of reversible and irreversible inhibitors of monoamine oxidase on tyramine deamination in dog intestine. In: Tipton, K.F., Dostert, P., and Strolin Benedetti, M. (Eds.), *Monoamine Oxidase and Disease.* Academic Press, London, pp. 443–448.

De Colibus, L., Li, M., Binda, C., Lustig, A., Edmondson, D.E., and Mattevi, A. (2005). Three-dimensional structure of human monoamine oxidase A (MAO A): relation to the structures of rat MAO A and human MAO B. *Proc. Natl. Acad. Sci. USA* **102**, 12684–12689.

Della Corte, L., and Tipton, K.F. (1980). The turnover of the A- and B-forms of monoamine oxidase in rat liver. *Biochem. Pharmacol.* **29**, 811–815.

Dixon, M. (1972). The graphical determination of K_m and K_i. *Biochem. J.* **129**, 197–202.

Edmondson, D.E., Bhattacharyya, A.K., and Walker, M.C. (1993). Spectral and kinetic studies of imine product formation in the oxidation of *p*-(*N,N*-dimethylamino)- benzylamine analogues by monoamine oxidase B. *Biochemistry* **32**, 5196–5202.

Erwin, V.G., and Deitrich, R.A. (1971). The labelling *in vivo* of monoamine oxidase by [^{14}C]-pargyline: a tool for studying the synthesis of the enzyme. *Mol. Pharmacol.* **7**, 219–228.

Felner, A.E., and Waldmeier, P.C. (1979). Cumulative effects of irreversible MAO inhibitors *in vivo*. *Biochem. Pharmacol.* **28**, 995–1002.

Fierro, A., Osorio-Olivares, M., Cassels, B.K., Edmondson, D.E., Sepúlveda-Boza, S., and Reyes-Parada, M. (2007). Human and rat monoamine oxidase-A are differentially inhibited by (*S*)-4-alkylthioamphetamine derivatives: insights from molecular modeling studies. *Bioorg. Med. Chem.* **15**, 5198–5206.

Fischer, A.G., Schulz, A.R., and Oliner, L. (1968). Thyroidal biosynthesis of iodothyronines. II. General characteristics and purification of mitochondrial monoamine oxidase.. *Biochim. Biophys. Acta* **159**, 460–471.

Fowler, J.S., Volkow, N.D., Logan, J., Wang, G.J., MacGregor, R.R., Schlyer, D., Wolf, A.P., Pappas, N., Alexoff, D., and Shea, C. (1994). Slow recovery of human brain MAO B after L-deprenyl (Selegiline) withdrawal. *Synapse* **18**, 86–93.

Frieden, C. (1964). Treatment of Enzyme Kinetic Data. I. The effect of modifiers on the kinetic parameters of single substrate enzymes. *J. Biol. Chem.* **239**, 3522–3531.

Garel, J.R., and Labouesse, B. (1970). Binding interactions between two ligands and a monomeric protein. Study on indole, protons and chymotrypsin. *J. Mol. Biol.* **47**, 41–56.

Garrick, N.A., and Murphy, D.L. (1982). Species differences in the deamination of dopamine and other substrates for monoamine oxidase in brain. *Psychopharmacology* **72**, 27–33.

Green, A.R., Mitchell, D., Todoff, A., and Youdim, M.B.H. (1977). Evidence for dopamine deamination by by both type A and type B monoamine oxidase in rat brain *in vivo* and for the degree of inhibition necessary for increased functional activity of dopamine and 5-hydroxytryptamine. *Br. J. Pharmacol.* **60**, 343–3429.

Hasan, F., McCrodden, J.M., Kennedy, N.P., and Tipton, K.F. (1988). The involvement of intestinal monoamine oxidase in the transport and metabolism of tyramine. *J. Neural Transm.* **26**(Suppl.), 1–9.

Hearon, J.Z., Bernhard, S.A., Friess, S.L., Botts, D.J., and Morales, M.F. (1959). Enzyme Kinetics. In: Boyer, P.D., Lardy, H.A., and Myrback, K. (Eds.), *The Enzymes*. 2nd edn., Vol. 1. Academic Press, New York, pp. 49–142.

Henderson, P.J.F. (1973). Steady-state enzyme kinetics with high-affinity substrates or inhibitors. A statistical treatment of dose-response curves.. *Biochem. J.* **135**, 101–107.

Houslay, M.D., and Tipton, K.F. (1973). The reaction pathway of membrane-bound rat liver mitochondrial monoamine oxidase. *Biochem. J.* **135**, 735–750.

Houslay, M.D., and Tipton, K.F. (1974). A kinetic evaluation of monoamine oxidase activity in rat liver mitochondrial outer membranes. *Biochem. J.* **139**, 645–652.

Houslay, M.D., and Tipton, K.F. (1975a). Amine competition for oxidation by rat liver mitochondrial monoamine oxidase. *Biochem. Pharmacol.* **24**, 627–631.

Houslay, M.D., and Tipton, K.F. (1975b). Rat liver mitochondrial monoamine oxidase. A change in the reaction mechanism on solubilization.. *Biochem. J.* **145**, 311–321.

Houslay, M.D., Garrett, N.J., and Tipton, K.F. (1974). Mixed substrate experiments with human brain monoamine oxidase. *Biochem. Pharmacol.* **23**, 1937–1944.

Husain, M., Edmondson, D.E., and Singer, T.P. (1982). Kinetic studies on the catalytic mechanism of liver monoamine oxidase. *Biochemistry* **21**, 595–600.

Inoue, H., Castagnoli, K., Van Der Schyf, C., Mabic, S., Igarashi, K., and Castagnoli, N. (1999). Species-dependent differences in monoamine oxidase A and B-catalyzed oxidation of various C4 substituted 1-methyl-4-phenyl-1,2,3, 6-tetrahydropyridinyl derivatives. *J. Pharmacol. Exp. Ther.* **291**, 856–864.

Krueger, M.J., Mazouz, F., Ramsay, R.R., Milcent, R., and Singer, T.P. (1995). Dramatic species differences in the susceptibility of monoamine oxidase B to a group of powerful inhibitors. *Biochem. Biophys. Res. Commun.* **206**, 556–562.

McDonald, A.G., and Tipton, K.F. (2002). Kinetics of catalyzed reactions—biological. In: Horváth, I.T. (Ed.), *Encyclopedia of Catalysis.* John Wiley & Sons, New York. 10.1002/0471227617.eoc127. http://www.mrw.interscience.wiley.com/enccat/.

McDonald, G.R., Olivieri, A.R., Ramsay, R.R., and Holt, A. (2010). On the formation and nature of the imidazoline I_2 binding site on human monoamine oxidase-B. *Pharmacol. Res.* **62**, 475–488.

Miller, J.R., and Edmondson, D.E. (1999). Structure-activity relationships in the oxidation of para-substituted benzylamine analogues by recombinant human liver monoamine oxidase A. *Biochemistry* **38**, 13670–13683.

Nair, N.P., Ahmed, S.K., and Kin, N.M. (1993). Biochemistry and pharmacology of reversible inhibitors of MAO-A agents: focus on moclobemide. *J. Psychiatry Neurosci.* **18**, 214–225.

Neff, N.H., and Goridis, C. (1972). Neuronal monoamine oxidase: specific enzyme types and their rates of formation. *Adv. Biochem. Psychopharmacol.* **5**, 307–323.

Nelson, D.L., Herbet, A., Glowinski, J., and Hamon, M.J. (1979). [^{3}H]Harmaline as a specific ligand of MAO A–II. Measurement of the turnover rates of MAO A during ontogenesis in the rat brain. *J. Neurochem.* **32**, 1829–1936.

O'Brien, E.M., Dostert, P., and Tipton, K.F. (1995). Species differences in the interactions of the anticonvulsant milacemide and some analogues with monoamine oxidase-B. *Biochem. Pharmacol.* **50**, 317–324.

Oi, S., Yasunobu, K.T., and Westley, J. (1971). The effect of pH on the kinetic parameters and mechanism of beef liver monoamine oxidase. *Arch. Biochem. Biophys.* **145**, 557–564.

Parkinson, D., Lyles, G.A., Browne, B.J., and Callingham, B.A. (1980). Some factors influencing the metabolism of benzylamine by type A and B monoamine oxidase in rat heart and liver. *J. Pharm. Pharmacol.* **32**, 844–850.

Pearce, L.B., and Roth, J.A. (1985). Human brain monoamine oxidase type B: mechanism of deamination as probed by steady-state methods. *Biochemistry* **24**, 1821–1826.

Pettersson, G. (1972). Dalziel rate behaviour in ternary-complex mechanisms for enzyme reactions involving two substrates. *Biochim. Biophys. Acta* **276**, 1–11.

Pollack, S.J., Atack, J.R., Knowles, M.R., McAllister, G., Ragan, C.I., Baker, R., Fletcher, S.R., Iverson, L.I., and Broughton, H.B. (1994). Mechanism of inositol monophosphatase, the putative target of lithium therapy. *Proc. Natl. Acad. Sci. USA* **91**, 5766–5770.

Ramsay, R.R. (1991). Kinetic mechanism of monoamine oxidase A. *Biochemistry* **30**, 4624–4629.

Reid, A.A., Hill, J.L., and Murphy, D.L. (1988). Interactions of tricyclic antidepressant drugs with human and rat monoamine oxidase type B. *Naunyn-Schmied. Arch. Pharmacol.* **388**, 678–683.

Riederer, P., and Lachenmayer, L. (2003). Selegiline's neuroprotective capacity revisited. *J. Neural Transm.* **110**, 1273–1278.

Saura, J., Kettler, R., Da Prada, M., and Richards, J.G. (1992). Quantitative enzyme radioautography with ^{3}H-Ro 41-1049 and ^{3}H-Ro 19-6327 *in vitro*: localization and abundance of MAO-A and MAO-B in rat CNS, peripheral organs, and human brain. *J. Neurosci.* **12**, 1977–1999.

Saura, J., Bleuel, Z., Ulrich, J., Mendelowitsch, A., Chen, K., Shih, J.C., Malherbe, P., Da Prada, M., and Richards, J.G. (1996). Molecular neuroanatomy of human monoamine oxidases A and B revealed by quantitative enzyme radioautography and *in situ* hybridization. *Histochemistry* **70**, 755–774.

Spears, G., Sneyd, J.G., and Loten, E.G. (1971). A method for deriving kinetic constants for two enzymes acting on the same substrate. *Biochem. J.* **125**, 1149–1151.

Strolin Benedetti, M., Boucher, T., Carlsson, A., and Fowler, C.J. (1983). Intestinal metabolism of tyramine by both forms of monoamine oxidase in the rat. *Biochem. Pharmacol.* **32**, 47–52.

Sullivan, J.P., and Tipton, K.F. (1992). Interactions of the neurotoxin MPTP and its demethylated derivative (PTP) with monoamine oxidase-B. *Neurochem. Res.* **8**, 791–796.

Szedlacsek, S.E., and Duggleby, R.G. (1995). Kinetics of slow and tight-binding inhibitors. *Methods Enzymol.* **249**, 144–181.

Tan, A.K., and Ramsay, R.R. (1993). Substrate-specific enhancement of the oxidative half-reaction of monoamine oxidase. *Biochemistry* **32**, 2137–2143.

Tipton, K.F. (1989a). Mechanism-based inhibitors. In: Sandler, M. and Smith, H.J. (Eds.), *Design of Enzyme Inhibitors as Drugs.* Oxford University Press, Oxford, pp. 70–93.

Tipton, K.F. (1989b). Monoamine oxidase inhibitors as antidepressants. In: Tipton, K.F. and Youdim, M.B.H. (Eds.), *Biochemical and Pharmacological Aspects of Depression.* Taylor & Francis, London, pp. 1–24.

Tipton, K.F. (1996). Patterns of enzyme inhibition. In: Engel, P.C. (Ed.), *EnzymologyLabFax.* Bios Scientific Publishers and Academic Press, Oxford, UK and San Diego, USA, pp. 115–174.

Tipton, K.F. (1997). Inhibitors of monoamine oxidase and the pressor response to dietary amines. *Vopr. Med. Khim.* **46**, 494–503.

Tipton, K.F. (2001). Enzymes: Irreversible Inhibition. In: Finazzi Agrò, A. (Ed.), *Encyclopedia of Life Sciences.* John Wiley & Sons, Ltd, Chichester. http://www.els.net/WileyCDA/[doi:10.1038/npg.els.0000601].

Tipton, K.F., and Fowler, C.J. (1984). The kinetics of monoamine oxidase inhibitors in relation to their clinical behaviour. In: Tipton, K.F., Dostert, P., and Strolin Benedetti, M. (Eds.), *Monoamine Oxidase and Disease.* Academic Press, London, pp. 27–40.

Tipton, K.F., and Singer, T.P. (1993). Advances in our understanding of the mechanisms of the neurotoxicity of MPTP and related compounds. *J. Neurochem.* **61**, 1191–1206.

Tipton, K.F., Balsa, D., and Unzeta, M. (1993). Newer approaches to the study of monoamine oxidase inhibition. In: Yashuhara, H., Parvez, S.H., Oguchi, K., Sandler, M., and Nagatsu, T. (Eds.), *Monoamine Oxidase: Basic and Clinical Aspects.* VSP BV, Utrecht, The Netherlands, pp. 1–13.

Waldmeier, P.C., Felner, A.E., and Tipton, K.F. (1983). The monoamine oxidase inhibiting properties of CGP 11305A. *Eur. J. Pharmacol.* **94**, 73–83.

Westley, A.M., and Westley, J. (1996). Enzyme inhibition in open systems. Superiority of uncompetitive agents. *J. Biol. Chem.* **271**, 5347–5352.

Williams, J.W., and Morrison, J.F. (1979). The kinetics of reversible tight-binding inhibition. *Methods Enzymol.* **63**, 437–467.

Yamada, M., and Yasuhara, H. (2004). Clinical pharmacology of MAO inhibitors: safety and future. *Neurotoxicology* **25**, 215–221.

Youdim, M.B.H., and Tipton, K.F. (2002). Rat striatal monoamine oxidase-B inhibition by l-deprenyl and rasagiline: its relationship to 2-phenylethylamine-induced stereotypy and Parkinson's disease. *Parkinsonism Relat. Disord.* **8**, 247–253.

Youdim, M.B.H., Edmondson, D.E., and Tipton, K.F. (2006). The therapeutic potential of monoamine oxidase inhibitors. *Nat. Rev. Neurosci.* **7**, 295–309.

Yu, P.H., and Tipton, K.F. (1989). Deuterium isotope effect of phenelzine on the inhibition of rat liver mitochondrial monoamine oxidase activity. *Biochem. Pharmacol.* **38**, 4245–4251.

THE PHARMACOLOGY OF SELEGILINE

Kálmán Magyar[1,2]

[1]Department of Pharmacodynamics, Semmelweis University, Budapest, Hungary
[2]Neurochemical Research Unit, Hungarian Academy of Sciences, Budapest, Hungary

Abstract

Selegiline, the *R*-optical enantiomer of deprenyl (phenyl-isopropyl-methyl-propargylamine), was almost exclusively used MAO-B inhibitor during the past decades to treat Parkinson's disease. Oral treatment prolongs the need of levodopa administration. Selegiline is rapidly metabolized by the microsomal enzymes to amphetamine, methamphetamine, and desmethyl-deprenyl. In addition, the flavin-containing monooxigenase is synthesizing deprenyl-*N*-oxide. Selegiline in rather low concentrations (10^{-9}–10^{-13} M), does not influence MAO-B, but it has an antiapoptotic activity in tissue culture. The neuroprotective effect of selegiline has a biphasic character. In higher concentrations than 10^{-7} M increases the rate of apoptosis (proapoptotic activity). The metabolites are also taking part in the complex pharmacological activity of selegiline. The simultaneous presence of the pro- and antiapoptotic effects of selegiline and its metabolites frequently hindered its clinical usage. During the past years rasagiline has been introduced to replace

INTERNATIONAL REVIEW OF
NEUROBIOLOGY, VOL. 100
DOI: 10.1016/B978-0-12-386467-3.00004-2

selegiline in clinical application. MAO-B inhibitors beside their effect on the enzyme MAO-B could hold different spectrum of pharmacological activities. Selegiline is administered orally and it possesses an intensive "first pass" metabolism. To circumvent the "first pass" metabolism, parenteral administration of the drug might lead to different distribution and pharmacological activity of selegiline.

ABBREVIATIONS

A	amphetamine
CNS	central nervous system
DNO	deprenyl-*N*-oxide
DD	desmethyl-deprenyl
DA	dopamine
FMO	flavin-containing monooxigenase
GIT	gastrointestinal tract
i.p.	intraperitoneally
i.v.	intravenously
MA	methamphetamine
MAO	monoamine oxidase
NA	noradrenaline
p.o.	orally
PD	Parkinson's disease
5-HT	serotonin
s.c.	subcutaneously

I. Historical Aspects of Selegiline

Early observations in the 1950s, namely, that the tuberculostatic iproniazid elevated the mood of patients infected with tuberculosis, made a major impact on the development of modern biological psychiatry. Zeller discovered the monoamine oxidase (MAO) inhibitory effect of iproniazid that is, as a consequence of enzyme inhibition, the concentration of the transmitter amines was elevated in the central nervous system (CNS), which regulates the mood of the patients (Zeller and Barsky, 1952). Due to this concept, a lot of MAO inhibitors have been synthesized (hydrazines, cyclopropylamines, and *N*-propargyl-derivates) to treat human depression, the most frequent psychiatric disorder. The hope to discover a clinically applicable antidepressant gave the impetus to the Chinoin Pharmaceutical Company in 1962 to synthesize a "me-too" drug on the name of E-250. This number indicated that deprenyl was the 250th in the series of compounds. The patent of deprenyl was issued in Hungary in 1962. It is worse to mention that neither in the patent nor in the first paper, in which the pharmacological properties of deprenyl were described by Knoll *et al.* (1965), a word was found about the Parkinson's disease, whereas that is the only accepted therapeutic indication of the drug nowadays (Parnham, 1993). The deprenyl story is rather colorful, as the important discoveries usually are; it contains subjective elements from the part of the contemporary participants. Nevertheless, deprenyl is an original Hungarian drug, synthesized by Z. Ecsery in Budapest and developed by Knoll and his group. It was synthesized as an antidepressant, psychostimulant agent, but its therapeutic application had been realized in totally different field, as it was originally planned.

Deprenyl is a propargylamine derivate; the propargyl moiety is simply attached to the amino group of methamphetamine (MA). It is closely similar to pargyline, which is also an *N*-propargyl derivate of benzylamine, synthesized earlier.

It was generally agreed that the MAO inhibitors elicit antidepressive activity, but the ingestion of foods, containing high concentration of tyramine (cheese, red vine, salted herrings), induce hypertensive crises, sometimes causing hemorrhage in the CNS (Blackwell *et al.*, 1967; Natoff, 1964). Because of the serious side effects of the nonselective (first generation) MAO inhibitors—named "cheese effect"—they were not the drugs of choice to treat depression (Jarrott and Vajda, 1987; Youdim and Finberg, 1987). Deprenyl, similarly to the other MAO inhibitors was neglected. In spite of the fact that Knoll and his group in rat vas-deferens preparation and on the blood pressure of cats gave experimental evidence, that deprenyl does not potentiate tyramine effect, neither *in vitro*, nor *in vivo* studies (Knoll *et al.*, 1968). In 1967 it was published that the levorotatory enantiomer of deprenyl is more potent inhibitor of MAO than the dextrorotatory form, and after 1967, all of the studies, including the clinical application of the drug, were conducted with the (−)-optical antipode of deprenyl, named selegiline (Magyar *et al.*, 1967).

Due to the discovery of the heterogenic nature of MAO (Johnston, 1968) and the selective enzyme inhibitory properties of selegiline, it became the only MAO inhibitor, which survived the shock of "cheese effect" (Magyar *et al.*, 2010).

II. The Multiplicity of Monoamine Oxidase

In 1968, on the basis of substrate specificity and inhibitor sensitivity, Johnston discovered that the MAO enzyme exists in multiple forms, being sensitive or not to the effect of clorgyline, using the substrate of 5-HT. The sensitive part of the enzyme was named MAO-A, while the other one MAO-B. We used ^{14}C-labeled tyramine as a substrate in our studies, and the enzyme inhibition curve obtained with selegiline has shown a double sigmoid character. After the discovery of Johnston, we immediately realized that the strange inhibition curve is due to the existence of MAO-A and MAO-B izoenzymes, when a mixed type of substrate is used, and the enzyme activity is inhibited with a selective inhibitor which cannot be anything else in our case, just selegiline (Knoll and Magyar, 1972). The enzyme inhibition curve unanimously showed that selegiline, on the contrary to clorgyline, is a potent, selective inhibitor of MAO-B (Magyar, 1993). The ratio of selectivity (IC_{50} of MAO-A/MAO-B) was ~500. Similar inhibition curve was obtained, when the metabolism of a mixed substrate was inhibited with clorgyline. Table I shows the substrate specificity and inhibitor sensitivity of the subtypes of MAO, in case of the most frequently used inhibitors and substrates. Our results were presented in the conference organized in Sardinia in 1971, devoted to the 70 years anniversary of H. Blaschko, who discovered, together with Hare, the MAO enzyme itself (Blaschko, 1974; Hare, 1928). The results were published a year later in "Advances in Biochemical Psychopharmacology" (Knoll and Magyar, 1972). The paper obtained international attention and became a "Science Citation Classic" in 1982. It was declared in the paper that selegiline is a potent B-type selective irreversible MAO inhibitor without "cheese effect." Initially, it forms a noncovalent complex with the flavin containing MAO enzyme, but its subsequent oxidation leads to the formation of a covalent bound with the enzyme complex.

The existence of MAO-A and MAO-B has received extensive attention from the biochemical and pharmaceutical communities to facilitate the development of more and more selective and effective drugs. The subtypes of MAO were differentiated not only on the basis of substrate specificity and inhibitor sensitivity but also with immunohystochemical methods. Molecular genetics was rarely used for this purpose (Shih and Chen, 2004).

Nevertheless, the localization of the genes was determined on the X-chromosomes. Initial studies on purified enzymes utilized liver mitochondria to have large

Table I
SUBSTRATES AND INHIBITORS OF MAO.

	MAO-A	MAO-A and -B	MAO-B
Substrate specificity	Serotonin (5-HT)	dopamine (D) noradrenaline (NA) tyramine (T)	β-Phenylethylamine (PEA)
Inhibitor sensitivity	Clorgyline[a]	Iproniazide Phenelsine Tranylcypromine	Selegiline[a]

[a] Selective inhibitors.

quantity of MAO-B, while human placental mitochondria served as a source for MAO-A.

Thanks to the excellent work of Edmondson and his group, the structure of MAO-A and MAO-B has been determined (Edmondson *et al.*, 2004). His studies on MAO will bring our understanding of both the structure and catalytic mechanism of this enzyme to a new level. Cloning and sequencing the respective genes, convincingly demonstrated that MAO-A and MAO-B are two separate enzymes that share many similar properties, for example, 70% sequence identity. It was also discovered on knock out rats of MAO-A that the animals become aggressive. It is rather interesting that the MAO-A can be inhibited totally by clorgyline, but the aggressive behavior of the animals did not appear (Chen *et al.*, 2007).

Interesting findings have been published by Oreland and his group, detecting association between the platelet MAO-B activity and personality traits such as sensation seeking, behavior, and impulsiveness. In human platelet, only the MAO-B is expressed which is readily accessible in venous blood. Recent results suggest an association between the platelet MAO-B activity and personality as well as vulnerability in the second type of alcoholism (Oreland *et al.*, 2004).

The existence of the subtypes of MAO offers acceptable explanation for the lack of the "cheese reaction" in case of selegiline treatment; mainly if we take into consideration the uneven distribution of the isoforms of MAO. High quantity of MAO-A can be found in the gastrointestinal tract (GIT), while more MAO-B are located in the CNS, compared to other organs. The food-derived amines are metabolized by the MAO-A. Due to the "first pass" metabolism, food-derived amines are inactivated in the GIT. MAO-A inhibitors protect the "first pass" metabolism of amines and increases the chance of the "cheese effect." On the contrary, the quantity of MAO-B is high in the CNS and plays an essential role in the metabolism of dopamine (DA). Nevertheless, DA and tyramine are substrates for both MAO subtypes. Consequently, the selective MAO-B inhibitors preferentially raise the concentration of DA in the CNS. The selectivity of the inhibitors is relative and concentration-dependent. Both selegiline and clorgyline inhibit the reversed subtypes of the enzyme in high molar concentrations. We are aware of the chemical structure of the purified MAO-A and MAO-B. The question is how these achievements can serve to find new better inhibitors which can over-exceed the clinical value of the presently available drugs.

It was firmly established that, in the serotonergic nerve endings, the MAO-B activity is high, while in dopaminergic nerves, the concentration of MAO-A is elevated. High MAO-B activity in human blood is localized in platelets, which can be determined without any ethical hindrance in the rest of the blood, which is normally taken during clinical examination of the patients. It is interesting that the generally used laboratory animals (rats, mice, cats, and rabbits) do not contain platelet MAO-B, only the pig contains in a certain but measurable amount (10–15%) of MAO-B activity compared to human platelets.

MAO activity is localized in the outer membrane of the mitochondria; therefore, drugs which inhibit MAO should reach the mitochondrial surface.

III. The Pharmacology of MAO-B Inhibition

The following reaction is catalyzed by MAO:

$$R - CH_2 - NH_2 \xrightarrow[H_2O+O_2]{MAO} R - COH + H_2O_2 + NH_3$$

The following pharmacological consequences are elicited by the treatment with selegiline, the selective inhibitor:

1. Selegiline treatment increases the concentration of DA in the CNS, the lack of which neurotransmitter plays the primary role in the pathogenesis of Parkinson's disease (PD) (Hornykiewicz, 2001, 2002).
2. Selegiline treatment decreases both the oxidative and nutritive stress by lowering the H_2O_2 overproduction as well as diminishes the amount of reactive oxygen and nitrogen species (ROS and RNS). Superoxide anion radicals can react with nitric oxide (NO) and form peroxynitrate, which spontaneously decomposes to $^{\circ}OH^-$ and NO_2 (Tipton *et al.*, 2004).
3. Selegiline inhibits the age-related increase of MAO-B activity; consequently, it can slow down the oxidative damage of the CNS, presumably prolonging life expectancy (Knoll, 1988).
4. Long-term treatment with selegiline increases the capacity of the protective mechanisms against oxidative damage. In addition, the long-term treatment also increases the activity of *superoxide dismutase* (SOD), mainly the soluble form of that (CuZn SOD). The catalase activity is also enhanced (Carrillo *et al.*, 1991).
5. MAO is localized partly on the outer membrane of the mitochondria inside the nerve endings and in glial cells. Synaptic transmission is increased by the inhibition of mitochondrial MAO, but MAO-B inhibition elevates the concentration of DA in the glia as well, which takes part in the extrasynaptic DA transmission.

IV. Structure–Activity Relationship Studies

Compounds from which selegiline has been selected gave us the possibility to analyze the structure–activity relationship. It became clear that even small alterations of the amphetaminergic structure—such as alkylation of the side chain (–methyl–isopropyl groups), saturation, or halogenation of the ring—resulted in a decrease of the MAO inhibitory potency. Nevertheless, replacing the phenyl ring with a furan or an indenyl group resulted in potent inhibitors of MAO-B (Magyar *et al.*, 1980). The indenyl derivate J-508 exceeded the MAO-B inhibitory potency of selegiline by one order of magnitude. In contrast to selegiline, the (+)-enantiomer of J-508 is a more potent inhibitor than the (−)-variant on MAO-B. In spite of some favorable properties of J-508, we did not continue the study on this compound, as we were highly satisfied with selegiline, on the other hand, since it became the "gold standard" of MAO inhibitors for decades. Selegiline was the drug of choice of MAO-B inhibitors to treat Parkinson's disease, without any important competitor on the market.

Presently, rasagiline—under the trade-name of Azilect—appeared on the market, which is structurally a nor-J-508 (Youdim *et al.*, 2001). Azilect dominates the market nowadays. Its privilege can partly be due to the novelty of the drug. Nevertheless, the complexity of the multifunctional effects of rasagiline might lead to its success. The studies of MAO-B inhibitors during the past decades have shown us that concentrating only the enzyme inhibitory potency led us to a myopic vision.

Among the new compounds, promising MAO-B inhibitors were found, although their further analysis had been stopped.

It is difficult to maintain the selective irreversible inhibition of selegiline *in vivo* during prolonged administration. Nevertheless, in a dose of 0.05–0.25 mg/day injected subcutaneously to rats, the selective spectrum can be maintained, but in a dose of 1 mg/kg, it was lost (Ekstedt *et al.*, 1979).

V. Effects of Selegiline Not Relating to MAO-B Inhibition

It is generally accepted that a selective transport mechanism exists in the synapse, localized on the membrane of the nerve endings. The physiological role of the transport (uptake) is to carry the released transmitter amines from the synaptic gap back to the nerve endings. This process preserves transmitter amines released by nerve impulses from metabolic inactivation by catechol-O-methyl-transferase (COMT). Moreover the transport is the most effective terminating mechanism of transmitter action. The (+)-optical isomer of deprenyl is more effective to inhibit transport proteins and facilitates the transmission than selegiline.

Our studies on selegiline and its metabolites unequivocally proved that the MAO-B enzyme inhibitory potency and the inhibition of transport are independent mechanisms. Some of the metabolites of deprenyl are strong inhibitors of uptake, without possessing enzyme inhibitory effect in structure–activity relationship studies, we found metabolites capable to inhibit DA, NA, and 5-HT uptake selectively with or without MAO inhibitory potency. The IC_{50} of MAO inhibition in μmol/l and the relative potency on the transport were presented in 2000 (Tekes *et al.*, 1900). The inhibitory potency of the compounds on DA uptake was measured in striatal synaptosomes of rats, while on NA and 5-HT uptake, that was determined in the hypothalamus and hippocampus, respectively.

The transporter protein is able to take up selective toxins into the nerve endings. MPTP (1-methyl-4-phenyl-1,2,3,6-tetrahydro-pyridine), a petidine analog, is the most extensively studied selective neurotoxin, which damages dopamin neurons in the substantia nigra pars compacta. The MPTP story strengthened the role of the neurotoxin hypothesis in the pathogenesis of PD. MPTP is a pretoxin, which is converted to MPDP by MAO-B, and without enzymatic interference, it leads to the formation of MPP^+ (1-methyl-4-phenyl-piperidine), a toxin, which is selectively taken up by the nigrostriatal system of the CNS. The reason of the high affinity of MPP^+ to the dopaminergic system in the CNS, but not in the periphery, are not known. The conversion of MPTP to MPDP can be inhibited by selegiline pretreatment, which prevents the toxicity. A comprehensive paper about the mechanism of MPTP toxicity was published by Glover *et al.* (1986). The protective effect of selegiline on MPTP toxicity, described by Langston, greatly enhanced the respect of selegiline and simultaneously strengthened the concept of the existence of neuroprotection (Langston, 1990).

There are other toxins, such as DSP-4, selective to noradrenergic- (Haberle *et al.*, 2002), 5,6-dihydroxy-triptamine to serotonergic- (Fowler and Tipton, 1982), and AF64A to cholinergic (Ricci *et al.*, 1992) nerves. Their effect can be prevented by selegiline administration before the insult caused by the toxins. It is interesting to note that selegiline seemed to be effective, when it was administered a few hours later following DSP-4 treatment, indicating some neuro-restoring property of selegiline.

Among the insecticides, there are selective dopaminergic toxins, such as rotenone and paraquate, which might play a certain role in the etiology of Parkinson's disease (Choi *et al.*, 2010). Against all of these toxins, selegiline shows up neuroprotective effect, which is mostly independent of its MAO-B inhibitory property.

It was published in several studies that selegiline has an antitumor effects (Thyaga Rajan and Felten, 2002; Thyaga Rajan and Quadri, 1999; Thyaga Rajan *et al.*, 1999, 2000). A novel effect of selegiline is to increase cell-to-cell adhesion (Jenei *et al.*, 2005) and the inhibition of hyperpermeability of vascular endothelial cells (Tharakan *et al.*, 2010; Whaley *et al.*, 2009).

VI. Antiapoptotic Effect of Selegiline

In 1994, Tatton and his group published an important discovery, namely, that selegiline reduces the rate of apoptosis in cell culture by inducing new protein synthesis, in a concentration range (10^{-9}–10^{-12} M) which is too low to inhibit the

MAO-B enzyme (Tatton *et al.*, 1994). They used PC12 cells of neuroectodermal origin, where the apoptosis was elicited by serum and/or NGF deprivation of the cell culture. It was postulated that the apoptotic mechanism is likely to be responsible, at least partly, for cell death in neurodegenerative diseases, such as the Parkinson's and the Alzheimer's diseases as well as in the amyotrophic lateral sclerosis. Drugs which are able to inhibit apoptosis could be promising candidates for neuroprotective therapy in the future. It was demonstrated that selegiline can prevent or diminish the apoptosis, induced by serum deprivation, glutation depletion and toxins, such as okadaic acid, nitric oxide, peroxynitrate, cytosine arabinoside, as well as peripheral nerve crush and axotomy (Tatton *et al.*, 1996). In our studies, we used the A2058 human melanoma cell line of neuroectodermal origin (Magyar and Szende, 2000, 2004; Magyar *et al.*, 1996, 1998; Szende, 2004; Szende *et al.*, 2001); however, our culture was not differentiated by nerve growth factor (NGF) deprivation, which might explain the differences of the results between the two research groups. Serum withdrawal for 5 days resulted in the appearance of a high number of apoptotic cells. The opposite antipode of (−)-deprenyl did not influence apoptosis.

High concentration of selegiline (10^{-3}–10^{-4} M) induces proapoptotic cell damage. Tatton described that SKF-525A (Proadiphene) pretreatment might prevent the antiapoptotic activity of selegiline. SKF-525A is an inhibitor of the microsomal drug metabolizing enzymes. Neither Tatton and Chalmers-Redman (1996) in PC12 cells nor our group in A2058 melanoma cell culture observed antiapoptotic effects in trophycally withdrawn culture, where the metabolites of selegiline, namely MA (methamphetamine) and A (amphetamine), were in low concentrations (10^{-9}–10^{-13} M) (Szende *et al.*, 2001; Tatton *et al.*, 1994).

DD ((−)-desmethyl-deprenyl) showed a dose-dependent inhibition of tumor growth *in vivo* on A2058 human melanoma stenographs (Szende *et al.*, 2000). These studies convincingly have shown that species variations of the cell cultures (origin of cells) differentiated or not differentiated cells, further, the neuronal or nonneuronal cultures behaved differently regarding the proapoptotic and antiapoptotic effects of selegiline. We proved that selegiline decreased the apoptotic effect of ischemia-reperfusion in rats, in a dose usually applied in human therapy (Toronyi *et al.*, 2002). Quin and his group reported an important finding, namely, that selegiline has antiapoptotic effect on nonneural cells, such as cardiac tissue (Qin *et al.*, 2003).

The mechanism of the established antiapoptotic effect of selegiline is most likely achieved through the modulation of gene expression, interfering with the apoptotic cascades, especially the mitochondrial pathway. Apoptosis is mediated by the loss of mitochondrial membrane potential and the opening of the mitochondrial membrane permeability transition pores, through which the proapoptotic factors are released, such as cytochrome *C*. Increased quantity of BCL2 helps to minimize the loss of mitochondrial membrane potential and closes the transition pores (Tatton and Chalmers-Redman, 1996; Wadia *et al.*, 1998). Selegiline, in a low concentration, might modulate the regulation of antiapoptotic BCL2 or the proapoptotic BAX concentrations.

The cytoprotective activity of selegiline, DD, and DNO (deprenyl-*N*-oxide) was studied against L-buthionine-(*S*,*R*)-sulfoximine (BSO) toxicity in serum deprived A-2058 melanoma cell culture. DD was the most effective compound in decreasing apoptotic activity, while DNO stabilized the cell number on the control level and increased the ratio of mitotic cells above the level measured in serum deprived control (Szende *et al.*, 2010).

VII. Pharmacokinetics of Selegiline

Selegiline is a highly lipid-soluble substance. Its distribution between hexane and phosphate buffer (pH 7.4) is very high in favor of hexane (82%) compared to the distribution of NA and A (0.01% and 0.32%, respectively). The fate of selegiline has been studied in many laboratories, including ours (Reynolds *et al.*, 1978; Riederer and Youdim, 1986; Rohatagi *et al.*, 1997a,b). In our early studies, we used ^{14}C-labeled selegiline; the label was positioned in the side chain or in the *N*-methyl or *N*-propargyl groups. We proved with others that selegiline is absorbed completely and rapidly from every routes of administration, but the bioavailability ($AUC_{0-\infty}$) of the drug highly depends on the type of administration (Barrett *et al.*, 1996a; Heinonen *et al.*, 1989, 1994; Magyar, 1994; Magyar *et al.*, 2004; Magyar and Tóthfalusi, 1984; Mahmood, 1997; Szebeni *et al.*, 1995). The $AUC_{0-\infty}$ values of selegiline, together with those of MA, DD, and A, were examined in our laboratory after oral (p.o.), subcutaneous (s.c.), intraperitoneal (i.p.), and intravenous (i.v.) administrations, and the results were presented in the paper of Magyar *et al.* (2004, 2010). Surprisingly, the $AUC_{0-\infty}$ values detected after p.o. administration of selegiline were rather low (25%) compared to those in the i.v. studies (100%). Azzaro and his coworkers registered only 4% $AUC_{0-\infty}$ in humans (Azzaro *et al.*, 2007). The highest plasma concentration of 0.5 μg/l was reached 1.5 h after p.o. administration (Heinonen *et al.*, 1994; Magyar and Tóthfalusi, 1984; Szebeni *et al.*, 1995). Food consumption surprisingly increased the efficiency of p.o. absorption of selegiline (Barrett *et al.*, 1996a). The low p.o. $AUC_{0-\infty}$ values compared to parenteral administration indicate an intensive "first pass" metabolism, thus a smaller amount of the parent drug reaches the systemic circulation (Barrett *et al.*, 1996a; Heinonen *et al.*, 1989; Magyar *et al.*, 2004, 2010; Mahmood *et al.*, 1994, 1995). In contrast to oral route, parenteral administration of selegiline shows much less "first pass" metabolism of the drug (~10%).

Parenteral treatment with selegiline results in relatively high inhibitor concentration in the brain, compared to other organs. It is due to the high concentration of selegiline in the arterial blood, the high perfusion rate of the brain, and the high lipid solubility of the inhibitor. After a usual daily dose of selegiline (10 mg/day in man), these factors can result in so high selegiline concentration in the brain, that is able to inhibit MAO-A, the enzyme known to play an important role in human depression. At the same time, MAO-A is not so efficiently inhibited in the GIT, mainly because of its lower perfusion rate. Table II shows this experimental finding demonstrated by an animal model. The special distribution of selegiline might explain the lack of "cheese effect." This concept is supported by the distribution of MAO, namely, that 80% of the total MAO activity localized in human brain (Tolosa and Katzenschlager, 2007).

This can explain why MAO-B inhibitors lack "cheese effect" (Barrett *et al.*, 1996b; Feiger *et al.*, 2006; Wecker *et al.*, 2003). It is worth to mention that during the preclinical studies of selegiline, the animals were treated s.c., while in human therapy; selegiline was almost exclusively administered orally. Because of the different distribution pattern as well as the different routes of administration in the former cases, it is difficult to draw valid conclusions from animal experiments to human studies.

The whole body autoradiography studies on mice have shown a rapid penetration of ^{14}C-selegiline into the brain and fatty tissues, such as brown-fat. The rapid rise of selegiline concentration was followed by a fast decrease after i.v. administration of the drug (Magyar, 1994). In some of these studies, doubly and alternatively labeled radio isomers of selegiline were used, when the ^{3}H-labeling

Table II
The Effect of Selegiline in Different Organs of MAO-A and -B After Chronic Treatments of Rats.

MAO	Selegiline			
	Orally		Subcutaneously	
	0.5 mg/kg	5 mg/kg	0.5 mg/kg	5 mg/kg
Brain MAO inhibition in %				
MAO-B	45.47	92.08	93.75	98.95
MAO-A	5.05	17.99	0.01	84.87
Liver MAO inhibition in %				
MAO-B	45.52	92.04	43.06	91.23
MAO-A	24.48	31.59	19.32	33.83
Bowel MAO inhibition in %				
MAO-B	68.55	84.92	68.29	80.31
MAO-A	36.49	84.13	24.67	61.49

Rats were treated orally or subcutaneous during 5 days.
The daily dose was 0.5 or 5.0 mg/kg.
Animals were decapitated 2 h after last injection.
Substrates: MAO-A, 5-HT; MAO-B, PEA.

in the ring indicated the total quantity of the parent compound and its metabolites, while ^{14}C-labeling in the propargyl group has shown the concentration of intact molecules. The evaluations of these results are complicated, but it has been shown that repeated administration of selegiline can lead to a stabilized long-lasting steady-state concentration of the inhibitor. That can be explained by the irreversible binding of selegiline to MAO-B.

From the pharmacokinetic studies of selegiline in dogs, the following important parameters were obtained: the absorption is fast (t_{max} 25 ± 5.8 min), the distribution volume is high (6.56 ± 0.56 l/kg), the elimination half-life is short ($t_{1/2}^{\beta}$ 60.24 ± 9.56 min), and the bioavailability depends on the route of administration (8.51 ± 3.31%). Parameters of the metabolism are published in the paper of Magyar *et al.* (2007). The metabolite concentrations of selegiline are rather low in the blood, because of the rapid tissue penetration and the intensive "first pass" metabolism. To determine the concentration of metabolites in tissues, a sensitive GC/MS analysis was performed.

VIII. Metabolism of Selegiline

Selegiline is readily metabolized in experimental animals and in humans, when mainly MA, DD, and A are formed and excreted into the urine. The metabolism of selegiline is stereospecific. Schacter was the first, who proved, that there is no racemization during metabolism of the selegiline (Schachter *et al.*, 1980). The metabolites are pharmacologically active. The *S*-enantiomers are more potent to inhibit the transporter proteins than the *R*-antipades. A chiral capillary electrophoresis method was developed in our laboratory for the separation of the enantiomers (Szökő and Magyar, 1995). CYP isoforms, which dealkylate selegiline, are the CYP206 (Grace *et al.*, 1994; Mahmood, 1997), the CYP3A4 (Dragoni *et al.*, 2003; Taavitsainen *et al.*, 2000), and the CYP2E1 (Valoti *et al.*, 2000). The flavin-containing monooxigenase (FMO) enzymes play a role in the formation of N-oxide (Szökő *et al.*, 2004). In addition to dealkylation, the *para*-hydroxylation can also be observed (Dragoni *et al.*, 2003; Kalász *et al.*, 1990; Lengyel *et al.*, 1997) in rats or in humans (Shin, 1997; Tarjányi *et al.*, 1998). Conjugates of the *para*-hydroxy-metabolites can be found in the urine. Ephedrine-type urinary metabolites were also produced by β-OH-formation in low amounts and excreted into the urine (Rohatagi *et al.*, 1997a). The formation of a new metabolite, selegiline-N-oxide (DNO), in a quantity less than 1% of the dose, was suggested by Japanese authors (Katagi *et al.*, 2001). In earlier studies with gas chromatography (GC), this metabolite was not detected because of its heat degradation. In the very early period, using thin-layer chromatography, we detected this metabolite although its chemical structure remained to be undetermined (Magyar and Szüts, 1982). The HPLC/MS method is suitable to the quantitative determination of DNO, which has quaternary nitrogen with an additional new chiral center. Consequently in contrast to selegiline, DNO has four stereoisomers.

DNO is excreted into the urine; its quantity during 24 h was about 1–5% of the dose administered, comparable to the quantity of MA and A (Katagi *et al.*, 2002). Para-hydroxy metabolites are more intensively formed. Their total urinary excretion can reach 50–60%, mostly conjugates. It worth to mention, that some publications described a retro-reduction of DNO (Sugiura and Kato, 1977). DNO is the ultimate metabolite having propargyl group, which is a pro-requisite of neuroprotection. Currently, an intensive study is going on in our lab to define the function of DNO. As a quaternary compound, its MAO inhibitory potency is lost.

Katagi and his group published in 2002 that DNO can be used as a reliable indicator of selegiline administration. They developed a sensitive method which is suitable to detect DNO, in cases of clinical and forensic toxicology, when selegiline administration should be distinguished from MA abuse (Katagi *et al.*, 2002).

IX. The Role of Birkmayer and His Group in the Introduction of Selegiline in the Therapy of Parkinson's Disease

The symposium organized in Sardinia in 1971 played an important role in the therapeutic application of selegiline. It was reported on the symposium by Knoll and Magyar, that (−)-deprenyl selectively inhibits the clorgyline insensitive part of MAO and (−)-deprenyl does not potentiate the hypertensive effect of exogenous catecholamines, neither *in vitro* nor *in vivo* studies. The lack of tyramine potentiation in clinical studies was also published by Varga and Tringer, however, Hungarian psychiatrists did not take part in selegiline research any further (Varga and Tringer, 1967). Being aware of the knowledge of the inhibitory spectrum of selegiline, P. Riederer and M. Youdim recognized the relevance of its potential benefit in the management of Parkinson's disease and suggested to W. Birkmayer in Vienna, the application of selegiline as an antiparkinson drug. Selegiline was found to be effective in parkinsonian patients in the clinical practice, and their results were published in the literature (Birkmayer *et al.*, 1975, 1977, 1985).

X. Future Perspectives

After three decades of unlimited use of selegiline as a gold standard in the treatment of PD, the question is arised whether it is reasonable or not to speak about the future perspectives of selegiline or, in a broader sense, the future perspectives of the MAO-B inhibitors. The answer is: yes! The introduction of rasagiline (*N*-propargylamine-inden, Azilect), recently, gives positive answer for the question. Rasagiline has 10-times higher potency to inhibit MAO-B, compared to selegiline. Nevertheless, this is not the main reason for its introduction in human therapy. The complexity of the spectrum of its activity should be considered, which makes it possible to choose the most suitable inhibitor. Recent findings firmly suggest that all of the biochemical and pharmacological properties of MAO-B inhibitors experienced in preclinical and clinical studies are not characteristic of all substances equally, but some characters are unique to a certain compound. Selegiline and rasagiline are more than simple MAO-B inhibitors.

The physicochemical properties of the inhibitors can be different which can modify the fate of drugs in the body (absorption, distribution, metabolism, elimination; ADME). The route of administration is an important factor, as it can alter the blood and the brain concentrations of the substances. It would be advisable to find the most suitable way of parenteral administration with a good efficiency and tolerability, mainly in the case of chronic administration.

To recite all of the possibilities would be impossible and useless, but it should be mentioned that the antiapoptotic effects of the MAO-B inhibitors offer a large scale of neuroprotective activity possessed by the generally used MAO-B inhibitors, and those that will be synthesized in the future.

XI. Overall Conclusions

Selegiline is an original Hungarian drug, which has been discovered by Knoll and his coworkers, and during the past three decades, it was almost exclusively used inhibitor of MAO-B to treat PD. PD is a motor and cognitive disorder, characterized by the progressive loss of dopaminergic neurons of the substantia nigra pars compacta and with the presence of Lewy bodies in the cytoplasm. Besides dopaminergic neurons, the noradrenergic neurons in the locus coeruleus are also degenerated. Selegiline, in a dose of 10 mg/day, delays the need of levodopa administration and decreases the dose applied. The neuroprotective effect of selegiline treatment was proved in the controlled, randomized, double blind DATATOP studies conducted on 800 parkinsonian patients. Nevertheless, because of the irreversible inhibition of MAO-B it is difficult to distinguish between the neuroprotective and symptomatic effects of the drug. The clinical effects of selegiline are summarized in the brilliant paper of Riederer *et al.* (2004).

In vitro studies on tissue cultures and *in vivo* observations in experimental animals proved that selegiline and some of its metabolites accelerate the degeneration of neurons in high concentrations, especially in supersensitivity of the treated subjects; while in a low concentrations, insufficient to inhibit MAO-B activity, the drug and some of the metabolites were found to be neuroprotective. Due to the proapoptotic effect of the high dose of selegiline, its administration is just like walking on the razor's edge. The debate between neurologists on the administration of MAO-B inhibitors or DA-agonists in the initial treatment of parkinsonian patients has not been settled yet, but if we stay on the ground of preclinical studies, MAO-B inhibitors, applied in a suitable dose, should be administered as early as the clinical signs appear. The therapeutic application of MAO-B inhibitors does not interfere with the effect of DA-agonists. The effectiveness of DA-agonists is not limited in time, in contrast to MAO-B inhibitors, which are effective as long as functioning neurons exist. The efficacy of DA-agonists is maintained even after the neurons had been totally lost. Large portion of MAO-B activity are localized in glial cells and its putative role is similar to DA-agonists.

Selegiline is readily metabolized by the microsomal enzymes (mixed function oxidases), which convert selegiline to A, MA, DD, *para*-hydroxy-metabolites, and their conjugates, all of which are excreted into the urine. In addition to the microsomal enzymes, the FMO synthesizes DNO. As a result of its quaternary character, DNO cannot penetrate passively through the membrane. The N-oxides DNO can also be reduced back to the parent compound, selegiline. DNO has two chiral centers and four optical enantiomers. We published recently that DNO increases the mitotic rate of melanoma cells in culture. Both selegiline and its metabolites can elicit diverse pharmacological activities, which all play a role in the complex effects of selegiline. In all concentration ranges, A and MA induced proapoptotic effect as well as damaged neurons. It is generally accepted that the metabolites possessing *N*-propargyl moiety can be neuroprotective (Youdim and Weinstock, 2002). Both DD and the DNO are metabolites suspected to possess neuroprotective property similarly to the parent compound, selegiline.

The dose–response curve of selegiline in neuroprotection is bell-shaped. High concentrations ($> 10^{-7}$ M) of the drug induce proapoptotic effect, while in small concentrations neuroprotection can be observed. It would be advisable to find a compound (metabolites), which would be neuroprotective in any concentration range, and would not have apoptosis inducing property. Currently, further studies are carried out in our laboratory to clear up the nature of neuroprotective activity of DNO. It is time to understand more precisely the complex molecular and pharmacological mechanism of neuroprotection induced by MAO-B inhibitors after three decades of their usage in the treatment of Parkinson's disease. In spite of the tremendous efforts presently there are only questions and no answers.

Acknowledgments

The author wishes to thank to Maria Knippel for her kind assistance in preparing the manuscript. This study was supported by the Neurochemical Research Group of Hungarian Academy of Sciences, Hungarian Academy of Sciences Foundation OTKA 63415 and ETT 141/2003 grants.

References

Azzaro, A.J., Ziemniak, J., Kemper, E., Campbell, B.J., and VanDenBerg, C. (2007). Pharmacokinetics and absolute bioavailability of selegiline following treatment of healthy subjects with the selegiline transdermal system (6 mg/24 h): a comparison with oral selegiline capsules. *J. Clin. Pharmacol.* **47**, 1256–1267.

Barrett, J.S., Rohatagi, S., DeWitt, K.E., Morales, R.J., and DiSanto, A.R. (1996a). The effect of dosing regimen and food on the bioavailability of the extensively metabolized, highly variable drug Eldepryl$^{(R)}$ (selegiline hydrochloride). *Am. J. Ther.* **3**, 298–313.

Barrett, J.S., Hochadel, T.J., Morales, R.J., Rohatagi, S., DeWitt, K.E., Watson, S.K., and DiSanto, A. R. (1996b). Pharmacokinetics and safety of a selegiline transdermal system relative to single-dose oral administration in the elderly. *Am. J. Ther.* **3**, 688–698.

Birkmayer, W., Riederer, P., Youdim, M.B., and Linauer, W. (1975). The potentiation of the anti akinetic effect after L-dopa treatment by an inhibitor of MAO-B, Deprenil. *J. Neural Transm.* **36**, 303–326.

Birkmayer, W., Riederer, P., Ambrozi, L., and Youdim, M.B. (1977). Implications of combined treatment with 'Madopar' and L-deprenil in Parkinson's disease. A long-term study. *Lancet* **1**, 439–443.

Birkmayer, W., Knoll, J., Riederer, P., Youdim, M.B., Hars, V., and Marton, J. (1985). Increased life expectancy resulting from addition of L-deprenyl to Madopar treatment in Parkinson's disease: a longterm study. *J. Neural Transm.* **64**, 113–127.

Blackwell, B., Marley, E., Price, J., and Taylor, D. (1967). Hypertensive interactions between monoamine oxidase inhibitors and foodstuffs. *Br. J. Psychiatry* **113**, 349–365.

Blaschko, H. (1974). The natural history of amine oxidases. *Rev. Physiol. Biochem. Pharmacol.* **70**, 83–148.

Carrillo, M.C., Kanai, S., Nokubo, M., and Kitani, K. (1991). (−)-Deprenyl induces activities of both superoxide dismutase and catalase but not of glutathione peroxidase in the striatum of young male rats. *Life Sci.* **48**, 517–521.

Chen, K., Cases, O., Rebrin, I., Wu, W., Gallaher, T.K., Seif, I., and Shih, J.C. (2007). Forebrain-specific expression of monoamine oxidase A reduces neurotransmitter levels, restores the brain structure, and rescues aggressive behavior in monoamine oxidase A-deficient mice. *J. Biol. Chem.* **282**, 115–123.

Choi, W.S., Abel, G., Klintworth, H., Flavell, R.A., and Xia, Z. (2010). JNK3 mediates paraquat- and rotenone-induced dopaminergic neuron death. *J. Neuropathol. Exp. Neurol.* **69**, 511–520.

Dragoni, S., Bellik, L., Frosini, M., Sgaragli, G., Marini, S., Gervasi, P.G., and Valoti, M. (2003). l-Deprenyl metabolism by the cytochrome P450 system in monkey (Cercopithecus aethiops) liver microsomes. *Xenobiotica* **33**, 181–195.

Edmondson, D.E., Mattevi, A., Binda, C., Li, M., and Hubalek, F. (2004). Structure and mechanism of monoamine oxidase. *Curr. Med. Chem.* **11**, 1983–1993.

Ekstedt, B., Magyar, K., and Knoll, J. (1979). Does the B form selective monoamine oxidase inhibitor lose selectivity by long term treatment? *Biochem. Pharmacol.* **28**, 919–923.

Feiger, A.D., Rickels, K., Rynn, M.A., Zimbroff, D.L., and Robinson, D.S. (2006). Selegiline transdermal system for the treatment of major depressive disorder: an 8-week, double-blind, placebo-controlled, flexible-dose titration trial. *J. Clin. Psychiatry* **67**, 1354–1361.

Fowler, C.J., and Tipton, K.F. (1982). Deamination of 5-hydroxytryptamine by both forms of monoamine oxidase in the rat brain. *J. Neurochem.* **38**, 733–736.

Glover, V., Gibb, C., and Sandler, M. (1986). The role of MAO in MPTP toxicity—a review. *J. Neural Transm. Suppl.* **20**, 65–76.

Grace, J.M., Kinter, M.T., and Macdonald, T.L. (1994). Atypical metabolism of deprenyl and its enantiomer, (S)-(+)-N, alpha-dimethyl-N-propynylphenethylamine, by cytochrome P450 2D6. *Chem. Res. Toxicol.* **7**, 286–290.

Haberle, D., Magyar, K., and Szoko, E. (2002). Determination of the norepinephrine level by high-performance liquid chromatography to assess the protective effect of MAO-B inhibitors against DSP-4 toxicity. *J. Chromatogr. Sci.* **40**, 495–499.

Hare, M.L. (1928). Tyramine oxidase: A new enzyme system in liver. *Biochem. J.* **22**, 968–979.

Heinonen, E.H., Myllyla, V., Sotaniemi, K., Lamintausta, R., Salonen, J.S., Anttila, M., Savijarvi, M., Kotila, M., and Rinne, U.K. (1989). Pharmacokinetics and metabolism of selegiline. *Acta Neurol. Scand. Suppl.* **126**, 93–99.

Heinonen, E.H., Anttila, M.I., and Lammintausta, R.A. (1994). Pharmacokinetic aspects of l-deprenyl (selegiline) and its metabolites. *Clin. Pharmacol. Ther.* **56**, 742–749.

Hornykiewicz, O. (2001). Dopamine and Parkinson's disease. A personal view of the past, the present, and the future. *Adv. Neurol.* **86**, 1–11.

Hornykiewicz, O. (2002). Dopamine miracle: from brain homogenate to dopamine replacement. *Mov. Disord.* **17**, 501–508.

Jarrott, B., and Vajda, F.J. (1987). The current status of monoamine oxidase and its inhibitors. *Med. J. Aust.* **146**, 634–638.

Jenei, V., Zor, K., Magyar, K., and Jakus, J. (2005). Increased cell-cell adhesion, a novel effect of R-(−)-deprenyl. *J. Neural Transm.* **112**, 1433–1445.

Johnston, J.P. (1968). Some observations upon a new inhibitor of monoamine oxidase in brain tissue. *Biochem. Pharmacol.* **17**, 1285–1297.

Kalász, H., Kerecsen, L., Knoll, J., and Pucsok, J. (1990). Chromatographic studies on the binding, action and metabolism of (−)-deprenyl. *J. Chromatogr.* **499**, 589–599.

Katagi, M., Tatsuno, M., Miki, A., Nishikawa, M., Nakajima, K., and Tsuchihashi, H. (2001). Simultaneous determination of selegiline-N-oxide, a new indicator for selegiline administration, and other metabolites in urine by high-performance liquid chromatography-electrospray ionization mass spectrometry. *J. Chromatogr. B Biomed. Sci. Appl.* **759**, 125–133.

Katagi, M., Tatsuno, M., Tsutsumi, H., Miki, A., Kamata, T., Nishioka, H., Nakajima, K., Nishikawa, M., and Tsuchihashi, H. (2002). Urinary excretion of selegiline N-oxide, a new indicator for selegiline administration in man. *Xenobiotica* **32**, 823–831.

Knoll, J. (1988). The striatal dopamine dependency of life span in male rats. Longevity study with (−) deprenyl. *Mech. Ageing Dev.* **46**, 237–262.

Knoll, J., and Magyar, K. (1972). Some puzzling pharmacological effects of monoamine oxidase inhibitors. *Adv. Biochem. Psychopharmacol.* **5**, 393–408.

Knoll, J., Ecseri, Z., Kelemen, K., Nievel, J., and Knoll, B. (1965). Phenylisopropylmethylpropinylamine (E-250), a new spectrum psychic energizer. *Arch. Int. Pharmacodyn. Ther.* **155**, 154–164.

Knoll, J., Vizi, E.S., and Somogyi, G. (1968). Phenylisopropylpropanylamine (E-250), a monoamine oxidase inhibitoe antagonizing the effects of tyramine. *Arzneimittelforschung* **18**, 109–112.

Langston, J.W. (1990). Selegiline as neuroprotective therapy in Parkinson's disease: concepts and controversies. *Neurology* **40**(Suppl.), 61–66discussion 67–69.

Lengyel, J., Magyar, K., Hollósi, I., Bartók, T., Báthori, M., Kalász, H., and Fürst, S. (1997). Urinary excretion of deprenyl metabolites. *J. Chromatogr. A* **762**, 321–326.

Magyar, K. (1993). Pharmacology of monoamine oxidase type-B inhibitors. In: I. Szelényi, I. (Ed.), *Inhibitors of Monoamine Oxidase B, Pharmacology and Clinical Use in Neurodegenerative Disorders.* Birkhauser Verlag, Basel, pp. 125–143.

Magyar, K. (1994). Behaviour of (−)-deprenyl and its analogues. *J. Neural Transm. Suppl.* **41**, 167–175.

Magyar, K., and Szende, B. (2000). The neuroprotective and neuronal rescue effect of (−)-deprenyl. In: R.G. Cameron, R.G. and G. Feuer, G. (Eds.), *Handbook of Experimental Pharmacology.* Springer-Verlag, Heidelberg, pp. 457–472.

Magyar, K., and Szende, B. (2004). (−)-Deprenyl, a selective MAO-B inhibitor, with apoptotic and anti-apoptotic properties. *Neurotoxicology* **25**, 233–242.

Magyar, K., and Szüts, T. (1982). The fate of (−)-deprenyl in the body—preclinical studies. In: B. Szebeni, B. (Ed.), *Proceedings of the International Symposium on (−)-Deprenyl, Jumex.* Chinoin, Budapest, pp. 25–31.

Magyar, K., and Tóthfalusi, L. (1984). Pharmacokinetic aspects of deprenyl effects. *Pol. J. Pharmacol. Pharm.* **36**, 373–384.

Magyar, K., Vizi, E.S., Ecseri, Z., and Knoll, J. (1967). Comparative pharmacological analysis of the optical isomers of phenyl-isopropyl-methyl-propinylamine (E-250). *Acta Physiol. Acad. Sci. Hung.* **32**, 377–387.

Magyar, K., Ecseri, Z., Bernáth, G., Sátory, É., and Knoll, J. (1980). Structure-activity relationship of selective inhibitors of MAO-B. In: K. Magyar, K. (Ed.), *Monoamine Oxidases and Their Selective Inhibitors.* Pergamon Press, Budapest, pp. 11–21.

Magyar, K., Szende, B., Lengyel, J., and Tekes, K. (1996). The pharmacology of B-type selective monoamine oxidase inhibitors; milestones in (−)-deprenyl research. *J. Neural Transm. Suppl.* **48**, 29–43.

Magyar, K., Szende, B., Lengyel, J., Tarczali, J., and Szatmáry, I. (1998). The neuroprotective and neuronal rescue effects of (−)-deprenyl. *J. Neural Transm. Suppl.* **52**, 109–123.

Magyar, K., Pálfi, M., Tábi, T., Kalász, H., Szende, B., and Szökő, É. (2004). Pharmacological aspects of (−)-deprenyl. *Curr. Med. Chem.* **11**, 2017–2031.

Magyar, K., Szatmáry, I., Szebeni, G., and Lengyel, J. (2007). Pharmacokinetic studies of (−)-deprenyl and some of its metabolites in mouse. *J. Neural Transm.* (Suppl.), 165–173.

Magyar, K., Szende, B., Jenei, V., Tabi, T., Palfi, M., and Szoko, E. (2010). R-deprenyl: pharmacological spectrum of its activity. *Neurochem. Res.* **35**, 1922–1932.

Mahmood, I. (1997). Clinical pharmacokinetics and pharmacodynamics of selegiline. An update. *Clin. Pharmacokinet.* **33**, 91–102.

Mahmood, I., Peters, D.K., and Mason, W.D. (1994). The pharmacokinetics and absolute bioavailability of selegiline in the dog. *Biopharm. Drug Dispos.* **15**, 653–664.

Mahmood, I., Marinac, J.S., Willsie, S., and Mason, W.D. (1995). Pharmacokinetics and relative bioavailability of selegiline in healthy volunteers. *Biopharm. Drug Dispos.* **16**, 535–545.

Natoff, I.L. (1964). Cheese and monoamine oxidase inhibitors. Interactions in anaesthetised cats. *Lancet* **1**, 532–533.

Oreland, L., Hallman, J., and Damberg, M. (2004). Platelet MAO and personality–function and dysfunction. *Curr. Med. Chem.* **11**, 2007–2016.

Parnham, M. (1993). The history of l-deprenyl. In: I. Szelényi, I. (Ed.), *Inhibitors of Monoamine Oxidase B, Pharmacology and Clinical Use in Neurodegenerative Disorders.* Birkhauser Verlag, Basel, pp. 237–251.

Qin, F., Shite, J., Mao, W., and Liang, C.S. (2003). Selegiline attenuates cardiac oxidative stress and apoptosis in heart failure: association with improvement of cardiac function. *Eur. J. Pharmacol.* **461**, 149–158.

Reynolds, G.P., Elsworth, J.D., Blau, K., Sandler, M., Lees, A.J., and Stern, G.M. (1978). Deprenyl is metabolized to methamphetamine and amphetamine in man. *Br. J. Clin. Pharmacol.* **6**, 542–544.

Ricci, A., Mancini, M., Strocchi, P., Bongrani, S., and Bronzetti, E. (1992). Deficits in cholinergic neurotransmission markers induced by ethylcholine mustard aziridinium (AF64A) in the rat hippocampus: sensitivity to treatment with the monoamine oxidase-B inhibitor L-deprenyl. *Drugs Exp. Clin. Res.* **18**, 163–171.

Riederer, P., and Youdim, M.B. (1986). Monoamine oxidase activity and monoamine metabolism in brains of parkinsonian patients treated with l-deprenyl. *J. Neurochem.* **46**, 1359–1365.

Riederer, P., Lachenmayer, L., and Laux, G. (2004). Clinical applications of MAO-inhibitors. *Curr. Med. Chem.* **11**, 2033–2043.

Rohatagi, S., Barrett, J.S., DeWitt, K.E., and Morales, R.J. (1997a). Integrated pharmacokinetic and metabolic modeling of selegiline and metabolites after transdermal administration. *Biopharm. Drug Dispos.* **18**, 567–584.

Rohatagi, S., Barrett, J.S., McDonald, L.J., Morris, E.M., Darnow, J., and DiSanto, A.R. (1997b). Selegiline percutaneous absorption in various species and metabolism by human skin. *Pharm. Res.* **14**, 50–55.

Schachter, M., Marsden, C.D., Parkes, J.D., Jenner, P., and Testa, B. (1980). Deprenyl in the management of response fluctuations in patients with Parkinson's disease on levodopa. *J. Neurol. Neurosurg. Psychiatry* **43**, 1016–1021.

Shih, J.C., and Chen, K. (2004). Regulation of MAO-A and MAO-B gene expression. *Curr. Med. Chem.* **11**, 1995–2005.

Shin, H.S. (1997). Metabolism of selegiline in humans. Identification, excretion, and stereochemistry of urine metabolites. *Drug Metab. Dispos.* **25**, 657–662.

Sugiura, M., and Kato, R. (1977). Reduction of tertiary amine N-oxides by rat liver mitochondria. *J. Pharmacol. Exp. Ther.* **200**, 25–32.

Szebeni, G., Lengyel, J., Székács, G., Magyar, K., Gaál, J., and Szatmári, I. (1995). Gas chromatographic procedure for simultaneous determination of selegiline metabolites, amphetamine, methamphetamine and demethyl-deprenyl in pig plasma. *Acta Physiol. Hung.* **83**, 135–141.

Szende, B. (2004). Cell proliferation and cell death in relation to dose of various chemical substances. In: T. Török, T. and I. Klebovich, I. (Eds.), *Monoamine Oxidase Inhibitors and Their Role in Neurotransmission.* Medicina, Budapest, pp. 133–140.

Szende, B., Magyar, K., and Szegedi, Z. (2000). Apoptotic and antiapoptotic effect of (−)-deprenyl and (−)-desmethyl-deprenyl on human cell lines. *Neurobiology (Bp.)* **8**, 249–255.

Szende, B., Bokonyi, G., Bocsi, J., Kéri, G., Timár, F., and Magyar, K. (2001). Anti-apoptotic and apoptotic action of (−)-deprenyl and its metabolites. *J. Neural Transm.* **108**, 25–33.

Szende, B., Barna, G., and Magyar, K. (2010). Cytoprotective effect of (−)-deprenyl, (−)desmethyl-deprenyl and (−)deprenyl-N-oxide on glutathione depleted A-2058 melanoma cells. *J. Neural Transm.* **117**, 695–698.

Szökő, É., and Magyar, K. (1995). Chiral separation of deprenyl and its major metabolites using cyclodextrine-modified capillary zone electrophoresis. *J. Chromatogr. A* **709**, 157–162.

Szökő, É., Tábi, T., Borbás, T., Dalmadi, B., Tihanyi, K., and Magyar, K. (2004). Assessment of the N-oxidation of deprenyl, methamphetamine, and amphetamine enantiomers by chiral capillary electrophoresis: an in vitro metabolism study. *Electrophoresis* **25**, 2866–2875.

Taavitsainen, P., Anttila, M., Nyman, L., Karnani, H., Salonen, J.S., and Pelkonen, O. (2000). Selegiline metabolism and cytochrome P450 enzymes: in vitro study in human liver microsomes. *Pharmacol. Toxicol.* **86**, 215–221.

Tarjányi, Z., Kalász, H., Szebeni, G., Hollosi, I., Báthori, M., and Fürst, S. (1998). Gas-chromatographic study on the stereoselectivity of deprenyl metabolism. *J. Pharm. Biomed. Anal.* **17**, 725–731.

Tatton, W.G., and Chalmers-Redman, R.M. (1996). Modulation of gene expression rather than monoamine oxidase inhibition: (−)-deprenyl-related compounds in controlling neurodegeneration. *Neurology* **47**, S171–S183.

Tatton, W.G., Ju, W.Y., Holland, D.P., Tai, C., and Kwan, M. (1994). (−)-Deprenyl reduces PC12 cell apoptosis by inducing new protein synthesis. *J. Neurochem.* **63**, 1572–1575.

Tatton, W.G., Wadia, J.S., Ju, W.Y., Chalmers-Redman, R.M., and Tatton, N.A. (1996). (−)-Deprenyl reduces neuronal apoptosis and facilitates neuronal outgrowth by altering protein synthesis without inhibiting monoamine oxidase. *J. Neural Transm. Suppl.* **48**, 45–59.

Tekes, K., Tóthfalusi, L., Gaál, J., and Magyar, K. (1988). Effect of MAO inhibitors on the uptake and metabolism of dopamine in rat and human brain. *Pol. J. Pharmacol. Pharm.* **40**, 653–658.

Tharakan, B., Whaley, J.G., Hunter, F.A., Smythe, W.R., and Childs, E.W. (2010). (−)-Deprenyl inhibits vascular hyperpermeability after hemorrhagic shock. *Shock* **33**, 56–63.

Thyaga Rajan, S., and Felten, D.L. (2002). Modulation of neuroendocrine–immune signaling by L-deprenyl and L-desmethyldeprenyl in aging and mammary cancer. *Mech. Ageing Dev.* **123**, 1065–1079.

Thyaga Rajan, S., and Quadri, S.K. (1999). L-Deprenyl inhibits tumor growth, reduces serum prolactin, and suppresses brain monoamine metabolism in rats with carcinogen-induced mammary tumors. *Endocrine* **10**, 225–232.

Thyaga Rajan, S., Madden, K.S., Stevens, S.Y., and Felten, D.L. (1999). Inhibition of tumor growth by L-deprenyl involves neural-immune interactions in rats with spontaneously developing mammary tumors. *Anticancer Res.* **19**, 5023–5028.

Thyaga Rajan, S., Madden, K.S., Stevens, S.Y., and Felten, D.L. (2000). Anti-tumor effect of L-deprenyl is associated with enhanced central and peripheral neurotransmission and immune reactivity in rats with carcinogen-induced mammary tumors. *J. Neuroimmunol.* **109**, 95–104.

Tipton, K.F., Boyce, S., O'Sullivan, J., Davey, G.P., and Healy, J. (2004). Monoamine oxidases: certainties and uncertainties. *Curr. Med. Chem.* **11**, 1965–1982.

Tolosa, E., and Katzenschlager, R. (2007). Pharmacologic management of Parkinson's disease. In: J. Jankovic, J. and E. Tolosa, E. (Eds.), *Parkinson's Disease and Movement Disorders.* Lippincott, Williams and Wilkins, Philadelphia, pp. 110–145.

Toronyi, E., Hamar, J., Magyar, K., and Szende, B. (2002). Antiapoptotic effect of (−)-deprenyl in rat kidney after ischemia-reperfusion. *Med. Sci. Monit.* **8**, BR65–BR68.

Valoti, M., Fusi, F., Frosini, M., Pessina, F., Tipton, K.F., and Sgaragli, G.P. (2000). Cytochrome P450-dependent N-dealkylation of L-deprenyl in C57BL mouse liver microsomes: effects of in vivo pretreatment with ethanol, phenobarbital, beta-naphthoflavone and L-deprenyl. *Eur. J. Pharmacol.* **391**, 199–206.

Varga, E., and Tringer, L. (1967). Clinical trial of a new type promptly acting psychoenergetic agent (phenyl-isopropyl-methylpropinyl-HCl, "E-250"). *Acta Med. Acad. Sci. Hung.* **23**, 289–295.

Wadia, J.S., Chalmers-Redman, R.M., Ju, W.J., Carlile, G.W., Phillips, J.L., Fraser, A.D., and Tatton, W.G. (1998). Mitochondrial membrane potential and nuclear changes in apoptosis caused by serum and nerve growth factor withdrawal: time course and modification by (−)-deprenyl. *J. Neurosci.* **18**, 932–947.

Wecker, L., James, S., Copeland, N., and Pacheco, M.A. (2003). Transdermal selegiline: targeted effects on monoamine oxidases in the brain. *Biol. Psychiatry* **54**, 1099–1104.

Whaley, J.G., Tharakan, B., Smith, B., Hunter, F.A., and Childs, E.W. (2009). (−)-Deprenyl inhibits thermal injury-induced apoptotic signaling and hyperpermeability in microvascular endothelial cells. *J. Burn Care Res.* **30**, 1018–1027.

Youdim, M.B., and Finberg, J.P. (1987). Monoamine oxidase B inhibition and the "cheese effect". *J. Neural Transm. Suppl.* **25**, 27–33.

Youdim, M.B., and Weinstock, M. (2002). Novel neuroprotective anti-Alzheimer drugs with anti-depressant activity derived from the anti-Parkinson drug, rasagiline. *Mech. Ageing Dev.* **123**, 1081–1086.

Youdim, M.B., Gross, A., and Finberg, J.P. (2001). Rasagiline [N-propargyl-1R(+)-aminoindan], a selective and potent inhibitor of mitochondrial monoamine oxidase B. *Br. J. Pharmacol.* **132**, 500–506.

Zeller, E.A., and Barsky, J. (1952). In vivo inhibition of liver and brain monoamine oxidase by 1-Isonicotinyl-2-isopropyl hydrazine. *Proc. Soc. Exp. Biol. Med.* **81**, 459–461.

TYPE A MONOAMINE OXIDASE REGULATES LIFE AND DEATH OF NEURONS IN NEURODEGENERATION AND NEUROPROTECTION

Makoto Naoi[1], Wakako Maruyama[2], Keiko Inaba-Hasegawa[1] and Yukihiro Akao[3]

[1]Department of Neurosciences, Gifu International Institute of Biotechnology, Kakamigahara, Gifu, Japan

[2]Department of Cognitive Brain Science, National Research Center for Geriatrics and Gerontology, Obu, Aichi, Japan

[3]United Graduate School of Drug Discovery and Medical Information Sciences, Gifu University, Gifu, Japan

Abstract

In Parkinson's disease, type B monoamine oxidase (MAO-B) is proposed to play an important role in the pathogenesis through production of reactive oxygen species and neurotoxins from protoxicants, such as 1-methyl-4-phenyl-1,2,3,6-tetrahydropyridine. In addition, inhibitors of MAO-B protect neurons in the cellular and animal models of Parkinson's and Alzheimer's diseases. However, the role of type A MAO (MAO-A) in neuronal death and neuroprotection by MAO-B inhibitors has been scarcely elucidated. This chapter presents our recent results on the involvement of MAO-A in the activation of mitochondrial death signal pathway and in the induction of prosurvival genes to prevent cell death with MAO-B inhibitors. The roles of MAO-A in the regulation of neuronal survival and death are discussed in concern to find a novel strategy to protect neurons in age-associated neurodegenerative disorders and depression.

INTERNATIONAL REVIEW OF NEUROBIOLOGY, VOL. 100
DOI: 10.1016/B978-0-12-386467-3.00005-4

I. Introduction

Monoamine oxidase [MAO; monoamine: oxygen oxidoreductase (deaminating) EC 1.3.3.4] is a key enzyme in the catabolism of monoamine neurotransmitters and xenobiotic amines. Type A MAO (MAO-A) and type B MAO (MAO-B) are designated according to the substrate specificity and sensitivity to distinct inhibitors. MAO-A preferentially oxidizes serotonin (5-hydroxytryptamine, 5-HT) and norepinephrine (NE) and is selectively inhibited by clorgyline. MAO-B oxidizes β-phenylethylamine and benzylamine, and selegiline [(−)deprenyl] is a typical MAO-B inhibitor. Dopamine (DA) is oxidized by both the isoenzymes. MAO-A and MAO-B are coded by different genes localized closely in the chromosome X (Xp11.23; Shih *et al.*, 1999). They are distributed differentially in distinct cells of the brain regions, indicating their specified role in neuronal functions. MAO is now proposed to play an important role in age-dependent neurodegenerative disorders, including Parkinson's and Alzheimer's diseases, and also in depression (Bortolato *et al.*, 2008; Riederer *et al.*, 2004; Youdim and Bakhle, 2006).

In Parkinson's disease (PD), MAO is considered to be associated with selective loss of DA neurons in the substantia nigra (Cohen *et al.*, 1997). MAO oxidizes DA and other monoamine transmitters and generates reactive oxygen species (ROS). MAO-B expression in the brain increases with aging (Fowler *et al.*, 1997). In PD, the results on MAO-B levels in the brain are conflicting, whereas in Alzheimer's disease (AD), MAO-B activity was reported to increase as a function of activated astrocytes, whereas MAO-A activity decreased (Kennedy *et al.*, 2003). Even subtle elevation of MAO-B activity impaired mitochondrial function (Kumar *et al.*, 2003); therefore, MAO-B may induce a malignant cycle among oxidative stress, mitochondrial dysfunction, impaired ubiquitin–proteasome system, excitotoxicity, and apoptotic signaling to cause neuronal death. On the other hand, selegiline is applied in patients with PD as a monotherapy or an adjunct of L-DOPA therapy to increase available DA, reduce ROS production, and alleviate oxidative stress (Birkmayer *et al.*, 1977). MAO-B oxidized MPTP, a protoxicant, into 1-methyl-4-phenylpyridinium (MPP^+), a parkinsonism-inducing neurotoxin (Heikkila *et al.*, 1984). Selegiline prevented MPTP oxidation and also neurotoxicity of MPP^+. These findings suggest that MAO-B may be a pathogenic factor in PD and MAO-B inhibitors (MAO-B-Is) may function as agents to halt or delay neuronal loss. The "MAO-B" hypothesis was extended to a clinical trial called "Deprenyl and tocopherol antioxidative therapy for parkinsonism" (DATATOP) study (Parkinson Study Group, 1993). However, this trial failed to prove the neuroprotection by selegiline, even though selegiline serves a symptomatic treatment in early PD and shows neuroprotective effects in animal and cellular models.

Recently, more direct role of MAO was found in neuronal death besides the production of ROS and neurotoxins. An endogenous dopaminergic neurotoxin,

N-methyl(*R*)salsolinol (*NMR*Sal; Maruyama *et al.*, 2001; Naoi *et al.*, 2004), bound to MAO-A and activated mitochondrial apoptosis signaling (Naoi *et al.*, 2006; Yi *et al.*, 2006a). MAO-A was upregulated in apoptosis induced by serum-withdrawal in SK-N-BE(2)-C cells, and the cell death was decreased in MAO-A-deficient mice, suggesting that MAO-A might be associated with apoptotic signal pathway (Ou *et al.*, 2006).

On the other hand, MAO-B-Is, selegiline and rasagiline, were proved to protect neuronal cells in animal and cellular models at a dose with little inhibition of MAO-B activity. MAO-B-Is suppress the activation of mitochondrial death signal pathway and prevent "intrinsic" apoptosis, and they increase prosurvival and antiapoptotic molecules, including *Bcl-2*, neurotrophic factors (NTFs), and antioxidant enzymes (Naoi and Maruyama, 2010). MAO-B-Is increased prosurvival genes in MAO-B-expressing astrocytes (Semkova *et al.*, 1996), and also in MAO-A-expressing human SH-SY5Y and rat pheochromocytoma PC12 cells (Akao *et al.*, 2002a,b; Naoi and Maruyama, 2009, 2010).

On the other hand, rasagiline and related propargylamine derivatives of MAO-B-Is require strict enantiomeric structure for the protective function (Maruyama *et al.*, 2002, 2003, 2004a,b), suggesting the occurrence of the binding site for MAO-B-Is. MAO-A was confirmed as one of the targets of MAO-B-Is (Naoi *et al.*, 2007; Naoi and Maruyama, 2009). In addition to irreversible MAO-B-Is, reversible MAO-A-Is, *N*-propargylamine (Yi *et al.*, 2006b), moclobemide (Chiou *et al.*, 2006), and befloxatone (Inaba-Hasegawa *et al.*, in preparation) protected neuronal cells. Therefore, it should be reexamined whether MAO-B-Is require MAO-A or MAO-B for their neuroprotective function.

In this chapter, we review our recent results on the role of MAO-A in regulation of death cascade, using cellular models expressing selectively MAO-A and MAO-B and those, where MAO expression was reduced by RNA interference with small interfering RNA (siRNA). MAO-A was confirmed to be the target protein of the endogenous neurotoxin and associated with mitochondria-dependent apoptosis. On the other hand, MAO-A was involved also in the suppression of apoptotic signal pathway of reversible and irreversible MAO-A- and MAO-B-Is. The "Janus"-faced roles of MAO-A are discussed in relation to the pathogenesis and treatment of age-dependent neurodegenerative disorders and depression. The future development of novel neuroprotective agents might be expected among compounds with affinity to MAO-A and potency to induce prosurvival genes.

II. Type A and B MAO in Cell Death of Neurons

Cell death of selective neurons in distinct brain regions is one of the pathological characteristics in neurodegenerative disorders. The postmortem studies suggest that programmed cell death (PCD) may occur in the brains of patients with

PD, AD, and other neurodegenerative disorders (Jellinger, 2001). Apoptosis, a form of PCD, is initiated by activation of mitochondrial apoptosis pathway or death receptor-mediated signal pathway (Kroemer *et al.*, 2007). The apoptosis cascade is well regulated and activated in a step-wise way, and proposed as a main target of neuroprotection therapy (Mattson *et al.*, 2008). Therefore, this review will be focused on the role of MAO in intrinsic apoptosis initiated by mitochondrial signal pathway.

MaoA and *MaoB* gene are proposed as pathogenic factors for neuronal loss in AD and PD. *MaoB*-gene polymorphisms were reported to be associated with PD (Bialecka *et al.*, 2007; Parsian *et al.*, 2004). In *MaoB*, specific G to A transition polymorphism in intron 13 (Singh *et al.*, 2008), a polymorphic GT repeat in intron 2, and longer repeat units than normal range (186 or 188 bp vs. 168–190 bp) were reported in a subgroup of older PD patients (Gao *et al.*, 2008). On the other hand, *MaoA* gene may be associated with the development, plasticity, and physiological function of the brain, and also with neuropsychiatric disorders and behaviors (Shih and Thompson, 1999). Association of various polymorphisms in *MaoA* was reported with AD (Wu *et al.*, 2007), major depressive disorder (Yu *et al.*, 2005), autism (Cohen *et al.*, 2003) attention-deficient hyperactivity disorder (ADHD; Roohi *et al.*, 2009), and frontotemporal dementia (Reif *et al.*, 2008).

MAO-B is considered to account for neuronal loss through ROS production by the enzymatic oxidation of monoamine substrates. Activated astrocytes expressing MAO-B were found to increase in the brain of patients with PD, AD, and postischemia or posttrauma brain. MAO-B knockout (KO) mice showed to resist MPTP toxicity through reduced enzymatic oxidation of this protoxicant (Grimsby *et al.*, 1997). However, few results are available to support the direct association of MAO-B with neuronal PCD in PD and AD.

On the other hand, MAO-A was confirmed to be directly associated with apoptosis induced by *N*M*R*Sal (Yi *et al.*, 2006a). This neurotoxin induces mitochondrial membrane permeability transition (MMPT), the release of molecules with less than 1.5 kDa, such as cytochrome *c* (Cyt *c*) and proapoptotic factors, the swelling of mitochondria, and the decline in mitochondrial membrane potential ($\Delta\Psi_m$; Akao *et al.*, 2002a; Maruyama *et al.*, 1997, 2001). Released Cyt *c* binds to AIF1 (apoptotic protease-activating factor-1), recruits caspase 9, and activates caspases 3, 6, and 7. Apoptosis is advanced through the cleavage of poly(ADP-ribose) polymerase (PARP) and other death substrates, such as caspase-activated DNase (CAD), the transfer of activated factors into nuclei, and finally the condensation and aggregation of nuclear DNA, accompanied with the translocation of glyceraldehydes-3-phosphate dehydrogenase (GAPDH), as shown in Fig. 1.

Inhibition of MAO-A expression with siRNA decreased *N*M*R*Sal-induced apoptosis in SH-SY5Y cells in accord to the reduction of the RNA, protein, and activity of MAO-A, whereas MAO-B overexpression did not induce MPPT and apoptosis. (Yi *et al.*, 2006a). Some ligands of imidazoline-binding site and the

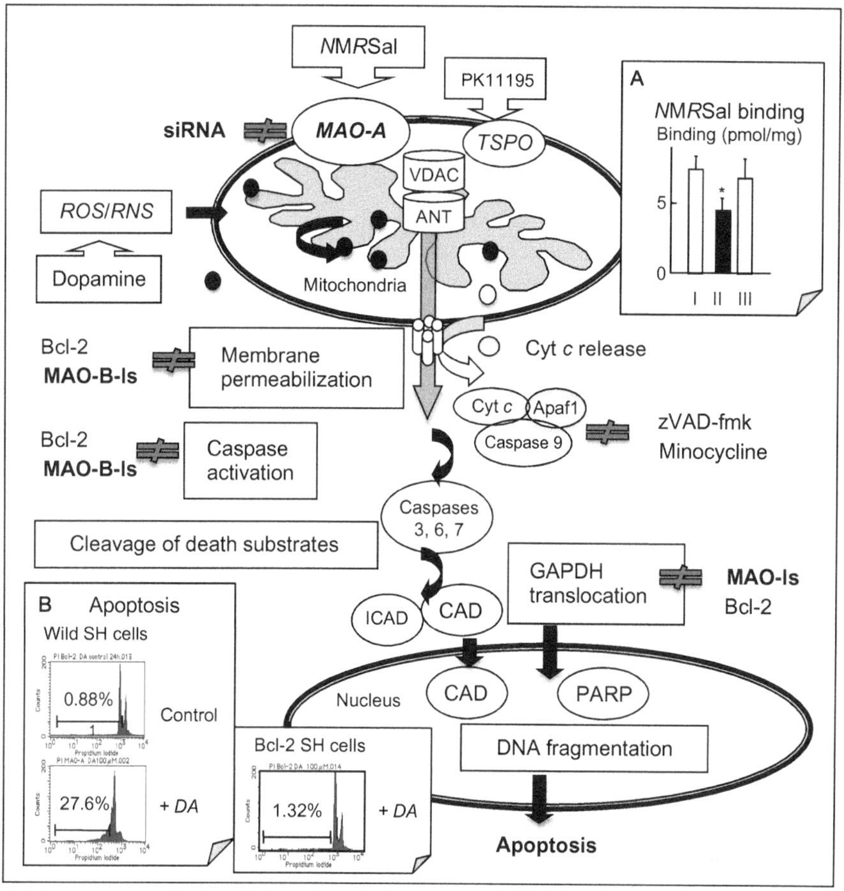

FIG. 1. The role of MAO-A in apoptosis pathway and the regulation of death signaling by Bcl-2 and rasagiline. An endogenous neurotoxin, *N*M*R*Sal, binds to MAO-A and opens MPP, followed by release of Cyt *c* and proapoptotic proteins in the cytoplasm, activation of caspases, and then activates death substrates (CAD) through cleavage of the inhibitory subunit (ICAD), which a caspase inhibitor, benzyloxycarbonylvalyl-alanyl-aspartyl fluoromethyl ketone (zVAD-fmk) and minocycline inhibit. These death substrates enter into nuclei, cleave DNA, and induce apoptosis. Bcl-2 and rasagiline prevent MMPT and apoptosis. (A) The treatment with siRNA reduces *N*M*R*Sal binding to mitochondria. The binding of *N*M*R*Sal was quantified in mitochondria from control (I), MAO-A-siRNA (II), or nonspecific siRNA-treated SH-SY5Y cells (III). (B) Bcl-2 overexpression in SH cells (Bcl-2 SH) reduces DA-treated apoptosis, as shown by FACS. GAPDH is incorporated in nuclei during apoptotic process, and irreversible MAO-B-Is and reversible MAO-A-Is prevent the translocation.

18-kDa-translocator protein [TSPO, formerly called "peripheral benzodiazepine receptor"] target MAO-A to initiate apoptosis (Naoi *et al.*, in preparation). In addition, *MaoA* gene was upregulated in apoptosis of serum-withdrawn PC12 cells, through activation of the p38 mitogen-activated protein kinase (MAPK) signal pathway (De Zutter and Davis, 2001). Also in SK-N-BE(2)-C cells, serum-withdrawal induced apoptosis with increased MAO-A expression through decrease of MAO-A-targeted transcription repressor R1 (Ou *et al.*, 2006). In apoptosis induced by staurosporine, MAO-A activity was increased by a posttranscriptional mechanism in SH-SY5Y cells (Fitzgerald *et al.*, 2007). The elevated MAO-A activity was proposed to increase oxidative stress leading to activate death signal pathway.

The role of MAO-A in neurodegeneration was indirectly suggested also by previous results: the bioactivation of a MPTP analogue, 1-methyl-(2′-methylphenyl)-1,2,3,4-tetrahydropyridine by MAO-A (Heikkila *et al.*, 1988) and the binding of MPP^+ to MAO-A (May, 1993). In nonneuronal cells, MAO-A mediated oxidative stress and apoptosis after ischemia-perfusion in renal cells (Kunduzova *et al.*, 2002) and isolated rat cardiomyocytes (Bianchi *et al.*, 2005).

III. MAO-A in Neuroprotection by MAO-Is

Rasagiline and related MAO-B-Is protect neuronal cells against cell death induced by neurotoxins, oxidative stress, and withdrawal of NTFs, and also in animal models of neurodegenerative disorders (Gelowitz and Paterson, 1999; Huang *et al.*, 1999; Stefanova *et al.*, 2008; Yu *et al.*, 1994). These MAO-B-Is protected DA (Finberg *et al.*, 1998), NE (Finnegan *et al.*, 1990), and cholinergic neurons (Ricci *et al.*, 1992; Speiser *et al.*, 1998) from cytotoxicity of cell type-specified neurotoxins. These results indicate that neuroprotective MAO-Is intervene common death pathway and protect cells, regardless of the cell type and kind of insults.

A. MAO-A in the Suppression of Mitochondrial Death Signal by MAO-B-Is

Rasagiline and selegiline prevent MMPT and protect neuronal cells from PCD. *N*M*R*Sal and a TSPO ligand, PK11195, cause MMPT and release apoptogenic protein in the cytoplasm. Bcl-2 overexpression and rasagiline suppress apoptosis through regulation of MPPT (Fig. 1). Rasagiline prevented cell death caused by peroxynitrite, a reactive nitrogen species (RNS) in SH-SY5Y cells, but could not in MAO-B-expressing U118MG cells, demonstrating the involvement of MAO-A. However, ethanol induced apoptosis in the same U118MG cells through upregulation of GAPDH and MAO-B, which was prevented by rasagiline, its metabolite 1(*R*)-aminoindan, and selegiline (Ou *et al.*, 2009). The conflicting results may be due to the different insults and apoptosis pathways mediated by MPPT or GAPDH translocation.

Rasagiline and selegiline protect neurons by inhibition of PCD at the other points of death pathway. MAO-B-Is scavenge ROS–RNS (Thomas *et al.*, 1997), prevent the nuclear translocation of GAPDH (Tatton *et al.*, 2003), and down-regulate PARP-1 protein (Brabeck *et al.*, 2003). In mitochondria, rasagiline prevents the oxidative stress-caused changes in redox state, regulated by levels of reduced glutathione (GSH) and *S*-glutathionylated protein. *S*-Glutathionylation preserves the higher structure of the mitochondrial electron transport chain (Naoi *et al.*, 2008) and regulates the opening of mitochondrial permeability pore (MPP) through adenine nucleotide translocator (ANT; Costantini *et al.*, 2000).

Not only MAO-B-Is but also MAO-A-specific inhibitors, *N*-propargylamine and clorgyline, can protect neuronal cells. *N*-Propergylamine protected SH-SY5Y cells from cell death induced by neurotoxins, serum-withdrawn, and oxidative stress (Yi *et al.*, 2006b; Yogev-Falach *et al.*, 2003). Reversible MAO-A inhibitors, pirlindole and dehydropirlindole, protected primary cultured rat brain cells against cell death induced by oxidative stress (Boland *et al.*, 2002). Clorgyline reduced malonate-induced striatal damage in mice (Maragos *et al.*, 2004). In N2a cells, clorgyline prevented cell death caused by glutamate (Maher and Davis, 1996) and MPTP (De Girolamo *et al.*, 2001). Clorgyline was considered to protect cells though the MAO inhibition and reduction of ROS production.

B. MAO-A Mediates Prosurvival Bcl-2 Induction by MAO-Is

Bcl-2 protein family regulates MMPT, either in suppressive or promoting way (Shacka and Roth, 2005). The overexpression of neuroprotective *Bcl-2* suppressed the activation of death signal pathway (Maruyama *et al.*, 2001). On the other hand, in lymphocytes from patients with PD, the negative correlation was reported between Bcl-2 level and the duration and severity of the disease (Blandini *et al.*, 2004). Upregulation of proapoptotic Bak, Bad, Bcl-2, and Bcl-xL was also reported by postmortem studies of AD and PD brains (Kitamura *et al.*, 1998). However, these results may not imply a causal relationship between Bcl-2 family protein and neurodegeneration.

Bcl-2 prevents apoptosis by suppressing MMPT and Cyt *c* release (Fig. 1). Bcl-2 inhibits formation of mitochondrial pore, directly by the complex formation with proapoptotic Bax and Bak, or by the interaction with voltage-dependent anion channel (VDAC) and other components of MPP (Tsujimoto and Shimizu, 2000). In addition, Bcl-2 and Bcl-xL stabilize mitochondrial contact site, scavenge ROS–RNS, increase antioxidant enzymes, and keep the mitochondrial function (Jang and Surh, 2004). In apoptosis induced by ceramide and etoposide, Bcl-2 prevented directly caspase-2 activation and suppressed caspase-2- and -9-dependent apoptosis (Lin *et al.*, 2005).

Rasagiline and selegiline induced the mRNA and protein of antiapoptotic Bcl-2 and Bcl-xL, but not proapoptotic Bax. Neither TV1022, (*S*)-enantiomer of rasagiline, or clorgyline affected the Bcl-2 expression (Akao *et al.*, 2002b). Figure 2 shows that MAO-A is required for the induction of Bcl-2 in SH-SY5Y cells (Naoi *et al.*, 2006; Inaba-Hasegawa *et al.*, in preparation). The inhibition of *MaoA* expression with siRNA in SH-SY5Y cells reduced Bcl-2 increase by rasagiline and selegiline. Befloxatone, a reversible MAO-A-I increased Bcl-2 levels, which is consistent with the previous result with moclobemide (Chiou *et al.*, 2006). On the other hand, in MAO-B-expressing Caco-2 cells and U118MG cells, rasagiline

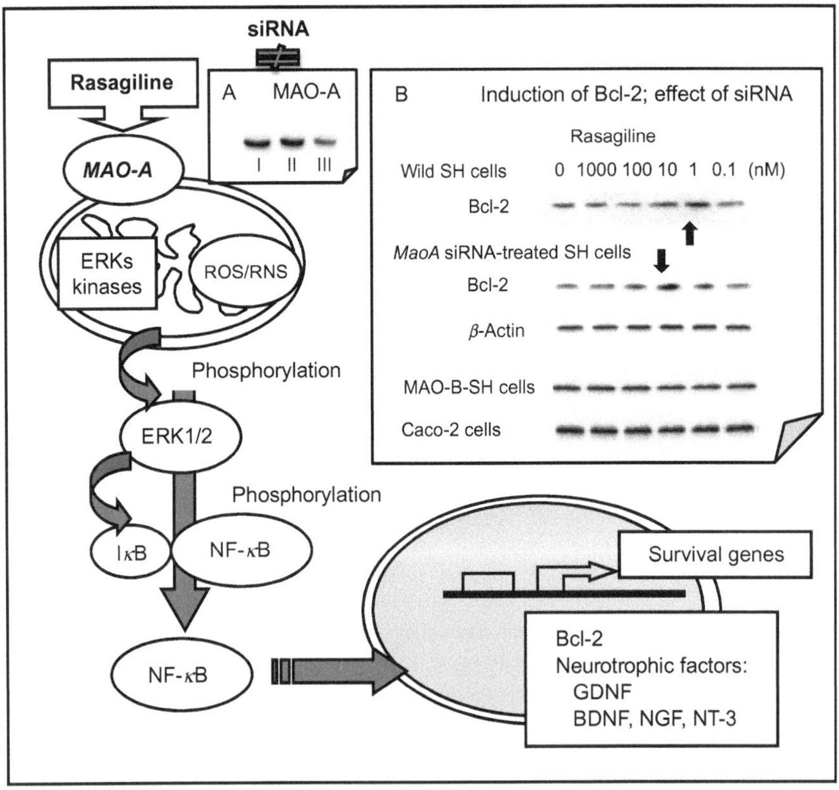

FIG. 2. Induction of Bcl-2 and prosurvival genes by rasagiline analogues and the effect of RNA interference of MAO-A. (A) MAO-A protein is decreased in cells treated with siRNA against *MaoA* (III) from wild (I) and nonspecific siRNA-treated cells (II). (B) Rasagiline increases Bcl-2 protein in wild SH-SY5Y cells, and siRNA treatment reduces the response to rasagiline, according to the reduction MAO-A protein. Rasagiline does not increase Bcl-2 in MAO-B-overexpressed SH cells or MAO-B-containing Caco-2 cells. The induction of prosurvival genes may be mediated through NF-κB activated by ROS–RNS or ERK1/2 or other kinases in mitochondria.

could not increase Bcl-2 expression. In *MaoB*-transfected SH-SY5Y cells, the response to rasagiline was markedly reduced, even though MAO-B protein was increased markedly. These results demonstrate that MAO-A is associated with increase of neuroprotective Bcl-2 protein by reversible and irreversible inhibitors of MAO-A and MAO-B.

IV. MAO-A in Induction of Neurotrophic Factors by MAO-Is

Selegiline and rasagiline increase *in vivo* and *in vitro* the expression of NTFs. NTFs are proposed as promising therapeutic tools for neurodegenerative disorders from their function in development, survival, and maintenance of specific neuronal subpopulations. Distinct NTF groups protect specified population of brain neurons, suggesting their beneficial application for different degenerative disorders and depression. NTFs comprise neurotrophin family, glial cell line-derived neurotrophic factors (GDNF) family ligands, neuropoietic cytokines, and a novel family consisting of mesencephalic astrocyte-derived neurotrophic factor (MANF; Petrova *et al.*, 2003) and cerebral dopamine neurotrophic factor (CDNF; Lindholm *et al.*, 2007). The role of MAO in the induction of neurotrophins and GDNF will be discussed in this review.

Neurotrophin family includes brain-derived neurotrophic factor (BDNF), nerve growth factor (NGF), and neurotrophin-3 and -4/5 (NT-3, NT-4/5). In AD and depression, reduced BDNF levels and Val66Met polymorphism of BDNF are proposed as the pathogenic factors for selective neuronal loss (Feher *et al.*, 2009). GDNF family ligands are a member of transforming growth factor-β (TGF-β), include GDNF, neurturin, artemin, and persephin, and support dopaminergic and motor neurons in the brain (Pascual *et al.*, 2008; Sariola and Saarma, 2003). GDNF is proposed to preserve the function of DA neurons in PD and protection of nigra-striatal DA neurons was confirmed in animal PD model (Eslamboli *et al.*, 2005). However, the clinical trials to treat parkinsonian patients with GDNF could not fully prove the neuroprotective effects (Hong *et al.*, 2008; Lang *et al.*, 2006). The technical difficulties to deliver GDNF effectively in the targeted brain nuclei may account for nonbeneficial results in clinical trials. The upregulation of endogenous levels of GDNF and BDNF with brain-permeable molecules might be a more promising approach to rescue dying neurons.

Rasagiline and related propargylamine derivatives of MAO-B-Is increase NTFs and neuroprotective fibroblast growth factors in animal models. Selegiline increased the expression of basic fibroblast growth factor (bFGF) mRNA in rat brain (Biagini *et al.*, 1994), GDNF mRNA in rat striatum (Tang *et al.*, 1998) and of BDNF protein, but not CDNF, in mouse cortex (Gyarfas *et al.*, 2010). Rasagiline increased the mRNA and protein of BDNF in the rat midbrain (Weinreb *et al.*,

2009). Harmine, a MAO-A inhibitor, increased BDNF levels in rat hippocampus (Fortunato *et al.*, 2010). Rasagiline increased GDNF most prominently, followed by NGF, BDNF, and NT-3 in the cerebrospinal fluid (CSF) from nonhuman primate after the systematical administration, whereas selegiline increased BDNF in the CSF from patients with PD (Maruyama *et al.*, in preparation).

As summarized in Table I, MAO-Is increase NTFs in cellular models. Rasagiline increased mRNA and protein of GDNF in SH-SY5Y cells (Maruyama *et al.*, 2004a) and mRNA of BDNF and GDNF in serum-deprived cells PC12 and SH-SY5Y (Bar-Am *et al.*, 2005; Weinreb *et al.*, 2004; Yogev-Falach *et al.*, 2002, 2003). Selegiline increased mRNA of a ciliary neurotrophic factor (CNTF) in cultured rat astrocytes (Seniuk *et al.*, 1994), NGF in cultured rat cortical astrocytes (Semkova *et al.*, 1996), and BDNF and NT-3 in embryonic stem cells (Esmaeli *et al.*, 2006). Selegiline and desmethylselegine increased mRNA of NGF, BDNF, and GDNF in cultured mouse astrocytes (Mizuta *et al.*, 2000). *R*-(−)-1-(Benzofuran-2-yl)-2-propyl-aminopentane [(−)-BPAP], a MAO-A-I (Knoll *et al.*, 1999) and an antiapoptotic agent (Maruyama *et al.*, 2004b), increased NGF, BDNF, and GDNF in cultured mouse astrocytes (Shimazu *et al.*, 2003). Rasagiline and selegiline preferentially increased the mRNA and protein of GDNF and neurotrophins (BDNF, NGF. 3-NT), respectively, in SH-SY5Y cells (Maruyama *et al.*, in preparation). Recently, we confirmed that MAO-A in SH-SY5Y cells is required for MAO-I-dependent synthesis of BDNA, NGF, and GDNF, but they are not increased in MAO-B-overexpressing cells, as in the case with Bcl-2 induction.

The effects of MAO-Is on NTFs have been studied *in vitro* using only limited types of cultured cell lines: SH-SY5Y and related neuroblastoma cells and astrocytes and related glioblastoma cells (Table I). These cell lines were used for the NTF induction by other small molecules. Antidepressants increased mRNA of neurotrophins and GDNF (Henkel *et al.*, 2008) and an antipsychotic olanzapine enhanced BDNF mRNA in SH-SY5Y cells (Lee *et al.*, 2010). In astrocytes, monoamine neurotransmitters increased NTFs (Mele *et al.*, 2010), and apomorphine (Ohta *et al.*, 2000) and cabergoline, a D2 agonist, simulated the synthesis of NGF, BDNF, and GDNF in cultured mouse astrocytes (Ohta *et al.*, 2004). At present, it remains to be clarified whether these compounds require MAO for the NFT induction.

V. Is MAO-A the Binding Site of MAO-Is for Induction of Neuroprotective Genes?

The binding site of rasagiline analogues in MAO-B has been intensively studied (Binda *et al.*, 2004), but it remains elusive how MAO-Is bind to MAO-A and induce prosurvival genes in nuclei. Rasagiline and selegiline inhibit MAO-A activity in a noncompetitive way to substrate and befloxatone and moclobemide are reversible MAO-A-Is, and all these MAO-Is induce Bcl-2 through binding to

Table I

Induction of Prosurvival Neurotrophic Factors by Rasagiline, Selegiline, and the Related Compounds.

MAO-Is	Induced gene product	Cell type	Type of MAO	Reference
Rasagiline	GDNF mRNA, protein	SH-SY5Y cells	MAO-A	Maruyama *et al.* (2004a)
	BDNF mRNA	PC12 cells	MAO-A	Yogev-Falach *et al.* (2002)
				Weinreb *et al.* (2004)
Rasagiline	BDNF, GDNF mRNA	SH-SY5Y cells	MAO-A	Bar-Am *et al.* (2005)
Rasagiline/selegiline	BDNF, NGF, NT-3, GDNF	SH-SY5Y cells	MAO-A	Maruyama *et al.* (in preparation)
	Protein, mRNA			Inaba-Hasegawa *et al.* (in preparation)
Selegiline	CTNF mRNA	Rat cortical astrocytes	MAO-B	Seniuk *et al.* (1994)
Selegiline	NGF mRNA, protein	Rat cortical astrocytes	MAO-B	Semkova *et al.* (1996)
Selegiline	NGF, BDNF, GDNF mRNA	Mouse astrocytes	MAO-B	Mizuta *et al.* (2000)
Selegiline	BDNF, NT-3	Embryonic stem cells	*	Esmaeli *et al.* (2006)
N-Propargylamine	GDNF, BDNF mRNA	SH-SY5Y cells, serum-deprived	MAO-A	Bar-Am *et al.* (2005)
(−)-BPAP, selegiline	NGF, BDNF, GDNF	Mouse astrocytes	MAO-B	Shimazu *et al.* (2003)
DA, NE, E	NT-3	Rat cortical astrocytes	*	Mele *et al.* (2010)

* MAO type was not mentioned.

MAO-A. These results suggest that the binding of neuroprotective MAO-Is to MAO-A may occur at the site other than the substrate-binding site. Imidazoline-binding domain is localized in MAO-B at the residues between K148 and M222 (Raddatz, *et al.*, 1997), and the imidazoline I_2-site ligands inhibit MAO-A and MAO-B activities in rat liver and adipocytes (Bour *et al.*, 2006; Ozaita *et al.*, 1997). However, an I_2 ligand guanabenze induced apoptosis in SH-SY5Y cells, which was inhibited by rasagiline, suggesting that rasagiline binds to MAO-A at the site different from the imidazoline-binding site (Inaba-Hasegawa *et al.*, in preparation). On the other hand, agmatine an endogenous I_2 ligand showed neuroprotective effects by *in vivo* and *in vitro* experiments (Qiu and Zheng, 2006), but the role of MAO was not studied in these studies.

TSPO is an MPP component, and PK11195 and other TSPO ligands inhibit the activity of MAO-A in preference to MAO-B and induce apoptosis in cells through MPP opening, which rasagiline can suppress completely. However, TSPO ligands do not increase Bcl-2 in SH-SY5Y cells and rasagiline does not bind to TSPO, suggesting that TSPO is not associated with the induction of prosurvival genes (Naoi *et al.*, in preparation). MAO-Is can bind to other proteins: D_2 DA receptor in rat striatal membrane (Levant, 2002), GAPDH (Ou *et al.*, 2009; Tatton *et al.*, 2003), and semicarbazide-sensitive amine oxidase (SSAO, EC 1.4.3.4; Holt *et al.*, 2004). However, it is not clear whether these proteins mediate MAO-I-dependent increase of prosurvival genes.

VI. Signal Pathway for Induction of Prosurvival Genes by MAO-Is

The signal pathway to increase the Bcl-2 and NTF transcription is another interesting issue. The constitutive Bcl-2 expression is regulated by extracellular signal-regulated kinases1/2 (ERK1/2)–nuclear factor-κB (NF-κB) pathway as proved by selective inhibitors of these kinases (Naoi *et al.*, 2007). Rasagiline increased NTFs and Bcl-2 through this ERK1/2–NF-κB pathway (Maruyama *et al.*, 2004a). Rasagiline and structurally related TV3326 [(*N*-propargyl-(3*R*)-aminoindan-5-yl)-ethyl methyl carbamate] and its (3*S*)-enantiomer TV3279 activated protein kinase C (PKC), MAPK, and tyrosine kinase-dependent pathway (Bar-Am *et al.*, 2004; Sagi *et al.*, 2003; Weinreb *et al.*, 2004; Yogev-Falach *et al.*, 2002). As alterative pathways, selegiline activated protein kinase A (PKA)-ERK1/2-c-Myc pathway to induce thioredoxin (Andoh *et al.*, 2005), and selegiline also activated phosphoinositol-3-kinase (PI3K)–nuclear factor-E2-related factor-2 (Nrf2) pathway to induce antioxidant enzymes in SH-SY5Y cells (Nakaso *et al.*, 2006). It has not been confirmed which signal pathway is primarily activated by MAO-Is, and the receptor or binding site of rasagiline and selegiline on the cell surface is not identified.

How does MAO on mitochondrial outer membrane mediate the signals to nuclei? Recently, several factors are proposed as the signals from mitochondria to nuclei. ROS and RNS activate mitochondrial kinases and mediate stress signaling from mitochondria to nuclei through activation of ERK1/2–NF-κB pathway (Biswas *et al.*, 2003; Dagda *et al.*, 2009; Horbinski and Chu, 2005). Activation of ERK1/2–NF-κB pathway in mitochondria functions either prosurvival or proapoptotic signaling, and NF-κB activation promotes cell survival through induction of antioxidant genes with a few exceptions (Morgan and Liu, 2011). It should be further studied whether rasagiline analogues activate prosurvival signal pathway in nuclei via regulation of redox state or through the activation of ERK–NF-κB signal pathway in mitochondria.

VII. Discussion

The association of MAO-A with neuronal death in PD is shown by the fact that MAO-A is a target of dopaminergic neurotoxins. However, we should be cautious to apply the results to neurodegenerative disorders and depression in general, since the cell death is regulated by various genetic and environmental factors. On the other hand, rasagiline and related MAO-B-Is are now proposed as neuroprotective agents in PD and AD, whereas selegiline and reversible MAO-A-Is are used as antidepressants. However, the role of MAO in promoting neuronal survival and ameliorating functional deterioration has not been well clarified. Recent advances in gene engineering to express *Mao* in specified neurons of brain region (Chen *et al.*, 2007) will bring us new insight on the function of MAO at the distinct phase of brain aging.

MAO-A may be associated with the fate of neuronal cells through regulation of 5-HT levels in the brain. 5-HT reduced apoptosis induced by *N*M*R*Sal through inhibiting the binding of the neurotoxin to MAO-A (Yi *et al.*, 2006a). In MAO-KO mice, proliferation of neuronal stem cells was diminished in the developing telencephalom and the deficit persisted in adult mice (Cheng *et al.*, 2010). Inhibition of MAO-A induced the differentiation of neural stem cells into serotonergic cells (Chiou *et al.*, 2006). These results show that MAO-A regulates neuronal development though controlling 5-HT level, which promotes proliferation and inhibits differentiation.

The induction of Bcl-2 and NTFs is clearly associated with MAO, as shown in Fig. 2. Considering that NTF signaling regulates Bcl-2 expression, the NTF induction may be the most vital for the neuroprotective activity of MAO-Is. BDNF binds to its tyrosine kinase receptor, TrkB, and activates prosurvival signal pathways, MAPK/ERK and PI3-kinase/Akt kinase cascade to increase neuroprotective genes (Almeida *et al.*, 2005). NGF downregulates Bim, the proapoptotic BH3-domain-only number of Bcl-2 family, by phosphorylation through MEK/MAP kinase pathway in PC12 cells (Biswas and Greene, 2002). Previous papers present that NTFs are increased by irreversible and reversible MAO Is only in MAO-A- or MAO-B-expressing cells, suggesting that MAO-A or MAO-B is required for the induction of NTFs. However, this issue also should be further studied in details.

In addition to propargylamine derivatives of MAO-B-Is, various reversible and irreversible inhibitors of MAO-A and MAO-B have been developed for the application as neuroprotective agents for AD and PD. VK-28 and M30, 8-hydroxyquinoline derivatives, are inhibitors of MAO-A and MAO-B and iron chelators, and they protected *in vivo* and *in vitro* neuronal cells from neurotoxins and oxidative stress (Gal *et al.*, 2005; Shachar *et al.*, 2004). A MAO-B-I, PF9601N [*N*-(2-propyl)-2-(5-benzyloxy-indolyl) methylamine] protected SH-SY5Y cells against MPP^+ toxicity (Sanz *et al.*, 2008). However, the association of MAO with their neuroprotective functions has not well been confirmed.

Upregulation of *MaoA* expression was observed in apoptosis, suggesting that *MaoA* may be an apoptogenic gene. NGF- or serum-withdrawal increased *MaoA* expression in PC12 or SK-N-BE(2)-C cells, and MAO-A was proposed as a target of proapoptotic p38 MAP kinase (De Zutter and Davis, 2001; Ou *et al.*, 2006). MAO-A synthesis is activated by Sp-1 and regulated by a repressor R1 (Zhu *et al.*, 1994; Chen *et al.*, 2005). However, we recently found that rasagiline increased *MaoA* expression through decrease of R1 in SH-SY5Y cells, whereas selegiline did not increase the expression, as reported in 1242 MG glioma cells (Ekblom *et al.*, 1996). The different effects of rasagiline and selegiline on *MaoA* expression are now studied.

As a conclusion, MAO-A regulates cellular signaling systems, especially ERK-NF-κB pathway, and plays a decisive role in the survival, death, and differentiation of cells in physiological conditions and neuropsychiatric disorders, PD and AD. In depression, MAO-A and BDNF may be not only associated with the etiology but also the response to the treatment (Balu *et al.*, 2008; Domschke *et al.*, 2008). MAO-A variants may determine the treatment responses, which may be ascribed to the difference in MAO-A-mediated BDNF induction by antidepressants. Our recent results on the different induction of neurotrophins and GDNF analogues by selegiline and rasagiline suggest the opportunities to increase endogenous NTFs specific for selective neuronal system: BDNF in depression and GDNF in PD. Compounds with high affinity to MAO-A should be screened for the induction of prosurvival genes and new therapeutic strategy will be developed for protection of neuronal cells in neurodegenerative disorders and depression.

Acknowledgment

This work was supported by the Research Grant for Longevity Sciences (21A-13) from the Ministry of Health, Labour, and Welfare (W. M. and M. N.).

References

Akao, Y., Maruyama, W., Shimizu, S., Yi, H., Shamoto-Nagai, M., Youdim, M.B., Tsujimoto, Y., and Naoi, M. (2002a). Mitochondrial permeability transition mediates apoptosis induced by *N*-methyl (*R*)salsolinol, an endogenous neurotoxin, and is inhibited by Bcl-2 and rasagiline, *N*-propargyl-1(*R*)-aminoindan. *J. Neurochem.* **82**, 913–923.

Akao, Y., Maruyama, W., Yi, H., Shamoto-Nagai, M., Youdim, M.B., and Naoi, M. (2002b). An anti-Parkinson's disease drug, *N*-propargyl-1(*R*)-aminoindan (rasagiline) enhances expression of antiapoptotic bcl-2 in human dopaminergic SH-SY5Y cells. *Neurosci. Lett.* **326**, 105–108.

Almeida, R.D., Manadas, B.J., Melo, C.V., *et al.* (2005). Neuroprotection by BDNF against glutamate-induced apoptotic cell death is mediated by ERK and PI3-kinase pathways. *Cell Death Differ.* **12**, 1329–1343.

Andoh, T., Chock, P.B., Murphy, D.L., and Chiueh, C.C. (2005). Role of the redox protein thioredoxin in cytoprotective mechanism evoked by (−)-deprenyl. *Mol. Pharmacol.* **68**, 1408–1414.

Balu, D.T., Hoshaw, B.A., Malberg, J.E., Rosenzweig-Lipson, S., Schechter, L.E., and Lucki, I. (2008). Differential regulation of central BDNF protein levels by antidepressant and non-antidepressant drug treatments. *Brain Res.* **1211**, 37–43.

Bar-Am, O., Yogev-Talach, M., Amit, T., Sagi, Y., and Youdim, M.B.H. (2004). Regulation of protein kinase C by the anti-parkinson drug, MAO-B inhibitor, rasagiline and its derivatives, *in vivo*. *J. Neurochem.* **89**, 1119–1125.

Bar-Am, O., Weinreb, O., Amit, T., and Youdim, M.B. (2005). Regulation of Bcl-2 protein family proteins, neurotrophic factors, and APP processing in the neurorescue activity of propargylamine. *FASEB J.* **19**, 1899–1901.

Biagini, G., Frasoldati, A., Fuxe, K., and Agnati, L.F. (1994). The concept of astrocyte-kinetic drug in the treatment of neurodegenerative diseases: evidence for L-Deprenyl-induced activation of reactive astrocytes. *Neurochem. Int.* **25**, 17–22.

Bialecka, M., Klodowska-Duda, G., Honczarenko, K., Gawrońska-Szklarz, B., Opala, G., Safranow, K., and Droździk, M. (2007). Polymorphisms of catechol-O-methyltransferase (COMT), monoamine oxidase B (MAOB), N-acetyltransferase 2 (NAT2) and cytochrome P450 2D6 (CYP2D6) gene in patients with early onset of Parkinson's disease. *Parkinsonism Relat. Disord.* **13**, 224–229.

Bianchi, P., Kunduzova, O., Masini, E., *et al.* (2005). Oxidative stress by monoamine oxidase mediates receptor-independent cardiomyocyte apoptosis by serotonin and postischemic myocardial injury. *Circulation* **112**, 3297–3305.

Binda, C., Hubalek, F., Li, M., Herzig, Y., Sterling, J., Edmondson, D.E., and Mattevi, A. (2004). Crystal structures of monoamine oxidase B in complex with four inhibitors of *N*-propargylaminoindan class. *J. Med. Chem.* **47**, 1767–1774.

Birkmayer, W., Riederer, P., Ambrozi, L., and Youdim, M.B.H. (1977). Implication of combined treatment with 'Madopar' and L-deprenyl in Parkinson's disease. *Lancet* **I**, 439–443.

Biswas, S.C., and Greene, L.A. (2002). Nerve growth factor (NGF) down-regulates the Bcl-2 homology 3 (BH3) domain-only protein Bim and suppresses its proapoptotic activity by phosphorylation. *J. Biol. Chem.* **277**, 40511–49516.

Biswas, G., Anandatheetthavarada, H.K., Zaidi, M., and Avadhani, N.G. (2003). Mitochondria to nucleus stress signaling: a distinctive mechanism of NFκB/Rel activation through calcineurin-mediated inactivation of IκBβ. *J. Cell Biol.* **161**, 507–519.

Blandini, F., Cosentino, M., Mangiagalli, A., *et al.* (2004). Modification of apoptosis-related protein levels in lymphocytes of patients with Parkinson's disease. The effect of dopaminergic treatment. *J. Neural Transm.* **111**, 1017–1030.

Boland, A., Gerardy, J., Mossay, D., Delapierrre, D., and Seutin, V. (2002). Pirlindole and dehydropirlindole protect rat cultured neuronal cells against oxidative stress-induced cell death through a mechanism unrelated to MAO-A inhibition. *Br. J. Pharmacol.* **135**, 713–720.

Bortolato, M., Chen, K., and Shih, J.C. (2008). Monoamine oxidase inactivation: from pathophysiology to therapeutics. *Adv. Drug Deliv. Rev.* **60**, 1527–1533.

Bour, S., Iglesias-Osman, M.-C., Marti, L., *et al.* (2006). The imidazoline I_2-site ligands BU224 and 2-BFI inhibit MAO-A and MAO-B activities, hydrogen peroxide production, and lipolysis in rodent and human adipocytes. *Eur. J. Pharmacol.* **552**, 20–30.

Brabeck, C., Pfeiffer, R., Leake, A., Beneke, S., Meyer, R., and Bürkle, A. (2003). L-Selegiline potentiates the cellular poly(ADP-ribosyl)lation response to ionizing radiation. *J. Pharmacol. Exp. Ther.* **306**, 973–979.

Chen, K., Ou, X.-M., Chen, G., Choi, S.H., and Chih, J.C. (2005). R1, a novel repressor of the human monoamine oxidase A. *J. Biol. Chem.* **280**, 11552–11559.

Chen, K., Cases, O., Rebrin, I., Wu, , Gallaher, T.K., Seif, I., and Shih, J.C. (2007). Forebrain-specific expression of monoamine oxidase A reduces neurotransmitter levels, restores the brain structure, and rescues aggressive behavior in monoamine oxidase A-deficient mice. *J. Biol. Chem.* **282**, 115–123.

Cheng, A., Scott, A.L., Ladenheim, B., *et al.* (2010). Monoamine oxidases regulate telencephalic neural progenitors in late embryonic and early postnatal development. *J. Neurosci.* **30**, 10752–10762.

Chiou, S.H., Ku, H.H., Tsai, T.H., Lin, H.L., Chen, L.H., Chien, C.S., Ho, L.L.-T., Lee, C.H., and Chang, Y.L. (2006). Moclobemide upregulated Bcl-2 expression and induce neural stem cell differentiation into serotoninergic neurons via extracellular-regulated kinase pathway. *Br. J. Pharmacol.* **148**, 587–598.

Cohen, G., Farooqui, P., and Kesler, N. (1997). Parkinson disease: a new link between monoamine oxidase and mitochondrial electron flow. *Proc. Natl. Acad. Sci. USA* **94**, 4890–4894.

Cohen, I.L., Liu, X., Schutz, C., White, B.N., Jenkins, E.C., Brown, W.T., and Holden, J.J.A. (2003). Association of autism severity with a monoamine oxidase A functional polymorphism. *Clin. Genet.* **64**, 190–197.

Costantini, J., Rees, H.D., Weintraub, S.T., *et al.* (2000). Oxidation of a critical thiol residue of the adenine nucleotide translocator enforces Bcl-2-independent permeability transition pore opening and apoptosis. *Oncogene* **19**, 307–314.

Dagda, R.K., Zhu, J., and Chu, C.T. (2009). Mitochondrial kinases in Parkinson's disease: converging insights from neurotoxin and genetic models. *Mitochondrion* **9**, 289–298.

De Girolamo, L.A., Hargreaves, A.J., and Billett, E.E. (2001). Protection from MPTP-induced neurotoxicity in differentiating mouse N2a neuroblastoma cells. *J. Neurochem.* **76**, 650–660.

De Zutter, G.S., and Davis, R.J. (2001). Pro-apoptotic gene expression mediated by the p38 mitogen-activated protein kinase signal transduction pathway. *Proc. Natl. Acad. Sci. USA* **98**, 6169–6173.

Domschke, K., Hohoff, C., Mortensen, L.S., Roehrs, T., Deckert, J., Arolt, V., and Baune, B.C. (2008). Monoamine oxidase A variant influences antidepressant treatment response in female patients with Major Depression. *Progr. Neuropsychopharmacol. Biol. Psychiatry* **32**, 224–228.

Ekblom, J., Zhu, Q.-S., Chen, K., and Shih, J.C. (1996). Monoamine oxidase gene transcription in human cell lines: treatment with psychoactive drugs and ethanol. *J. Neural Transm.* **103**, 681–692.

Eslamboli, A., Georgievska, B., Ridley, R.M., *et al.* (2005). Continuous low-level glial cell line-derived neurotrophic factor delivery using recombinant adeno-associated viral vectors provides neuroprotection and induces behavioral recovery in a primate model of Parkinson's disease. *J. Neurosci.* **25**, 769–777.

Esmaeli, F., Tiraihi, T., Movahedin, M., and Mowla, S.J. (2006). Selegiline induces neuronal phenotype and neurotrophins expression in embryonic stem cells. *Rejuvenation Res.* **9**, 475–484.

Feher, A., Juhasz, A., Rimanoczy, A., Kalman, J., and Janka, Z. (2009). Association between BDNF Val66Met polymorphism and Alzheimer disease, dementia with Lewy bodies, and Pick disease. *Alzheimer Dis. Assoc. Disord.* **23**, 224–228.

Finberg, J.P.M., Takeshima, T., Johnston, J.M., and Commissiong, J.W. (1998). Increased survival of dopaminergic neurons by rasagiline, a monoamine oxidase B inhibitor. *Neuroreport* **9**, 703–707.

Finnegan, K.T., Skratt, J.J., Irwin, I., DeLanney, L.E., and Langston, J.W. (1990). Protection against DSP-4-induced neurotoxicity by deprenyl is not relayed to its inhibition of MAO B. *Eur. J. Pharmacol.* **184**, 119–126.

Fitzgerald, J.C., Ufer, C., De Girolamo, L.A., Kuhn, H., and Billet, E.E. (2007). Monoamine oxidase-A modulates apoptotic cell death induced by staurosporine in human neuroblastoma cells. *J. Neurochem.* **103**, 2189–2199.

Fortunato, J.J., Reus, G.Z., Kirsch, T.R., *et al.* (2010). Chronic administration of harmine elicits antidepressant-like effects and increases BDNF levels in rat hippocampus. *J. Neural Transm.* **117**, 1131–1137.

Fowler, J.S., Volkow, N.D., Wang, G.J., Logan, J., Pappas, N., Shea, C., and MacGregor, R. (1997). Age-related increases in brain monoamine oxidase B in living healthy human subjects. *Neurobiol. Aging* **198**, 431–435.

Gal, S., Zheng, H., Fridkin, M., and Youdim, M.B.H. (2005). Novel multifunctional neuroprotective iron chelator-monoamine oxidase inhibitor drugs for neurodegenerative diseases. *In vivo* selective brain monoamine oxidase inhibition and prevention of MPTP-induced striatal dopamine depletion. *J. Neurochem.* **95**, 79–88.

Gao, X., Scott, W.K., Wang, G., Mayhew, G., Li, Y.J., Vance, J.M., and Martin, E.R. (2008). Gene-gene interaction between FGF20 and MAOB in Parkinson disease. *Ann. Hum. Genet.* **72**(Pt 2), 157–162.

Gelowitz, D.L., and Paterson, I.A. (1999). Neuronal sparing and behavioral effects of the antiapoptotic dug, (−)deprenyl, following kainic acid administration. *Pharmacol. Biochem. Behav.* **62**, 255–262.

Grimsby, J., Toth, M., Chen, K., *et al.* (1997). Increased stress response and β-phenylethylamine in MAOB-deficient mice. *Nat. Genet.* **17**, 206–210.

Gyarfas, T., Knuuttila, J., Lindholm, P., Rantamaki, T., and Castren, E. (2010). Regulation of brain-derived neurotrophic factors (BDNF) and cerebral dopamine neurotrophic factor (CDNF) by anti-parkinsonian drug therapy in vivo. *Cell. Mol. Neurobiol.* **30**, 361–368.

Heikkila, R.E., Manzino, L., Cabbat, F.S., and Duvoisin, R.N. (1984). Protection against dopaminergic neurotoxicity of 1-methyl-4-phenyl-1,2,3,6-tetrahydropyridine by monoamine oxidase inhibitors. *Nature* **311**, 467–469.

Heikkila, R.E., Kindt, M.V., Sonsalla, P.K., Giovanni, A., Youngster, S.K., McKeown, K.A., and Singer, T.P. (1988). Importance of monoamine oxidase A in the bioactivation of neurotoxic analogs of 1-methyl-4-phenyl-1,2,3,6-tetrahydroptridine. *Proc. Natl. Acad. Sci. USA* **85**, 6172–6176.

Henkel, A.W., Sperling, W., Rotter, A., *et al.* (2008). Antidepressant drugs modulates growth factors in cultured cells. *BMC Pharmacol.* **8**, 6.

Holt, A., Wieland, B., and Boulton, A.A. (2004). On the binding site of monoamine oxidase inhibitors to some sites distinct from the MAO active site, and effects thereby elicited. *Neurotoxicology* **25**, 251–266.

Hong, M., Mukhida, K., and Mendez, I. (2008). GDNF therapy for Parkinson's disease. *Expert Rev. Neurother.* **8**, 11125–11139.

Horbinski, C., and Chu, C.T. (2005). Kinase signaling cascades in the mitochondria: a matter of life or death. *Free Radic. Biol. Med.* **38**, 2–11.

Huang, W., Chen, Y., Shohami, E., and Weinstock, M. (1999). Neuroprotective effect of rasagiline, a selective monoamine oxidase-B inhibitor, against closed head injury in the mouse. *Eur. J. Pharmacol.* **366**, 127–135.

Inaba-Hasegawa, K., Akao, Y., Maruyama, W., and Naoi, M. Type A monoamine oxidase is associated with the induction of neuroprotective Bcl-2 by rasagiline, an inhibitor of type B monoamine oxidase. (in preparation).

Jang, J.-H., and Surh, Y.-J. (2004). Bcl-2 attenuation of oxidative cell death is associated with up-regulation of γ-glutamyl cysteine ligase via constitutive NF-κB activation. *J. Biol. Chem.* **279**, 38779–38786.

Jellinger, K.A. (2001). Cell death mechanisms in neurodegeneration. *J. Cell. Mol. Med.* **5**, 1–17.

Kennedy, B.P., Ziegler, M.G., Alford, M., Hansen, L.A., Thal, L.J., and Masliah, E. (2003). Early and persistent alterations in frontal cortex MAO A and B in Alzheimer's disease. *J. Neural Transm.* **110**, 789–801.

Kitamura, Y., Shimohama, S., Kamoshima, W., *et al.* (1998). Alteration of proteins regulating apoptosis, Bcl-2, Bcl-x, Bax, Bak, Bad, ICH-1 and CPP32 in Alzheimer's disease. *Brain Res.* **780**, 260–269.

Knoll, J., Yoneda, F., Knoll, B., Ohde, H., and Miklya, I. (1999). (−)1-(Benzofuran-2-yl)-2-propylaminopentane, [(−)BPAP], a selective enhancer of the impulse propagation mediated release of catecholamines and serotonin in the brain. *Br. J. Pharmacol.* **128**, 1723–1732.

Kroemer, G., Galluzzi, L., and Brenner, C. (2007). Mitochondrial membrane permeabilization in cell death. *Physiol. Rev.* **87**, 99–163.

Kumar, M.J., Nicholls, D.G., and Andersen, J.A. (2003). Oxidative α-ketoglutarate dehydrogenase inhibition via subtle elevation in monoamine oxidase B levels results in loss of spare respiratory capacity. *J. Biol. Chem.* **278**, 46432–46439.

Kunduzova, O.R., Bianchi, P., Pizzinat, N., Escourrou, G., Seguelas, M.-H., Parini, A., and Cambon, G. (2002). Regulation of JNK/ERK activation, cell apoptosis, and tissue regeneration by monoamine oxidases after renal ischemia-reperfusion. *FASEB J.* **16**, 1129–1131.

Lang, A.E., Gill, S., Patel, N.K., *et al.* (2006). Randomized controlled trial of intraputmental glial cell line-derived neurotrophic factor infusion in Parkinson's disease. *Ann. Neurol.* **59**, 459–466.

Lee, J.G., Cho, H.Y., Park, S.W., Seo, M.K., and Kim, Y.H. (2010). Effects of olanzapine on brain-derived neurotrophic factor gene promotor activity in SH-SY5Y neuroblastoma cells. *Prog. Neuropsychopharm. Biol. Psychiatry* **34**, 1001–1006.

Levant, B. (2002). Novel drug interactions at D_2 dopamine receptors: modulation of [^{3}H]quinpirole binding by monoamine oxidase inhibitors. *Life Sci.* **71**, 2691–2700.

Lin, C.-F., Chen, C.-L., Chang, W.-T., *et al.* (2005). Bcl-2 rescues ceramide- and etoposide-induced mitochondrial apoptosis through blockage of caspase-2 activation. *J. Biol. Chem.* **280**, 23758–23765.

Lindholm, P., Voutilainen, M.H., Lauren, J., *et al.* (2007). Novel neurotrophic factor CDNF protects and rescues midbrain dopamine neurons *in vivo*. *Nature* **448**, 73–77.

Maher, P., and Davis, J.B. (1996). The role of monoamine metabolism in oxidative glutamate toxicity. *J. Neurosci.* **16**, 6394–6401.

Maragos, W.F., Young, K.L., Altman, C.S., *et al.* (2004). Striatal damage and oxidative stress induced by the mitochondrial toxin malonate are reduced in clorgyline and MAO-A deficient mice. *Neurochem. Res.* **29**, 741–746.

Maruyama, W., Naoi, M., Kasamatsu, T., Hashizume, Y., Takahashi, T., Kohda, K., and Dostert, P. (1997). An endogenous dopaminergic neurotoxin, *N*-methyl-(*R*)-salsolinol, induces DNA damage in human dopaminergic neuroblastoma SH-SY5Y cells. *J. Neurochem.* **69**, 322–329.

Maruyama, W., Akao, Y., Youdim, B.B.H., Davis, B.A., and Naoi, M. (2001). Transfection-enforced Bcl-2 overexpression and an anti-Parkinson drug, rasagiline, prevent nuclear accumulation of glyceraldehyde-3-phosphate dehydrogenase induced by an endogenous dopaminergic neurotoxin, *N*-methyl(*R*)salsolinol. *J. Neurochem.* **68**, 727–735.

Maruyama, W., Takahashi, T., Youdim, M., and Naoi, M. (2002). The anti-parkinson drug, rasagiline, prevents apoptotic DNA damage induced by peroxynitrite in human dopaminergic neuroblastoma SH-SY5Y cells. *J. Neural Transm.* **109**, 467–481.

Maruyama, W., Weinstock, M., Youdim, M.B.H., Nagai, M., and Naoi, M. (2003). Anti-apoptotic action of anti-Alzheimer drug, TV3326 [(*N*-propargyl)-(3*R*)-amnoindan-5-yl]-ethyl methyl carbamate, a novel cholinesterase-monoamine oxidase inhibitor. *Neurosci. Lett.* **341**, 233–236.

Maruyama, W., Nitta, A., Shamoto-Nagai, M., Hirata, Y., Akao, Y., Youdim, M., Furukawa, S., Nabeshima, T., and Naoi, M. (2004a). *N*-Propargyl-1-(*R*)-aminoindan, rasagiline, increases glial cell line-derived neurotrophic factor (GDNF) in neuroblastoma SH-SY5Y cells through activation of NF-κB transcription factor. *Neurochem. Int.* **44**, 293–400.

Maruyama, W., Yi, H., Takahashi, T., Shimazu, S., Ohde, H., Yoneda, F., Iwasa, K., and Naoi, M. (2004b). Neuroprotective function of *R*-(−)1-(benzofuran-2-yl)-2-propyl-aminopentane, [*R*-(−)-BPAP], against apoptosis induced by *N*-methyl(*R*)salsolinol, an endogenous dopaminergic neurotoxin, in human dopaminergic neuroblastoma SH-SY5Y cells. *Life Sci.* **75**, 107–117.

Maruyama, W., Nitta, A., Suzuki, J., and Naoi, M. Induction of glial cell line-derived neurotrophic factor and neurotrophin in non-human primate by rasagiline and selegiline. (in preparation).

Mattson, M.P., Gleichmann, M., and Cheng, A. (2008). Mitochondria in neuroplasticity and neurological disorders. *Neuron* **60**, 748–766.

May, T. (1993). 1-Methyl-4-phenylpyridinium (MPP^+) binds with high affinity to a β-carboline binding site localized on monoamine oxidase type A in rat brain. *Neurosci. Lett.* **162**, 55–58.

Mele, T., Carman-Krzan, M., and Juric, D.M. (2010). Regulatory role of monoamine neurotransmitters in astrocytic NT-3 synthesis. *Int. J. Dev. Neurosci.* **28**, 13–19.

Mizuta, I., Ohta, M., Ohta, K., Nishimura, M., Mizuta, E., Hayashi, K., and Kuno, S. (2000). Selegiline and desmethylselegiline stimulate NGF, BDNF, and GDNF synthesis in cultured mouse astrocytes. *Biochem. Biophys. Res. Commun.* **279**, 751–755.

Morgan, M.J., and Liu, Z.-G. (2011). Crosstalk of reactive oxygen species and NF-κB signaling. *Cell Res.* **21**, 103–115.

Nakaso, K., Nakamura, C., Sato, H., Imamura, K., Takesima, T., and Nakashima, K. (2006). Novel cytoprotective mechanism of anti-parkinsonian drug deprenyl: PI3K and Nrf2-derived induction of antioxidative proteins. *Biochem. Biophys. Res. Commun.* **339**, 915–922.

Naoi, M., and Maruyama, W. (2009). Functional mechanism of neuroprotection by inhibitors of type B monoamine oxidase in Parkinson's disease. *Expert Rev. Neurother.* **9**, 1233–1250.

Naoi, M., and Maruyama, W. (2010). Monoamine oxidase inhibitors as neuroprotective agents in age-dependent neurodegenerative disorders. *Curr. Pharmaceut. Design* **16**, 2799–2817.

Naoi, M., Maruyama, W., and Nagy, G.M. (2004). Dopamine-derived salsolinol derivatives as endogenous monoamine oxidase inhibitors: occurrence, metabolism and function in human brain. *Neurotoxicology* **25**, 193–204.

Naoi, M., Maruyama, W., Akao, Y., Yi, H., and Yamaoka, T. (2006). Involvement of type A monoamine oxidase in neurodegeneration: regulation of mitochondrial signaling leading cell death or neuroprotection. *J. Neural Transm. Suppl.* **71**, 67–77.

Naoi, M., Maruyama, W., Yi, H., Akao, Y., Yamaoka, Y., and Shamoto-Nagai, M. (2007). Neuroprotection by propargylamines in Parkinson's disease: intracellular mechanism underlying the anti-apoptotic function and search for clinical markers. *J. Neural Transm. Suppl.* **72**, 121–131.

Naoi, M., Maruyama, W., Yi, H., *et al.* (2008). Neuromelanin selectively induces apoptosis in dopaminergic SH-SY5Y cells by deglutathionylation in mitochondria: involvement of the protein and melanin component. *J. Neurochem.* **105**, 2489–2500.

Naoi, M., Maruyama, W., and Yi, H. Role of monoamine oxidase in apoptosis induced by imidazolines and ligands for 18-kDA-translocator protein. (in preparation).

Ohta, M., Mizuta, I., Ohta, K., Nishimura, M., Mizuta, E., Hayashi, K., and Kuno, S. (2000). Apomorphine up-regulates NGF and GDNF synthesis in cultured mouse astrocytes. *Biochem Biophys. Res. Commun.* **272**, 18–22.

Ohta, K., Fujinami, A., Kuo, S., Sakakimoto, A., Matsui, H., Kawahara, Y., and Ohta, M. (2004). Cabergoline stimulates synthesis and secretion of nerve growth factor, brain-derived neurotrophic factor and glial cell line-derived neurotrophic factor by mouse astrocytes in primary culture. *Pharmacology* **71**, 162–168.

Ou, X.-M., Chen, K., and Shih, J.C. (2006). Monoamine oxidase A and repressor R1 are involved in apoptotic signaling pathway. *Proc. Natl. Acad. Sci. USA* **103**, 10923–10928.

Ou, X.-M., Johnson, C., Chen, K., Youdim, M.B., Rajkowska, G., and Shih, J.C. (2009). Glyceraldehyde-3-phosphare dehydrogenase-monoamine oxidase B-mediated cell death-induced by ethanol is prevented by rasagiline and 1-R-aminoindan. *Neurotox. Res.* **16**, 148–159.

Ozaita, A., Olmos, G., Boronat, A., Lizcano, J.M., Unzeta, M., and Garcia-Sevilla, J.A. (1997). Inhibition of monoamine oxidase A and B activities by imidazol(ine)/guanidine dugs, nature of the interaction and distinction from I_2-imidazoline receptors in rat liver. *Br. J. Pharmacol.* **121**, 901–912.

Parkinson Study Group (1993). Effects of tocopherol and deprenyl on progression of disability in early Parkinson's disease. *N. Engl. J. Med.* **328**, 176–183.

Parsian, A., Racette, B., Zhang, Z.H., Rundle, M., and Perlmutter, J.S. (2004). Association of variations in monoamine oxidases A and B with Parkinson's disease subgroups. *Genomics* **83**, 454–460.

Pascual, A., Hidalgo-Figuerosa, M., Piruat, J., Pintado, C.O., Gomez-Diaz, R., and Lopez-Barneo, J. (2008). Absolute requirement of GDNF for adult catecholaminergic neuron survival. *Nat. Neurosci.* **11**, 755–761.

Petrova, P., Raibekas, A., Pevsner, J., *et al.* (2003). MANF: a new mesencephalic, astrocyte-derived neurotrophic factor with selectivity for dopaminergic neurons. *J. Mol. Neurosci.* **20**, 173–188.

Qiu, W.-W., and Zheng, R.-Y. (2006). Neuroprotective effects of receptor imidazoline 2 and its endogenous ligand agmatine. *Neurosci. Bull.* **22**, 187–191.

Raddatz, R., Parini, A., and Lanier, S.M. (1997). Localization of the imidazoline binding domain on monoamine oxidase B. *Mol. Pharmacol.* **52**, 549–553.

Reif, A., Scapini, E., Venturelli, E., Töpner, T., Fenoglio, C., Lesch, K.P., and Galimberti, D. (2008). The functional MAOA-uVNTR promoter polymorphism in patients with frontotemporal dementia. *Eur. J. Neurol.* **15**, 637–639.

Ricci, A., Mancini, M., Strocchi, P., Bongrani, S., and Bronzetti, E. (1992). Deficits in cholinergic neurotransmission markers induced by ethylcholine mustard aziridinium (AF64A) in the rat hippocampus: sensitivity to treatment with the monoamine oxidase-B inhibitor L-deprenyl. *Drugs Exp. Clin. Res.* **18**, 163–171.

Riederer, P., Lachenmayer, L., and Laux, G. (2004). Clinical applications of MAO-inhibitors. *Curr. Med. Chem.* **11**, 2033–2043.

Roohi, J., DeVincent, C.J., Hatchwell, E., and Gadow, K.D. (2009). Association of a monoamine oxidase-a promoter polymorphism with ADHD and anxiety in boys with autism spectrum disorder. *J. Autism Dev. Disord.* **39**, 67–74.

Sagi, Y., Mandel, S., Amit, T., and Youdim, M.B. (2003). Activation of tyrosine kinase receptor signaling pathway by rasagiline facilitates neurorescue and restoration of nigrostriatal dopamine neurons in post-MPTP-induced parkinsonism. *Neurobiol. Dis.* **25**, 35–44.

Sanz, E., Quintana, A., Battaglia, V., *et al.* (2008). Anti-apoptotic effect of Mao-B inhibitor PF9601N [N-(2-propynyl)-2-(5-benzyloxy-indol) methylamine] is mediated by p53 pathway inhibition in MPP^+-treated SH-SY5Y human dopaminergic cells. *J. Neurochem.* **105**, 2404–2417.

Sariola, H., and Saarma, M. (2003). Novel functions and signaling pathways for GDNF. *J. Cell Sci.* **116**, 3853–3862.

Semkova, I., Wolz, P., Schilling, M., and Krieglstein, J. (1996). Selegiline enhances NGF synthesis and protects central nervous system neurons from excitotoxic and ischemic damage. *Eur. J. Pharmacol.* **315**, 19–30.

Seniuk, N.A., Henderson, J.T., Tatton, W.G., and Roder, J.C. (1994). Increased CNTF gene expression in process-bearing astrocytes following injury is augmented by R(−)-deprenyl. *J. Neurosci. Res.* **37**, 278–286.

Shachar, D.B., Kahana, N., Kampel, V., Warshawsky, A., and Youdim, M.B. (2004). Neuroprotection by a novel brain permeable iron chelator, VK-28, against 6-hydroy dopamine lesion in rats. *Neurophamacology* **46**, 254–263.

Shacka, J.J., and Roth, K.A. (2005). Regulation of neuronal cell death and neurodegeneration by members of the Bcl-2 family: therapeutic implications. *Curr. Drug Targets—CNS Neurol. Dis.* **4**, 25–39.

Shih, J.C., and Thompson, R.F. (1999). Monoamine oxidase in neuropsychiatry and behavior. *Am. J. Hum. Genet.* **65**, 593–598.

Shih, J.C., Chen, K., and Ridd, M.J. (1999). Monoamine oxidase: from genes to behavior. *Annu. Rev. Neurosci.* **22**, 197–217.

Shimazu, S., Tanigawa, A., Sato, N., Yoneda, F., Hayashi, K., and Knoll, J. (2003). Enhancer substrates: selegiline and R-(−)-1-(benzofuran-2-yl9-2-propylaminopentane [(−)-BPAP] enhances the neurotrophic factor synthesis on cultured mouse astrocytes. *Life Sci.* **72**, 2785–2792.

Singh, M., Khan, A.J., Shah, P.P., Shukla, R., Khanna, V.K., and Parmar, D. (2008). Polymorphism in environment responsive genes and association with Parkinson disease. *Mol. Cell. Biochem.* **312**, 131–138.

Speiser, Z., Katzir, O., Rehavi, M., Zabarski, T., and Cohen, S. (1998). Sparing by rasagiline (TVP-1012) of cholinergic functions and behavior in the postnatal anoxia rat. *Pharmacol. Biochem. Behav.* **60**, 378–393.

Stefanova, N., Poewe, W., and Wenning, G.K. (2008). Rasagiline is neuroprotective in a transgenic model of multiple system atrophy. *Exp. Neurol.* **210**, 421–427.

Tang, Y.P., Ma, Y.L., Chao, C.C., Chen, K.Y., and Lee, E.H.Y. (1998). Enhanced glial cell line-derived neurotrophic factor mRNA expression upon (−)-deprenyl and melatonin treatment. *J. Neurosci. Res.* **53**, 593–604.

Tatton, W., Chalmers-Redman, R., and Tatton, N. (2003). Neuroprotection by deprenyl and other propargylamines: glyceraldehydes-3-phosphate dehydrogenase rather than monoamine oxidase B. *J. Neural Transm.* **110**, 509–515.

Thomas, C.E., Huber, E.W., and Ohlweiler, D.F. (1997). Hydroxyl and peroxyl radical trapping by the monoamine oxidase-B inhibitors deprenyl and MDL 72, 974A: implications for protection of biological substrates. *Free Radic. Biol. Med.* **22**, 733–737.

Tsujimoto, Y., and Shimizu, S. (2000). VDAC regulation by the Bcl-2 family proteins. *Cell Death Differ.* **7**, 1174–1181.

Weinreb, O., Bar-Am, O., Amit, T., Chillag-Talmor, O., and Youdim, M.B.H. (2004). Neuroprotection via pro-survival protein kinase C isoforms associated with Bcl-2 family members. *FASEB J.* **18**, 1471–1473.

Weinreb, O., Amit, T., Sagi, Y., Drigues, N., and Youdim, M.B.H. (2009). Genomic and proteomic study to survey the mechanism of action of the anti-Parkinson's disease drug, rasagiline compared with selegiline, in the rat brain. *J. Neural Transm.* **116**, 1456–1472.

Wu, Y.H., Fisher, D.F., and Swaab, D.F. (2007). A promotor polymorphism in the monoamine oxidase A gene is associated with the pineal MAOA activity in Alzheimer's disease patients. *Brain Res.* **116**, 13–19.

Yi, H., Akao, Y., Maruyama, W., Chen, K., Shih, J.C., and Naoi, M. (2006a). Type A monoamine oxidase is the target of an endogenous dopaminergic neurotoxin, *N*-methyl(*R*)salsolinol, leading to apoptosis in SH-SY5Y cells. *J. Neurochem.* **96**, 541–549.

Yi, H., Maruyama, W., Akao, Y., Takahashi, T., Iwasa, K., Youdim, M.B., and Naoi, M. (2006b). *N*-Propargylamine protects SH-SY5Y cells from apoptosis induced by an endogenous neurotoxin, *N*-methyl(*R*)salsolinol, through stabilization of mitochondrial membrane and induction of anti-apoptotic Bcl-2. *J. Neural Transm.* **113**, 21–32.

Yogev-Falach, M., Amit, T., Bar-Am, O., Weinstock, M., and Youdim, M.B.H. (2002). The involvement of mitogen-activated protein (MAP) kinase in the regulation of amyloid precursor protein processing by novel cholinesterase inhibitors derived from rasagiline. *FASEB J.* **16**, 1674–1676.

Yogev-Falach, M., Amit, T., Bar-Am, O., and Youdim, M.B.H. (2003). The importance of propargyl moiety in the anti-Parkinson drug rasagiline and its derivatives for MAP-dependent amyloid precursor protein processing. *FASEB J.* **16**, 2325–2327.

Youdim, M.B.H., and Bakhle, Y.S. (2006). Monoamine oxidase: isoforms and inhibitors in Parkinson's disease and depressive illness. *Br. J. Pharmacol.* **147**, S287–S296.

Yu, P.H., Davis, B.A., Fang, J., and Boulton, A.A. (1994). Neuroprotective effects of some monoamine oxidase-B inhibitors against DSP-4-induced noradrenaline depletion in the mouse hippocampus. *J. Neurochem.* **63**, 1820–1828.

Yu, Y., W, Y., Tsai, S.J., Hong, C.J., Chen, T.J., Chen, M.C., and Yang, C.W. (2005). Association study of a monoamine oxidase A gene promoter polymorphism with major depressive disorder and antidepressant response. *Neuropharmacology* **30**, 1719–1723.

Zhu, Q., Chen, K., and Shih, J.C. (1994). Bidirectional promotor of human monoamine oxidase A (MAO A) controlled by transcription factor Sp1. *J. Neurosci.* **14**, 7393–7403.

MULTIMODAL DRUGS AND THEIR FUTURE FOR ALZHEIMER'S AND PARKINSON'S DISEASE

Cornelis J. Van der Schyf and Werner J. Geldenhuys

Department of Pharmaceutical Sciences, College of Pharmacy, Northeast Ohio Medical University, Rootstown, Ohio, USA

Abstract

This chapter discusses the rationale for developing multimodal or multifunctional drugs (also called designed multiple ligands or DMLs) aimed at disease-modifying treatment strategies for the most common neurodegenerative diseases Alzheimer's and Parkinson's disease (AD and PD). Both the prevalence and incidence of AD and PD have seen consistent and dramatic increases, a disconcerting phenomenon which, ironically, has been attributed to extended life expectancy brought about by better health care globally. In spite of these statistics, the development and introduction to the clinic of new therapies proven to prevent or delay the onset of AD and PD have been disappointing. Evidence has accumulated to suggest that the etiopathology of these diseases is extremely complex, with an array of potential drug targets located within a number of deleterious

INTERNATIONAL REVIEW OF NEUROBIOLOGY, VOL. 100
DOI: 10.1016/B978-0-12-386467-3.00006-6

0074-7742/11 $35.00

biochemical pathways. Therefore, in these diseases, it is unlikely that the complex pathoetiological cascade leading to disease initiation or progression will be mitigated by any one drug acting on a single pathway or target. The pursuit of novel DMLs may offer far better outcomes. Although certainly not the only, and perhaps not even the best, approach but farthest along the drug development pipeline in the DML paradigm are drugs that combine inhibition of monoamine oxidase with associated etiological targets unique to either AD or PD. These compounds will constitute the major focus of this chapter, which will also explore radically new paradigms that seek to combine cognitive enhancers with proneurogenesis compounds.

I. Introduction

Alzheimer's disease (AD) and Parkinson's disease (PD), the most prevalent neurodegenerative diseases, have an escalating impact on society as a result of a larger percentage of the total population reaching old age. The pathoetiological complexity of AD and PD makes the treatment of these diseases one of the greatest therapeutic challenges in medicine today. Both AD and PD are characterized by a diminished number of cells in certain neuronal populations due to cell death, a situation that is clinically reflected in the appearance of disease-specific symptoms. From a pharmacological point of view, several approaches toward therapy are being explored. First, the etiopathogenic approach is considered to be "disease-modifying" and is aimed at halting cell death and/or promoting the recovery of injured cell populations. The pathophysiological approach is currently the most widely applied in AD and PD therapy and seeks to prevent or palliate the appearance of symptoms resulting as a consequence of the alteration in levels of neurotransmitters due to the disease process. Therefore, the primary objective of the pathophysiological approach is to maintain normal function in the presence of progressive disease insult: in the case of AD and PD, increased neurodegeneration. A cohort of drugs falling in the latter category is currently used in the clinic and forms the mainstay for approaches aimed at controlling the symptoms of AD and PD sufferers. Increasing evidence from recent studies suggests that the number of therapeutic possibilities uncovered in the discovery of new drug targets is likely to rise sharply in the future but investigative new drug candidates entering the market that are either pathophysiological or etiopathogenic agents are lacking. Perhaps the most striking evidence underscoring the complexity of these diseases is the fact that, to date, no single etiopathogenic drug has been developed and brought to the clinic that possesses the ability to prevent the degenerative process or restore dying neurons. At most, pharmaceutical research has only been able to bring medicines to market that can, at best, pathophysiologically modulate the symptoms in patients suffering from these diseases.

As pointed out by Morphy's group (Morphy *et al.*, 2004; Morphy and Rankovic, 2005), this gap in the development pipeline of neuroprotective/neurorestorative medications for AD and PD can at least in part be attributed to a "one-disease-one-target" approach that has been followed in most drug discovery laboratories in the pharmaceutical industry and in academia. The quest to develop highly selective compounds has been a mantra of Big Pharma for decades, driven by the observation that there likely exists an inverse relationship between drug target selectivity and an acceptable side-effect profile. In this paradigm, it is assumed that the design of a highly selective compound that targets only one receptor or enzyme will mitigate the likelihood of side effects resulting from cross-reactivity at other biological sites in the body. Recently, however, a multi-drug approach to therapy has gained increasing acceptance by seeking to address complex disease etiologies through a multitargeted-single-ligand approach (Zimmermann *et al.*, 2007). In a 2005 review, Morphy and Rankovic identified over 300 research articles that have been published in primary medicinal chemistry journals in the preceding decade and a half. In these, the structure and activity of a multitude of compounds designed as multiple-mechanistic drugs were reported (Morphy and Rankovic, 2005). The nomenclature for these drugs is still in a process of evolution, and terms such as "dual mechanism," "dual ligand," "bifunctional," "multifunctional," "multi-mechanistic," "multimodal," "pan-agonist," or "hybrid" drugs have all gained some measure of popularity and are often used interchangeably (Youdim and Buccafusco, 2005a,b). With so many different labels to describe a similar concept, it has been suggested by Morphy and Rankovic (2005) that the terminology should be unified, and that such compounds should be named "designed multiple ligands" (DMLs)—that is compounds that are specifically and rationally designed to target the multiple mechanisms underlying the etiology of a specific disease. In essence, a significant shift in focus has been seen away from "magic bullets" designed to be highly selective and potent, to a more promiscuous "dirty drug" approach (Zimmermann *et al.*, 2007) which argues for single-molecule modulators of several key components (i.e., "dirty targets") in a particular disease pathway. Thus, an emerging drug-design concept for AD and PD therapeutics is ligands that modulate the multitude of drug targets in these disorders. In this chapter, we present examples of novel multimodal drugs with various CNS targets in development for symptomatic and possibly neuroprotective–neurorescue treatment of AD and PD. Notably, we focus on drugs that combine inhibition of monoamine oxidase (MAO), a strong common denominator in both AD and PD, with associated etiological targets unique to either AD or PD.

1997 report of The Alzheimer's Disease Cooperative Study also showed that progression of disease in AD patients with moderately severe impairment was slowed (Sano *et al.*, 1997).

In PD, the substantia nigra—the brain area primarily affected—contains high numbers of MAO-B-containing astrocytes. Hydrogen peroxide produced by MAO-B within these astrocytes may be either broken down to water by the glutathione system (thereby detoxifying hydrogen peroxide) or diffuse into nearby vulnerable dopaminergic neurons where it may elicit toxic effects due to oxidative stress, mitochondrial DNA damage, and impairment of mitochondrial function (Shadrina *et al.*, 2010). The majority of PD cases are idiopathic with reduced levels of mitochondrial complex I activity generally associated with the disease (Navarro and Boveris, 2009). In addition, xenobiotic toxins such as rotenone and MPTP also cause an acute and irreversible parkinsonian syndrome that involves (in the case of MPTP) MAO-B activity in the conversion of a nontoxic to a toxic species that specifically inhibits complex I (Langston and Ballard, 1983). These findings suggest that one of the principal causes in a complicated array of events leading to neuronal loss and consequential motor impairment in PD is toxin-induced and MAO-associated mitochondrial stress in dopaminergic neurons (Di Monte, 2003). Transcriptomics, proteomic profiling, and genetic analyses also indicate that PD is polyetiological in origin (Youdim and Van der Schyf, 2007). Thus, an emerging drug-design concept for AD and PD therapeutics is ligands that modulate the multitude of drug targets in these disorders. In this chapter, we present examples of novel multimodal drugs with various CNS targets in development for symptomatic and possibly neuroprotective–neurorescue treatment of AD and PD.

IV. Multimodal Drugs That Target MAO (1)

A. Combining MAO Inhibition with Actions Against Iron and Cholinesterases

1. *Rasagiline, Ladostigil, and Others*

Rasagiline (*N*-propargyl-1-(*R*)-aminoindan; Teva Pharmaceutical Co., Israel and USA) is an example of an MAO-inhibiting drug initially intended for use in PD. PD, due to its inherently complex etiology that almost always involves a role for MAO in conjunction with other biotoxic mechanisms, appears to be particularly amenable to drug therapy where multiple pathoetiological targets are addressed, rather than focusing on a single target. Rasagiline was approved by the US Food and Drug Administration (FDA) in 2006 as the first once-daily oral treatment for PD under the brand name Azilect®. Its initial approved use was stated as a monotherapy in early PD and as an adjunct therapy to levodopa in moderate to advanced PD. It was originally designed as an MAO-B inhibitor to increase levels of dopamine in the brain, specifically in the striatum, thereby mitigating the characteristic movement disorders seen in PD patients (Chen *et al.*, 2007). Several clinical studies have shown benefits attributed to rasagiline in improving motor outcome in PD patients (Schapira and Olanow, 2008), and in particular led to a decreased requirement to add levodopa early in the treatment regimen. The latter characteristic led to a decreased risk for the appearance of dyskinesia in these patients (Haycox *et al.*, 2009). Notably, rasagiline appears to have beneficial effects on disease progression above and beyond mere symptomatic relief. It has been shown that patients, who were treated with rasagiline early in the disease progression, did better than their control counterparts after 9 months of therapy. Of significance is that a dose of 1 mg per day was reported to have an apparent disease-modifying effect in a double-blind, delayed-start trial of rasagiline in PD (Olanow *et al.*, 2009).

The success of rasagiline led to the designed development of a unique DML ladostigil (TV3326) from the same group of researchers (Weinreb *et al.*, 2008a). Ladostigil [(*N*-propargyl-(3*R*) aminoindan-5yl)-ethyl methyl carbamate] is a dual ACh-butyrylcholine (BuCh)-esterase and brain-selective MAO-A and -B inhibitor *in vivo* (with little or no peripheral MAO inhibitory effect). Ladostigil is currently in Phase II studies and anticipated to be beneficial for the treatment of dementia comorbid with extrapyramidal disorders and depression. The identification of the propargylamine moiety as a key element that confers neuroprotective activity to MAO-B inhibitors played into the rationale that led to the development of this unique molecule (Van der Schyf *et al.*, 2007; Weinreb *et al.*, 2008a,b, 2009). As a classical DML, ladostigil was rationally designed to be a multi-AChE-BuChE/MAO-B inhibitor. The rationale for the design of ladostigil was to integrate the neuroprotective propargylamine moiety of rasagiline with the carbamate ChE inhibitory moiety of the anti-AD drug, rivastigmine (Exelon®; (*S*)-3-[1-(dimethylamino)ethyl]phenyl ethylmethylcarbamate; Novartis Pharmaceuticals, NJ). Ladostigil has also been shown to attenuate 1-methyl-4-phenyl-1,2,3,6-tetrahydropyridine (MPTP) toxicity in mice (Sagi *et al.*, 2003). ACh levels are shown to be depressed in AD patients compared to age-matched controls. Therefore, bolstering ACh levels in AD brain has been a strategy in antidementia drug development for many years (Amenta and Tayebati, 2008). Although a poor MAO-B inhibitor, the *S*-isomer of ladostigil, TV3279, has shown neuroprotective activity similar to ladostigil in laboratory animals. Recent findings also demonstrate that the major metabolite of ladostigil, hydroxy-1-(*R*)-aminoindan, has neuroprotective activity and thus may also contribute to the activity profile of its parent compound. Ladostigil has also been shown, similar to both its parent compounds rasagiline and rivastigmine, to possess characteristics that are both neuroprotective and antisymptomatic in AD and PD (Weinreb *et al.*, 2008a). From published work on ladostigil, it was shown that this drug is capable of modulating BCL-2/Bax, GDNF/BDNF, and PKC-MAPK kinases similar to rasagiline and thereby promoting neuronal survival (Weinreb *et al.*, 2008a; Yogev-Falach *et al.*, 2002). Additionally, ladostigil was also shown to be able to decrease age-related changes in glial activation (that lead to associated age-related increases in brain MAO activity in humans), and attenuate memory deficits in old, but not in young rats (Weinstock *et al.*, 2011).

The recent news that Eli Lilly and Company will terminate the development of the gamma secretase inhibitor semagacestat has raised concerns about the importance of the Aβ hypothesis in AD. In spite of this ostensibly disappointing outcome from the semagacestat clinical trial, the finding that AChE protein colocalizes with Aβ to promote and accelerate Aβ aggregation has renewed interest in the development of new DML-based AChE inhibitors for AD therapy. In this regard, a new class of selective AChE inhibitors with site-activated chelating, neurorescuing, and antioxidant activities is currently being developed. The lead compound in these studies, HLA20A, exhibits poor affinity for divalent metal ions, including Fe^{2+}, Cu^{2+}, and Zn^{2+}, but can be activated through inhibition of AChE to liberate the active chelator HLA20 (see above). HLA20A inhibits AChE in a time- and concentration-dependent manner, with both the HLA20A–AChE-complex and second-order-rate constants comparable to those of rivastigmine. The "prodrug" HLA20A in isolation shows little Fe^{2+}-binding capacity or activity against Fe^{2+}-induced lipid peroxidation (Zheng *et al.*, 2009). In contrast, the active compound HLA20 exhibits significant potency in Fe^{2+}-binding and inhibits Fe^{2+}-induced lipid peroxidation. These data suggest that HLA20A represents a promising candidate for further development of AD therapy (Zheng *et al.*, 2010a,b).

2. *Potential MAO-Inhibitors/Antiaggregation/Metal-Chelating Agents*

Both AD and PD are characterized by the presence of abnormal protein components (Butterfield and Kanski, 2001; Shastry, 2003) that accumulate in the brain and lead to age-dependent selective loss of neurons in AD (through Aβ-amyloid neuritic plaques and neurofibrillary tangles) and in PD (through α-synuclein, Lewy bodies). Interestingly, metals such as iron and copper appear not only to play a role in ROS generation and subsequent oxidative stress but also appear to play an important role in deleterious protein aggregation. Therefore, metals—in particular divalent cations—are likely to provide a link between the two pathological processes in AD and PD: those of abnormal protein aggregation and oxidative damage (Gaeta and Hider, 2005). Above, it was illustrated that multifunctional drugs under development for AD include dual inhibitors of acetylcholinesterase (AChE)–MAO, or AChE–Aβ aggregation, as well as metal chelators with antioxidant/amyloid binding properties. More recently, compounds with structures that clearly—based on known structure–activity relationship experience in MAO inhibition (Chimenti *et al.*, 2006)—may have MAO inhibitory properties, have emerged as bifunctional candidate small molecules that chelate Aβ-associated metal species, and regulate metal-induced Aβ aggregation and oxidative stress-induced neurotoxicity. The salutary influence of the bidentate ligand clioquinol on metal-involved Aβ aggregation has been explained through its metal chelation, which assists in part to disaggregate Aβ clusters. However, clioquinol does not completely hinder the progression of Aβ aggregation (Mancino *et al.*, 2009).

Recently, Choi *et al.* (2010) reported the synthesis of two bifunctional small molecules, *N*-(pyridin-2-ylmethyl)aniline and *N*(1),*N*(1)-dimethyl-*N*(4)-(pyridin-2-ylmethyl)benzene-1,4-diamine, and demonstrated that these compounds interact both with metal ions and Aβ species. *N*(1),*N*(1)-dimethyl-*N*(4)-(pyridin-2-ylmethyl) benzene-1,4-diamine modulated metal-induced Aβ aggregation and neurotoxicity *in vitro* as well as in human neuroblastoma cells while simultaneously disassembling Aβ aggregates in human AD brain tissue homogenates containing metal ions and Aβ protein. These findings follow an earlier study by the same group with two different compounds: 2-[4-(dimethylamino)phenyl]imidazo[1,2-*a*]pyridine-8-ol and *N*(1),*N*(1)-dimethyl-*N*(4)-(pyridin-2-ylmethylene)benzene-1,4-diamine, which were found to react with Cu(II)-associated Aβ aggregates more effectively than known metal chelating agents such as clioquinol and EDTA (Hindo *et al.*, 2009).

Of importance is the fact that these compounds show great promise as potential MAO inhibitors, an action that will bolster the multimodal promise of these investigational new drugs as potential therapeutics in AD and PD. Experiments are currently underway and results of these studies will be published soon.

V. Multimodal Drugs That Target MAO (2)

A. Combining MAO Inhibition with Actions Against the Adenosine A_{2A} Receptor: Istradefylline (KW-6002)

Adenosine receptor antagonists that are selective for the A_{2A} receptor subtype (A_{2A} receptor antagonists) are also currently being investigated as therapeutic agents for the symptomatic treatment of motor deficits associated with PD. A_{2A} receptor antagonists (but not A_1 receptor antagonists) have been demonstrated to protect against drug-induced neuronal injury as well as stimulate motor function in experimental animal models of PD. The nonselective A_1/A_{2A} receptor antagonist caffeine has been reported to protect mice against MPTP and 6-hydroxydopamine-induced nigrostriatal neurotoxicity, while caffeine consumption also is associated with a reduced risk of developing PD. A compound, which is of interest due to its multitude of activities that include antagonism of the A_{2A} receptor subtype, istradefylline/KW6002 (8-[(*E*)-2-(3,4-dimethoxyphenyl)vinyl]-1,3-diethyl-7-methyl-3,7-dihydro-1*H*-purine-2,6-dione), is being evaluated as PD therapy (Fernandez *et al.*, 2009; Jenner *et al.*, 2009; LeWitt *et al.*, 2008; Morelli *et al.*, 2009). Recently, disappointing data from the 6002-US-051 trial suggested that although istradefylline is safe and well tolerated as monotherapy in patients with PD, its efficacy in improving motor symptoms in early PD could not be statistically demonstrated. The drug is also facing significant challenges for FDA approval, and initial approval for its use in PD has been denied in 2008. However, pending submission of additional data from ongoing clinical trials suggests that istradefylline may still be considered for future approval. This drug targets both the MAO-B enzyme and the adenosine A_{2A} receptor. The combination of these two drug targets has been suggested to potentially enhance therapeutic outcomes in the treatment of PD (Hauser *et al.*, 2008; Jenner *et al.*, 2009; Morelli *et al.*, 2009). As earlier alluded to, the impetus for the development of dual A_{2A}/MAO-B inhibitors came from epidemiological data that suggested that caffeine appeared to counteract age-related cognitive decline. This pragmatic approach is supported by strong evidence indicating that critical changes in adenosine-related neurotransmission occur with aging and may be counteracted by adenosine receptor antagonism (Chen *et al.*, 2001; Dall'Igna *et al.*, 2003, 2004; Joghataie *et al.*, 2004; Maia and de Mendonca, 2002). Of particular interest are additional indications that caffeine may protect against β-amyloid neurotoxicity, while acute treatment with caffeine and the A_{2A} receptor antagonist ZM241385 was found to reverse age-related olfactory deficits and memory decline in rats (Prediger *et al.*, 2005a,b). The former effects are theoretically also linked to caffeine's antagonistic action on A_{2A} receptors in the CNS (Jacobson and Gao, 2006). Even though initial attempts at gaining approval for the clinical use of these compounds in PD treatment were unsuccessful, pathoetiological research strongly suggests involvement of A_{2A} receptors in cognitive decline and possibly also in neurodegenerative processes.

VI. Multimodal Drugs That Target MAO (3)

A. Combining MAO Inhibition with Actions Against MitoNEET—Targeting Inflammation: Pioglitazone and Rosiglitazone

With the rise in diabetes in the USA, several new drugs have been developed to combat this disease. Of interest to the multifunctional arena are peroxisome proliferator activated receptor gamma (PPAR-γ) agonists. PPAR-γ agonists containing the thiazolidinedione or TZD moiety have become a new focus group of compounds in the treatment of ischemic stroke and PD (Culman *et al.*, 2007). PPAR-γ belongs to the orphan receptor group of ligand-activated transcription factors. When a ligand activates PPAR-γ, it moves to the DNA where it interacts with response elements to regulate gene expression. Most well known in the TZD class are the rather controversial (due to purported severe toxicity and deaths among patients treated) drugs pioglitazone [(*RS*)-5-(4-[2-(5-ethylpyridin-2-yl)ethoxy]benzyl)thiazolidine-2,4-dione; Actos®, USA and UK; Glustin®, Europe; Glizone®/Pioz®, India; Zactos®, Mexico by Takeda Pharmaceuticals] and rosiglitazone [(*RS*)-5-[4-(2-[methyl(pyridin-2-yl)amino]ethoxy)benzyl]thiazolidine-2,4-dione; Avandia®, USA, GlaxoSmithKline] which are clinically used to treat type II diabetes. Another drug in this class, troglitazone, was withdrawn from the market due to severe hepatic side effects (Herrine and Choudhary, 1999). Several studies have now been published that suggest that PPAR-γ agonists might be neuroprotective in stroke. Stroke's poor prognosis (up to 30% mortality rate) and significant resultant loss of quality of life has made it necessary to develop agents which would protect or restore the brain circuitry following brain ischemia (Culman *et al.*, 2007). Both TZDs which are currently on the market have been shown in animal models of stroke to be neuroprotective (Honda *et al.*, 2008; Ji *et al.*, 2009; Kaundal *et al.*, 2009; Lee *et al.*, 2009; Li *et al.*, 2008; Patzer *et al.*, 2008; Rennings *et al.*, 2010; Wu *et al.*, 2009; Yi *et al.*, 2008). Cerebral ischemia is associated with inflammatory responses that can injure the tissue. PPAR-γ agonists appear able to modulate this inflammatory response and reduce the size of the infarcted area (Bordet *et al.*, 2006; Collino *et al.*, 2006; Gautier *et al.*, 2003). Studies in both heart and brain tissue suggest that these compounds may attenuate reperfusion injury after an ischemic event by conditioning the tissue to better tolerate the oxidative burst which accompanies reperfusion of the injured tissue. Additionally, they seem to interact with mechanisms of inflammation as well as with pathways that lead to ROS, for example, inhibition of iNOS expression and inhibition of matrix metallopeptidase 9 (MMP-9). Also, it has been shown that CuZn-superoxide dismutase (SOD) levels are increased in cell culture following TZD treatment of the cells, which may bolster TZD's neuroprotective activity (Shimazu *et al.*, 2005). Pioglitazone also has been shown to reduce cytokine release when cells are treated with lipopolysaccharide (LPS) in culture media (Luna-Medina *et al.*, 2005). TZD's salutary effects in neuroinflammation have also led to the suggestion that AD patients may benefit from these compounds (Kummer and Heneka, 2008). The PPAR-γ agonists appear able to inhibit beta-amyloid-stimulated secretion of inflammatory mediators, as well as deposition of beta-amyloid in the brain, suggesting a role in AD neuroprotection (Combs *et al.*, 2000; Heneka *et al.*, 2005; Yan *et al.*, 2003). Several groups (Barbiero *et al.*, 2011; Breidert *et al.*, 2002; Quinn *et al.*, 2008) have shown that pioglitazone and rosiglitazone are protective in the MPTP parkinsonian mouse model by mitigating DA depletion in the striatum, reduce glial activation, and protect the substantia nigra. The neuroprotective mechanism was shown to be, at least in part, through inhibition of MAO-B (Quinn *et al.*, 2008) which prevented the conversion of MPTP to its toxic metabolite MPP^+. Recent findings also show that pioglitazone prevents iron-induced apoptosis via both endoplasmic reticular and mitochondrial pathways (Yu *et al.*, 2010) Further, inhibition of α-synuclein aggregation and neuroinflammation may contribute to pioglitazone-induced neuroprotection in the CNS. We have recently used the TZD moiety as a scaffold to search for therapeutic leads toward the treatment of PD by identifying structures that would prove to be appropriate DMLs combining MAOI with other salutary effects reported to be inherent in the TZD structure (Geldenhuys *et al.*, 2010a).

Recently, an exciting novel mitochondrial protein was identified as a target for the thiazolidinedione PPAR-γ agonist pioglitazone. This protein, referred to as mitoNEET (CISC1), belongs to a small group of proteins in the zinc-finger class that has been linked to accelerated aging (Chen *et al.*, 2009). It has been shown that pioglitazone stabilizes mitoNEET and acts as a neuroprotective drug by altering mitochondrial function. Ligands to mitoNEET have been shown to influence the oxidative state of mitochondria, depolarize the mitochondrial membrane, and prevent rotenone-induced cell death. Data from these studies suggest that the beneficial therapeutic activities seen following treatment by pioglitazone may be attributable to its multiple mechanisms that include interaction with mitoNEET and the nuclear receptor pathway (Colca *et al.*, 2004; Geldenhuys *et al.*, 2010a,b; Lin *et al.*, 2007; Paddock *et al.*, 2007; Wiley *et al.*, 2007a,b). At least one study has been published in which structure-based design techniques have been applied to develop mitoNEET ligands (Geldenhuys *et al.*, 2010b).

VII. Multimodal Drugs That Target Memory and Neurogenesis

A. Combining 5-HT_6 Receptor Antagonism with Neurogenesis: Dimebon and P7C3

A major problem with both AD and PD not only includes the loss of neurons as the disease progresses but also consists of deficits in memory and learning. Although these memory problems are usually associated with AD, PD patients also exhibit memory and cognition deficits, specifically later in the disease state. One method to combat this would be to combine a compound which would modulate cognition as well as potentially lead to new neuron formation (neurogenesis). Interest in such compounds was realized with the recent discovery of the serotonin-6 (5-HT_6) receptor. Great excitement and anticipation accompanied the disclosure of the

$5\text{-}HT_6$ receptor, based on the potential to develop compounds that could interact with these receptors (Kohen *et al.*, 1996). Several *in vivo* animal models suggested that the $5\text{-}HT_6$ receptor plays a role in cognition and memory. Largely inspired by industry and the academic group of Glennon *et al.* (2010), there was a surge in the development of molecular probes for the $5\text{-}HT_6$ receptor. One of these was the antihistamine dimebon, which early in its development had shown promise in the treatment of AD (Lermontova *et al.*, 2000). Early work characterized dimebon as both a cognitive enhancer and a neuroprotective agent (Bachurin *et al.*, 2001). Unfortunately, initial data from a recently completed clinical trial in the United States did not bode well for dimebon, as it had not met the required outcomes of the trail on AD patients (Bezprozvanny, 2010). However, a recent study by the Pieper and McKnight groups showed that both dimebon and its derivative P7C3 were potent proneurogenesis compounds. Mice missing the gene that encodes neuronal PAS domain protein 3 (NPAS3) lack hippocampal neurogenesis and exhibit other deficits that were corrected by P7C3 through normalization of apoptosis in newborn hippocampal neurons. Additionally, the cognitive ability of mice treated with these compounds was shown to be improved (Pieper *et al.*, 2010). As other chemically different types of 5-HT6 receptor antagonists do not seem to share the proneurogenesis ability of P7C3 and dimebon (Foley *et al.*, 2008), these finding suggest that future drug discovery efforts may need to focus on these indole compounds which could both enhance cognition and increase the development of new neurons.

VIII. Conclusion

In spite of technological advances in the drug research industry, such as combinatorial chemistry and HTS, the complexity of certain disease conditions has not been met with successful drugs to treat or prevent the progression of neurodegenerative diseases such as AD or PD. For instance, to date no compound has been approved as a disease-modifying drug, although some promising compounds such as rasagiline and ladostigil have been shown to possess such properties in their pharmacological profile in *in vitro* and *in vivo* studies. Based on these preclinical observations, compounds such as rasagiline also caught the attention of clinicians by demonstrating their abililty to modify disease progression. Correlation of such observations with experiments designed to uncover the structural requirements for targeting multiple pathological loci increases the attractiveness of DMLs as an attractive option both pharmacologically and as a viable medicinal chemistry design paradigm. This state of affairs for AD and PD in particular may derive from two factors, both of which are challenging drug-design obstacles: (1) the models employed may not be fully appropriate and homologous to the disease in question; (2) transcriptomics and proteomic profiling studies with animal models and human brain obtained at autopsy demonstrate a global picture of molecular events at neurodegenerative sites in the brain. These events are indicative of a cascade of "domino" events, each of which can initiate the neurodegenerative process. Thus, even though one pathway may be adequately blocked, others will continue to progress. The new paradigm shift toward the design of multimodal ligands that block several such pathways simultaneously may offer a higher rate of success, even if only to improve patient compliance by cutting down medication costs and ease of use of such medications. The conversion from a "magic bullet" to a "magic shotgun" approach may thus yet prove to be key in future treatment regimens for complex conditions such as AD and PD. We predict that several new DMLs now in the pipeline will successfully make the transition from preclinical testing to Phase I through III clinical trials, suggesting exciting prospects for the future control of AD and PD.

References

Amenta, F., and Tayebati, S.K. (2008). Pathways of acetylcholine synthesis, transport and release as targets for treatment of adult-onset cognitive dysfunction. *Curr. Med. Chem.* **15**, 488–498.

Bachurin, S., Bukatina, E., Lermontova, N., Tkachenko, S., Afanasiev, A., Grigoriev, V., Grigorieva, I., Ivanov, Y., Sablin, S., and Zefirov, N. (2001). Antihistamine agent Dimebon as a novel neuroprotector and a cognition enhancer. *Ann. N. Y. Acad. Sci.* **939**, 425–435.

Barbiero, J.K., Santiago, R.M., Lima, M.M., Ariza, D., Morais, L.H., Andreatini, R., and Vital, M.A. (2011). Acute but not chronic administration of pioglitazone promoted behavioral and neurochemical protective effects in the MPTP model of Parkinson's disease. *Behav. Brain Res.* **216**, 186–192.

Bezprozvanny, I. (2010). The rise and fall of Dimebon. *Drug News Perspect.* **23**, 518–523.

Bordet, R., Ouk, T., Petrault, O., Gele, P., Gautier, S., Laprais, M., Deplanque, D., Duriez, P., Staels, B., Fruchart, J.C., and Bastide, M. (2006). PPAR: a new pharmacological target for neuroprotection in stroke and neurodegenerative diseases. *Biochem. Soc. Trans.* **34**, 1341–1346.

Breidert, T., Callebert, J., Heneka, M.T., Landreth, G., Launay, J.M., and Hirsch, E.C. (2002). Protective action of the peroxisome proliferator-activated receptor-gamma agonist pioglitazone in a mouse model of Parkinson's disease. *J. Neurochem.* **82**, 615–624.

Butterfield, D.A., and Kanski, J. (2001). Brain oxidation in age-related neurodegenerative disorders that are associated with aggregated proteins. *Mech. Ageing Dev.* **122**, 945–962.

Chen, J.F., Xu, K., Petzer, J.P., Staal, R., Xu, Y.H., Beilstein, M., Sonsalla, P.K., Castagnoli, K., Castagnoli, N. Jr., and Schwarzschild, M.A. (2001). Neuroprotection by caffeine and A(2A) adenosine receptor inactivation in a model of Parkinson's disease. *J. Neurosci.* **21**, RC143.

Chen, J.J., Swope, D.M., and Dashtipour, K. (2007). Comprehensive review of rasagiline, a second-generation monoamine oxidase inhibitor, for the treatment of Parkinson's disease. *Clin. Ther.* **29**, 1825–1849.

Chen, Y.F., Kao, C.H., Kirby, R., and Tsai, T.F. (2009). Cisd2 mediates mitochondrial integrity and life span in mammals. *Autophagy* **5**, 1043–1045.

Chimenti, F., Bolasco, A., Manna, F., Secci, D., Chimenti, P., Granese, A., Befani, O., Turini, P., Cirilli, R., La Torre, F., Alcaro, S., Ortuso, F., *et al.* (2006). Synthesis, biological evaluation and 3D-QSAR of 1,3,5-trisubstituted-4,5-dihydro-(1 H)-pyrazole derivatives as potent and highly selective monoamine oxidase A inhibitors. *Curr. Med. Chem.* **13**, 1411–1428.

Choi, J.S., Braymer, J.J., Nanga, R.P., Ramamoorthy, A., and Lim, M.H. (2010). Design of small molecules that target metal-Aβ species and regulate metal-induced Aβ aggregation and neurotoxicity. *Proc. Natl. Acad. Sci. USA* **107**, 21990–21995.

Colca, J.R., McDonald, W.G., Waldon, D.J., Leone, J.W., Lull, J.M., Bannow, C.A., Lund, E.T., and Mathews, W.R. (2004). Identification of a novel mitochondrial protein ("mitoNEET") cross-linked specifically by a thiazolidinedione photoprobe. *Am. J. Physiol. Endocrinol. Metab.* **286**, E252–E260.

Collino, M., Aragno, M., Mastrocola, R., Gallicchio, M., Rosa, A.C., Dianzani, C., Danni, O., Thiemermann, C., and Fantozzi, R. (2006). Modulation of the oxidative stress and inflammatory response by PPAR-gamma agonists in the hippocampus of rats exposed to cerebral ischemia/reperfusion. *Eur. J. Pharmacol.* **530**, 70–80.

Combs, C.K., Johnson, D.E., Karlo, J.C., Cannady, S.B., and Landreth, G.E. (2000). Inflammatory mechanisms in Alzheimer's disease: inhibition of beta-amyloid-stimulated proinflammatory responses and neurotoxicity by PPARgamma agonists. *J. Neurosci.* **20**, 558–567.

Culman, J., Zhao, Y., Gohlke, P., and Herdegen, T. (2007). PPAR-gamma: therapeutic target for ischemic stroke. *Trends Pharmacol. Sci.* **28**, 244–249.

Dall'Igna, O.P., Porciuncula, L.O., Souza, D.O., Cunha, R.A., and Lara, D.R. (2003). Neuroprotection by caffeine and adenosine A2A receptor blockade of beta-amyloid neurotoxicity. *Br. J. Pharmacol.* **138**, 1207–1209.

Dall'Igna, O.P., Souza, D.O., and Lara, D.R. (2004). Caffeine as a neuroprotective adenosine receptor antagonist. *Ann. Pharmacother.* **38**, 717–718.

Di Monte, D.A. (2003). The environment and Parkinson's disease: is the nigrostriatal system preferentially targeted by neurotoxins? *Lancet Neurol.* **2**, 531–538.

Fahn, S., and Cohen, G. (1992). The oxidant stress hypothesis in Parkinson's disease: evidence supporting it. *Ann. Neurol.* **32**, 804–812.

Fernandez, H.H., Greeley, D.R., Zweig, R.M., Wojcieszek, J., Mori, A., and Sussman, N.M. (2009). Istradefylline as monotherapy for Parkinson disease: results of the 6002-US-051 trial. *Parkinsonism Relat. Disord.* **16**, 16–20.

Foley, A.G., Hirst, W.D., Gallagher, H.C., Barry, C., Hagan, J.J., Upton, N., Walsh, F.S., Hunter, A.J., and Regan, C.M. (2008). The selective 5-HT6 receptor antagonists SB-271046 and SB-399885 potentiate NCAM PSA immunolabeling of dentate granule cells, but not neurogenesis, in the hippocampal formation of mature Wistar rats. *Neuropharmacology* **54**, 1166–1174.

Fowler, C.J., Wiberg, A., Oreland, L., Marcusson, J., and Winblad, B. (1980). The effect of age on the activity and molecular properties of human brain monoamine oxidase. *J. Neural Transm.* **49**, 1–20.

Fowler, J.S., Volkow, N.D., Wang, G.J., Logan, J., Pappas, N., Shea, C., and MacGregor, R. (1997). Age-related increases in brain monoamine oxidase B in living healthy human subjects. *Neurobiol. Aging* **18**, 431–435.

Gaeta, A., and Hider, R.C. (2005). The crucial role of metal ions in neurodegeneration: the basis for a promising therapeutic strategy. *Br. J. Pharmacol.* **146**, 1041–1059.

Gautier, S., Petrault, O., Gele, P., Laprais, M., Bastide, M., Bauters, A., Deplanque, D., Jude, B., Caron, J., and Bordet, R. (2003). Involvement of thrombolysis in recombinant tissue plasminogen activator-induced cerebral hemorrhages and effect on infarct volume and postischemic endothelial function. *Stroke* **34**, 2975–2979.

Geldenhuys, W.J., Darvesh, A.S., Funk, M.O., Van der Schyf, C.J., and Carroll, R.T. (2010a). Identification of novel monoamine oxidase B inhibitors by structure-based virtual screening. *Bioorg. Med. Chem. Lett.* **20**, 5295–5298.

Geldenhuys, W.J., Funk, M.O., Barnes, K.F., and Carroll, R.T. (2010b). Structure-based design of a thiazolidinedione which targets the mitochondrial protein mitoNEET. *Bioorg. Med. Chem. Lett.* **20**, 819–823.

Glennon, R.A., Siripurapu, U., Roth, B.L., Kolanos, R., Bondarev, M.L., Sikazwe, D., Lee, M., and Dukat, M. (2010). The medicinal chemistry of 5-HT6 receptor ligands with a focus on arylsulfonyltryptamine analogs. *Curr. Top. Med. Chem.* **10**, 579–595.

Hauptmann, N., Grimsby, J., Shih, J.C., and Cadenas, E. (1996). The metabolism of tyramine by monoamine oxidase A/B causes oxidative damage to mitochondrial DNA. *Arch. Biochem. Biophys.* **335**, 295–304.

Hauser, R.A., Shulman, L.M., Trugman, J.M., Roberts, J.W., Mori, A., Ballerini, R., and Sussman, N.M. (2008). Study of istradefylline in patients with Parkinson's disease on levodopa with motor fluctuations. *Mov. Disord.* **23**, 2177–2185.

Haycox, A., Armand, C., Murteira, S., Cochran, J., and Francois, C. (2009). Cost effectiveness of rasagiline and pramipexole as treatment strategies in early Parkinson's disease in the UK setting: an economic Markov model evaluation. *Drugs Aging* **26**, 791–801.

Heneka, M.T., Sastre, M., Dumitrescu-Ozimek, L., Hanke, A., Dewachter, I., Kuiperi, C., O'Banion, K., Klockgether, T., Van Leuven, F., and Landreth, G.E. (2005). Acute treatment with the PPARgamma agonist pioglitazone and ibuprofen reduces glial inflammation and Abeta1-42 levels in APPV717I transgenic mice. *Brain* **128**, 1442–1453.

Herrine, S.K., and Choudhary, C. (1999). Severe hepatotoxicity associated with troglitazone. *Ann. Intern. Med.* **130**, 163–164.

Hindo, S.S., Mancino, A.M., Braymer, J.J., Liu, Y., Vivekanandan, S., Ramamoorthy, A., and Lim, M. H. (2009). Small molecule modulators of copper-induced Abeta aggregation. *J. Am. Chem. Soc.* **131**, 16663–16665.

Honda, T., Kaikita, K., Tsujita, K., Hayasaki, T., Matsukawa, M., Fuchigami, S., Sugiyama, S., Sakashita, N., Ogawa, H., and Takeya, M. (2008). Pioglitazone, a peroxisome proliferator-activated receptor-gamma agonist, attenuates myocardial ischemia-reperfusion injury in mice with metabolic disorders. *J. Mol. Cell. Cardiol.* **44**, 915–926.

Jacobson, K.A., and Gao, Z.G. (2006). Adenosine receptors as therapeutic targets. *Nat. Rev. Drug Discov.* **5**, 247–264.

Jenner, P., Mori, A., Hauser, R., Morelli, M., Fredholm, B.B., and Chen, J.F. (2009). Adenosine, adenosine A 2A antagonists, and Parkinson's disease. *Parkinsonism Relat. Disord.* **15**, 406–413.

Ji, S., Kronenberg, G., Balkaya, M., Farber, K., Gertz, K., Kettenmann, H., and Endres, M. (2009). Acute neuroprotection by pioglitazone after mild brain ischemia without effect on long-term outcome. *Exp. Neurol.* **216**, 321–328.

Joghataie, M.T., Roghani, M., Negahdar, F., and Hashemi, L. (2004). Protective effect of caffeine against neurodegeneration in a model of Parkinson's disease in rat: behavioral and histochemical evidence. *Parkinsonism Relat. Disord.* **10**, 465–468.

Kaundal, R.K., Iyer, S., Kumar, A., and Sharma, S.S. (2009). Protective effects of pioglitazone against global cerebral ischemic-reperfusion injury in gerbils. *J. Pharmacol. Sci.* **109**, 361–367.

Kohen, R., Metcalf, M.A., Khan, N., Druck, T., Huebner, K., Lachowicz, J.E., Meltzer, H.Y., Sibley, D.R., Roth, B.L., and Hamblin, M.W. (1996). Cloning, characterization, and chromosomal localization of a human 5-HT_6 serotonin receptor. *J. Neurochem.* **66**, 47–56.

Kummer, M.P., and Heneka, M.T. (2008). PPARs in Alzheimer's disease. *PPAR Res.* **2008**, 403896.

Langston, J.W., and Ballard, P.A. Jr. (1983). Parkinson's disease in a chemist working with 1-methyl-4-phenyl-1,2,5,6-tetrahydropyridine. *N. Engl. J. Med.* **309**, 310.

Lee, S.R., Kim, H.Y., Hong, J.S., Baek, W.K., and Park, J.W. (2009). PPARgamma agonist pioglitazone reduces matrix metalloproteinase-9 activity and neuronal damage after focal cerebral ischemia. *Biochem. Biophys. Res. Commun.* **380**, 17–21.

Lermontova, N.N., Lukoyanov, N.V., Serkova, T.P., Lukoyanova, E.A., and Bachurin, S.O. (2000). Dimebon improves learning in animals with experimental Alzheimer's disease. *Bull. Exp. Biol. Med.* **129**, 544–546.

LeWitt, P.A., Guttman, M., Tetrud, J.W., Tuite, P.J., Mori, A., Chaikin, P., and Sussman, N.M. (2008). Adenosine A2A receptor antagonist istradefylline (KW-6002) reduces "off" time in Parkinson's disease: a double-blind, randomized, multicenter clinical trial (6002-US-005). *Ann. Neurol.* **63**, 295–302.

Li, J., Lang, M.J., Mao, X.B., Tian, L., and Feng, Y.B. (2008). Antiapoptosis and mitochondrial effect of pioglitazone preconditioning in the ischemic/reperfused heart of rat. *Cardiovasc. Drugs Ther.* **22**, 283–291.

Lin, J., Zhou, T., Ye, K., and Wang, J. (2007). Crystal structure of human mitoNEET reveals distinct groups of iron sulfur proteins. *Proc. Natl. Acad. Sci. USA* **104**, 14640–14645.

Luna-Medina, R., Cortes-Canteli, M., Alonso, M., Santos, A., Martinez, A., and Perez-Castillo, A. (2005). Regulation of inflammatory response in neural cells in vitro by thiadiazolidinones derivatives through peroxisome proliferator-activated receptor gamma activation. *J. Biol. Chem.* **280**, 21453–21462.

Maia, L., and de Mendonca, A. (2002). Does caffeine intake protect from Alzheimer's disease? *Eur. J. Neurol.* **9**, 377–382.

Mancino, A.M., Hindo, S.S., Kochi, A., and Lim, M.H. (2009). Effects of clioquinol on metal-triggered amyloid-beta aggregation revisited. *Inorg. Chem.* **48**, 9596–9598.

Morelli, M., Carta, A.R., and Jenner, P. (2009). Adenosine A2A receptors and Parkinson's disease. *Handb. Exp. Pharmacol.* **193**, 589–615.

Morphy, R., and Rankovic, Z. (2005). Designed multiple ligands. An emerging drug discovery paradigm. *J. Med. Chem.* **48**, 6523–6543.

Morphy, R., Kay, C., and Rankovic, Z. (2004). From magic bullets to designed multiple ligands. *Drug Discov. Today* **9**, 641–651.

Navarro, A., and Boveris, A. (2009). Brain mitochondrial dysfunction and oxidative damage in Parkinson's disease. *J. Bioenerg. Biomembr.* **41**, 517–521.

Nowakowska, E., Kus, K., Chodera, A., and Rybakowski, J. (2001). Investigating potential anxiolytic, antidepressant and memory enhancing activity of deprenyl. *J. Physiol. Pharmacol.* **52**, 863–873.

Nunomura, A., Perry, G., Aliev, G., Hirai, K., Takeda, A., Balraj, E.K., Jones, P.K., Ghanbari, H., Wataya, T., Shimohama, S., Chiba, S., Atwood, C.S., *et al.* (2001). Oxidative damage is the earliest event in Alzheimer disease. *J. Neuropathol. Exp. Neurol.* **60**, 759–767.

Olanow, C.W., Rascol, O., Hauser, R., Feigin, P.D., Jankovic, J., Lang, A., Langston, W., Melamed, E., Poewe, W., Stocchi, F., and Tolosa, E. (2009). A double-blind, delayed-start trial of rasagiline in Parkinson's disease. *N. Engl. J. Med.* **361**, 1268–1278.

Paddock, M.L., Wiley, S.E., Axelrod, H.L., Cohen, A.E., Roy, M., Abresch, E.C., Capraro, D., Murphy, A.N., Nechushtai, R., Dixon, J.E., and Jennings, P.A. (2007). MitoNEET is a uniquely folded 2Fe 2 S outer mitochondrial membrane protein stabilized by pioglitazone. *Proc. Natl. Acad. Sci. USA* **104**, 14342–14347.

Patzer, A., Zhao, Y., Stock, I., Gohlke, P., Herdegen, T., and Culman, J. (2008). Peroxisome proliferator-activated receptorsgamma (PPARgamma) differently modulate the interleukin-6 expression in the peri-infarct cortical tissue in the acute and delayed phases of cerebral ischaemia. *Eur. J. Neurosci.* **28**, 1786–1794.

Pieper, A.A., Xie, S., Capota, E., Estill, S.J., Zhong, J., Long, J.M., Becker, G.L., Huntington, P., Goldman, S.E., Shen, C.H., Capota, M., Britt, J.K., *et al.* (2010). Discovery of a proneurogenic, neuroprotective chemical. *Cell* **142**, 39–51.

Prediger, R.D., Batista, L.C., and Takahashi, R.N. (2005a). Caffeine reverses age-related deficits in olfactory discrimination and social recognition memory in rats. Involvement of adenosine A1 and A2A receptors. *Neurobiol. Aging* **26**, 957–964.

Prediger, R.D., Pamplona, F.A., Fernandes, D., and Takahashi, R.N. (2005b). Caffeine improves spatial learning deficits in an animal model of attention deficit hyperactivity disorder (ADHD)—the spontaneously hypertensive rat (SHR). *Int. J. Neuropsychopharmacol.* **8**, 583–594.

Quinn, L.P., Crook, B., Hows, M.E., Vidgeon-Hart, M., Chapman, H., Upton, N., Medhurst, A.D., and Virley, D.J. (2008). The PPARgamma agonist pioglitazone is effective in the MPTP mouse model of Parkinson's disease through inhibition of monoamine oxidase B. *Br. J. Pharmacol.* **154**, 226–233.

Reinikainen, K.J., Soininen, H., and Riekkinen, P.J. (1990). Neurotransmitter changes in Alzheimer's disease: implications to diagnostics and therapy. *J. Neurosci. Res.* **27**, 576–586.

Rennings, A.J., Meijer, P., van Uden, D.J., Tack, C.J., Smits, P., Boerman, O.C., Oyen, W.J., and Rongen, G.A. (2010). Rosiglitazone reduces ischaemia-reperfusion injury in patients with the metabolic syndrome. *Eur. Heart J.* **31**, 983.

Sagi, Y., Weinstock, M., and Youdim, M.B. (2003). Attenuation of MPTP-induced dopaminergic neurotoxicity by TV3326, a cholinesterase-monoamine oxidase inhibitor. *J. Neurochem.* **86**, 290–297.

Sano, M., Ernesto, C., Thomas, R.G., Klauber, M.R., Schafer, K., Grundman, M., Woodbury, P., Growdon, J., Cotman, C.W., Pfeiffer, E., Schneider, L.S., and Thal, L.J. (1997). A controlled trial of selegiline, α-tocopherol, or both as treatment for Alzheimer's disease. The Alzheimer's Disease Cooperative Study. *N. Engl. J. Med.* **336**, 1216–1222.

Saura, J., Luque, J.M., Cesura, A.M., Da Prada, M., Chan-Palay, V., Huber, G., Loffler, J., and Richards, J.G. (1994). Increased monoamine oxidase B activity in plaque-associated astrocytes of Alzheimer brains revealed by quantitative enzyme radioautography. *Neuroscience* **62**, 15–30.

Sayre, L.M., Perry, G., and Smith, M.A. (2008). Oxidative stress and neurotoxicity. *Chem. Res. Toxicol.* **21**, 172–188.

Schapira, A.H., and Olanow, C.W. (2008). Drug selection and timing of initiation of treatment in early Parkinson's disease. *Ann. Neurol.* **64**(Suppl. 2), S47–S55.

Shadrina, M.I., Slominsky, P.A., and Limborska, S.A. (2010). Molecular mechanisms of pathogenesis of Parkinson's disease. *Int. Rev. Cell Mol. Biol.* **281**, 229–266.

Shastry, B.S. (2003). Neurodegenerative disorders of protein aggregation. *Neurochem. Int.* **43**, 1–7.

Shimazu, T., Inoue, I., Araki, N., Asano, Y., Sawada, M., Furuya, D., Nagoya, H., and Greenberg, J.H. (2005). A peroxisome proliferator-activated receptor-gamma agonist reduces infarct size in transient but not in permanent ischemia. *Stroke* **36**, 353–359.

Soong, N.W., Hinton, D.R., Cortopassi, G., and Arnheim, N. (1992). Mosaicism for a specific somatic mitochondrial DNA mutation in adult human brain. *Nat. Genet.* **2**, 318–323.

Strolin Benedetti, M., and Dostert, P. (1989). Monoamine oxidase, brain ageing and degenerative diseases. *Biochem. Pharmacol.* **38**, 555–561.

Terry, R.D., DeTeresa, R., and Hansen, L.A. (1987). Neocortical cell counts in normal human adult aging. *Ann. Neurol.* **21**, 530–539.

Tsunekawa, H., Noda, Y., Mouri, A., Yoneda, F., and Nabeshima, T. (2008). Synergistic effects of selegiline and donepezil on cognitive impairment induced by amyloid beta (25–35). *Behav. Brain Res.* **190**, 224–232.

Van der Schyf, C.J., Mandel, S., Geldenhuys, W.J., Amit, T., Avramovich, Y., Zheng, H., Fridkin, M., Gal, S., Weinreb, O., Bar Am, O., Sagi, Y., and Youdim, M.B. (2007). Novel multifunctional anti-Alzheimer drugs with various CNS neurotransmitter targets and neuroprotective moieties. *Curr. Alzheimer Res.* **4**, 522–536.

Weinreb, O., Amit, T., Bar-Am, O., Yogev-Falach, M., and Youdim, M.B. (2008a). The neuroprotective mechanism of action of the multimodal drug ladostigil. *Front. Biosci.* **13**, 5131–5137.

Weinreb, O., Bar-Am, O., Amit, T., Drigues, N., Sagi, Y., and Youdim, M.B. (2008b). The neuroprotective effect of ladostigil against hydrogen peroxide-mediated cytotoxicity. *Chem. Biol. Interact.* **175**, 318–326.

Weinreb, O., Mandel, S., Bar-Am, O., Yogev-Falach, M., Avramovich-Tirosh, Y., Amit, T., and Youdim, M.B. (2009). Multifunctional neuroprotective derivatives of rasagiline as anti-Alzheimer's disease drugs. *Neurotherapeutics* **6**, 163–174.

Weinstock, M., Luques, L., Poltyrev, T., Bejar, C., and Shoham, S. (2011). Ladostigil prevents age-related glial activation and spatial memory deficits in rats. *Neurobiol. Aging* **32**, 1069–1078.

Wiley, S.E., Murphy, A.N., Ross, S.A., van der Geer, P., and Dixon, J.E. (2007a). MitoNEET is an iron-containing outer mitochondrial membrane protein that regulates oxidative capacity. *Proc. Natl. Acad. Sci. USA* **104**, 5318–5323.

Wiley, S.E., Paddock, M.L., Abresch, E.C., Gross, L., van der Geer, P., Nechushtai, R., Murphy, A.N., Jennings, P.A., and Dixon, J.E. (2007b). The outer mitochondrial membrane protein mitoNEET contains a novel redox-active 2Fe-2S cluster. *J. Biol. Chem.* **282**, 23745–23749.

Wu, J.S., Cheung, W.M., Tsai, Y.S., Chen, Y.T., Fong, W.H., Tsai, H.D., Chen, Y.C., Liou, J.Y., Shyue, S.K., Chen, J.J., Chen, Y.E., Maeda, N., *et al.* (2009). Ligand-activated peroxisome proliferator-activated receptor-gamma protects against ischemic cerebral infarction and neuronal apoptosis by 14-3-3 epsilon upregulation. *Circulation* **119**, 1124–1134.

Yan, Q., Zhang, J., Liu, H., Babu-Khan, S., Vassar, R., Biere, A.L., Citron, M., and Landreth, G. (2003). Anti-inflammatory drug therapy alters beta-amyloid processing and deposition in an animal model of Alzheimer's disease. *J. Neurosci.* **23**, 7504–7509.

Yi, J.H., Park, S.W., Brooks, N., Lang, B.T., and Vemuganti, R. (2008). PPARgamma agonist rosiglitazone is neuroprotective after traumatic brain injury via anti-inflammatory and anti-oxidative mechanisms. *Brain Res.* **1244**, 164–172.

Yogev-Falach, M., Amit, T., Bar-Am, O., Weinstock, M., and Youdim, M.B. (2002). Involvement of MAP kinase in the regulation of amyloid precursor protein processing by novel cholinesterase inhibitors derived from rasagiline. *FASEB J.* **16**, 1674–1676.

Youdim, M.B., and Buccafusco, J.J. (2005a). CNS Targets for multi-functional drugs in the treatment of Alzheimer's and Parkinson's diseases. *J. Neural Transm.* **112**, 519–537.

Youdim, M.B., and Buccafusco, J.J. (2005b). Multi-functional drugs for various CNS targets in the treatment of neurodegenerative disorders. *Trends Pharmacol. Sci.* **26**, 27–35.

Youdim, M.B., and Van der Schyf, C.J. (2007). Magic bullets or novel multimodal drugs with various targets for Parkinson's Disease? *Nat. Rev. Drug Discov.* **6**, iii–vi.

Youdim, M.B., Edmondson, D., and Tipton, K.F. (2006). The therapeutic potential of monoamine oxidase inhibitors. *Nat. Rev. Neurosci.* **7**, 295–309.

Yu, H.C., Feng, S.F., Chao, P.L., and Lin, A.M. (2010). Anti-inflammatory effects of pioglitazone on iron-induced oxidative injury in the nigrostriatal dopaminergic system. *Neuropathol. Appl. Neurobiol.* **36**, 612–622.

Zheng, H., Youdim, M.B., and Fridkin, M. (2009). Site-Activated Multifunctional Chelator with Acetylcholinesterase and Neuroprotective-Neurorestorative Moieties for Alzheimer's Therapy. *J. Med. Chem.* **52**, 4095–4098.

Zheng, H., Youdim, M.B.H., and Fridkin, M. (2010a). New selective acetylcholinesterase inhibitor with site-activated chelating, neurorescuing and anti-amyloid activities for Alzheimer's therapy. *ACS Chem. Neurosci.* **1**, 737–746.

Zheng, H., Youdim, M.B.H., and Fridkin, M. (2010b). Site-activated chelators targeting acetylcholinesterase and monoamine oxidase for Alzheimer's therapy. *ACS Chem. Biol.* **5**, 603–610.

Zimmermann, G.R., Lehar, J., and Keith, C.T. (2007). Multi-target therapeutics: when the whole is greater than the sum of the parts. *Drug Discov. Today* **12**, 34–42.

NEUROPROTECTIVE PROFILE OF THE MULTITARGET DRUG RASAGILINE IN PARKINSON'S DISEASE

Orly Weinreb[1], Tamar Amit[1], Peter Riederer[2], Moussa B.H. Youdim[1] and Silvia A. Mandel[1]

[1]Eve Topf Centre of Excellence for Neurodegenerative Diseases and Department of Molecular Pharmacology, Technion-Rappaport Faculty of Medicine, Efron Street, P.O. Box 9697, Haifa, Israel

[2]Clinical Neurochemistry, National Parkinson Foundation Centre of Excellence Laboratories, Clinic and Polyclinic for Psychiatry, Psychosomatic, and Psychotherapy, Medical School, University of Würzburg, Würzburg, Germany

Abstract

Monoamine oxidase (MAO) inhibitors were reported to have therapeutic value in several common neurodegenerative conditions owed to their diverse pharmacological functions in neuron survival. Rasagiline (*N*-propargyl-1-(*R*)-aminoindan) is a novel, highly potent irreversible MAO-B inhibitor in the treatment of Parkinson's disease (PD). It has been demonstrated to be neuroprotective in PD model systems by preventing the formation of reactive oxygen species derived from prevention of derived from oxidation of dopamine by MAO-B and via an antiapoptotic action, which appears to be independent of MAO-B inhibition and related to its embedded *N*-propargyl moiety. This review reflects on earlier and present evidence supporting a role for rasagiline as a neuroprotective molecule in the treatment of PD.

DOI: 10.1016/B978-0-12-386467-3.00007-8

0074-7742/11 $35.00

I. Introduction

Monoamine oxidase (MAO)-B inhibitors are in clinical use for the treatment of Parkinson's disease (PD) since 1960. It was soon after the seminal findings of Carlsson *et al.* (1958) that iproniazid potentiated the effects of DL-DOPA in antagonizing the sedative actions of reserpine in rabbits and rats that Sano (1960) described the first use of iproniazid and other MAO inhibitors in PD and their mild potentiation of DL-DOPA's short-lasting effects. This more anecdotical description was followed in 1961 and 1962 by comprehensive clinical trials describing both beneficial effects of a variation of MAO inhibitors including the potentiation of L-DOPA and side effects including such related to cardiovascular functions (Birkmayer and Hornykiewicz, 1961, 1962; Bernheimer *et al.*, 1962). It was the pharmacological distinction of two subtypes of MAO, MAO-A, and MAO-B with different substrate—as well as inhibitor properties (Johnson, 1968), which led to a renaissance of MAO-inhibitor developments. This finding was confirmed by Collins *et al.* (1970) and O'Carroll *et al.* (1983) showing different isoenzymes of MAO in human brain tissue.

Meanwhile, Hungarian pharmacologists had synthesized the compound L-deprenyl (selegiline, E-250; phenyl-isopropyl-methyl-propinylamine; Knoll *et al.*, 1965; Magyar *et al.*, 1967; Fig. 1). This compound is a perfect irreversible MAO-B type inhibitor and free of the cheese-effect (Birkmayer *et al.*, 1975a, 1977; Knoll and Magyar, 1972; Sandler *et al.*, 1978; see Chapter 'Selective inhibitors of monoamine oxidase type B and the "cheese effect"' by Finberg and Gillman).

Such ideal side effect/adverse reaction profile led to the pioneer introduction of selegiline as an adjuvant to L-DOPA (Birkmayer *et al.*, 1975a, 1977) and later as monotherapy (Parkinson Study Group, 1989, 1996) in PD. Further, a retrospective analysis of Parkinsonian patients treated with L-DOPA plus the peripheral acting decarboxylase inhibitor benserazide alone or combined with selegiline came to the conclusion that selegiline might have neuroprotective actions (Birkmayer *et al.*, 1983, 1985). All this was the basis of nowadays developments of other safe MAO-B inhibitors, like rasagiline and the concept of "neuroprotection" as well as "disease modification".

Studies in Parkinsonian patients challenged with high doses of L-DOPA after receiving selegiline confirmed these findings (Birkmayer *et al.*, 1975a, 1977). The beneficial effects of selegiline might result from some of its metabolites, namely (−)-desmethyl-deprenyl and (−)-deprenyl-*N*-oxide (Magyar *et al.*, 2010; Szende *et al.*, 2010). However, selegiline is metabolized *in vivo* to L-amphetamine and L-methamphetamine, which may interfere with or neutralize its reported neuroprotective effects (Bar-Am *et al.*, 2004; Reynolds *et al.*, 1978a). Indeed, in pharmacological preparations and *in vivo*, selegiline possesses amphetamine-like sympathomimetic actions, which may result at high doses in increased blood

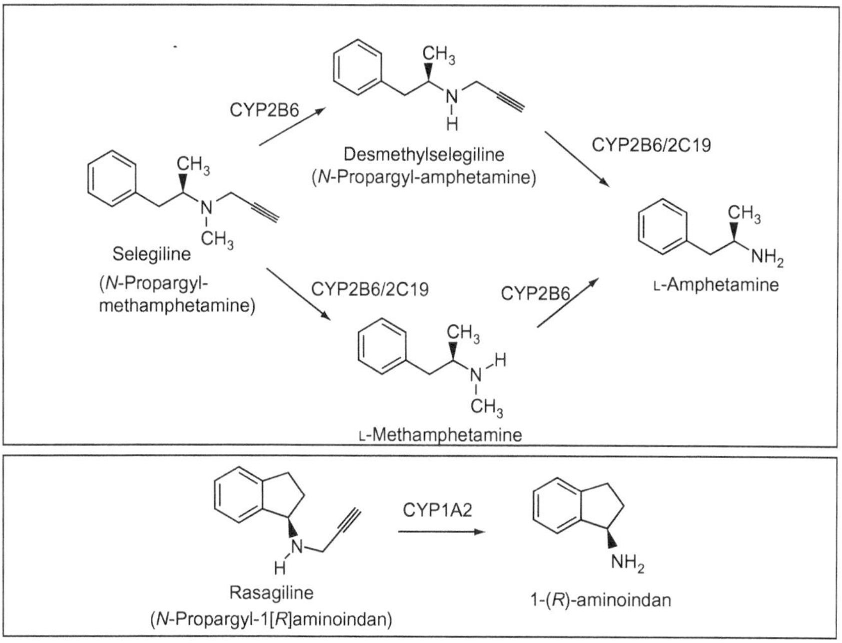

FIG. 1. Chemical structures of the MAO-B inhibitors, selegiline and rasagiline, and their respective major metabolites. Reproduced from Chen *et al.* (2007).

pressure and heart rate (Finberg *et al.*, 1981a; Simpson, 1978), although in clinical practice, it may not have a significant role compared to all other anti-Parkinsonian drug (Birkmayer *et al.*, 1975b, 1977; see Chapter 'Selective inhibitors of monoamine oxidase type B and the "cheese effect"' by Finberg and Gillman). Given that the therapeutically effective dose of selegiline in combination with L-DOPA in Parkinsonism is no more than 10–15 mg (Birkmayer *et al.*, 1977; Lees *et al.*, 1977), it is unlikely to produce a central amphetamine-like action, although a contribution of amphetamine to the improvement seen in patients on L-DOPA plus amphetamine cannot be ruled out (Reynolds *et al.*, 1978b).

Rasagiline (*N*-propargyl-1-(*R*)-aminoindan; TVP-1012; Azilect; Fig. 1) is a highly potent irreversible selective inhibitor of MAO-B that is not metabolized to l-amphetamine. It was developed as an anti-Parkinsonian drug (Finberg *et al.*, 1996, 1999; Kalir *et al.*, 1981; Sabbagh and Youdim., 1978; Sterling *et al.*, 1998; Youdim and Bakhle, 2006; Youdim *et al.*, 2001a). Basic and clinical studies with rasagiline have shown that it has very similar pharmacological action to that of selegiline (Chen *et al.*, 2007). Both MAO-B inhibitors were reported to protect neuronal cells against cell death induced in cellular and animal models of neurodegenerative diseases (Naoi and Maruyama, 2010). Their main differences are related to MAO-B inhibitory potency, metabolism, interaction with hepatic CYP-enzymes, and modulation of biological processes (Riederer and Laux, 2011).

Rasagiline selectively inhibits MAO-B with a potency of ~10 times greater than that of selegiline (Finberg and Youdim, 1985; Sterling *et al.*, 1998; Youdim *et al.*, 2001a). Rasagiline is the *R*-isomer of the selective MAO-B inhibitor, the racemic compound AGN1135 (Kalir *et al.*, 1981; Sabbagh and Youdim, 1978) being the active component, while the *S*-isomer is less than 1000-folds active as MAO-B inhibitory compound (Youdim *et al.*, 2001a). Similar to selegiline, rasagiline does not potentiate pressor responses to oral tyramine by single oral doses, up to 5 mg/kg, or following 21 days of chronic treatment at doses up to 2 mg/kg daily. At 10 mg/kg of rasagiline, both MAO-A and -B are inhibited, and the irreversible inhibition of MAO-A induces the "cheese reaction" (Finberg and Youdim, 1985; Finberg *et al.*, 1980, 1981a,b; Youdim, 2003). Thus, no cheese-effect is observed at selective efficient doses of rasagiline (deMarcaida *et al.*, 2006; Goren *et al.*, 2010).

Parkinsonian subjects respond to low doses of rasagiline (0.5–2 mg daily) in controlled monotherapy and adjunct L-DOPA therapy, as shown in early and late PD studies (Parkinson Study Group, 2002, 2004, 2005). These values are significantly lower than the dosage employed with selegiline (5–10 mg daily; Parkinson Study Group, 1993, 1996, 1998). In conclusion, the lack of tyramine sympathomimetic potentiation at its selective MAO-B inhibitory dosage, the lack of amphetamine-like metabolites, and the low dose in the clinics of PD point to a therapeutic advantage for rasagiline over selegiline.

In preclinical studies, rasagiline was shown to possess neuroprotective-anti-apoptotic activity (Tatton *et al.*, 2002; Weinreb *et al.*, 2004). More prominently, rasagiline induced an *in vivo* neurorestorative activity in the substantia nigra pars compacta (SNpC) neurons, when given post-*N*-methyl-4-phenyl-1,2,3,6-tetrahydropyridine (MPTP) or postlactacystin (proteasome inhibitor; Sagi *et al.*, 2007; Zhu *et al.*, 2008). This property was shown to be associated with activation of tyrosine kinase (TK) receptor and stimulation of various cell signaling transduction pathways (Sagi *et al.*, 2007). An additional neuroprotective effect, shared by rasagiline and other propargyl derivative compounds, is related to their ability to regulate the nonamyloidogenic processing of the Alzheimer's disease (AD) amyloid precursor protein (APP; Bar-Am *et al.*, 2010b; Hu *et al.*, 2008; Yang *et al.*, 2007; Yogev-Falach *et al.*, 2002, 2003). Thus, rasagiline may induce neuroprotective, neurorescue, and long-term potentiating effects. The results of the TEMPO clinical trial with rasagiline (Parkinson Study Group, 2002) and the recent Phase III delayed-start clinical study in Parkinsonian subjects, ADAGIO (attenuation of disease progression with Azilect given once-daily), were consistent with the possibility that rasagiline might be the first drug with the benefits of slowing the disease progression in PD (Olanow *et al.*, 2008, 2009). The ADAGIO trial was one of the first tests in PD of the delayed-start paradigm, which was designed to show benefits beyond symptoms improvements. The trial randomized a massive 1176 people with PD, not yet received anti-Parkinsonian medication, to receive 1 or 2 mg of rasagiline or placebo control. An early-start group was on active drug for the full 72-week duration of the trial, while a delayed-start group was initially on placebo and then switched to active drug at week 36 for the reminder of the trial. Olanow *et al.* (2009) reported that 1 mg of rasagiline successfully achieved all three prespecific trial efficacy endpoints (a combination of slope and difference measures), suggesting that rasagiline may have a disease-modifying effect, although this has been questioned recently (Ahlskog and Uitti, 2010; de la Fuente-Fernandez *et al.*, 2010).

Unlike selegiline, rasagiline is primarily metabolized by hepatic cytochrome P-450 isoenzyme 1A2 (CYP1A2)-mediated N-dealkylation to form the nontoxic compound 1-(*R*)-aminoindan (Fig. 1; Chen and Swope, 2005; Chen *et al.*, 2007). A kinetic and crystallographic analysis revealed that 1-(*R*)-aminoindan is not a substrate for MAO oxidation, but conversely, a weak reversible inhibitor (Binda *et al.*, 2005; Sterling *et al.*, 1998). Recent studies indicated that 1-(*R*)-aminoindan possesses beneficial pharmacological effects in animal and cell culture models of PD (Bar-Am *et al.*, 2007; Speiser *et al.*, 1998; Weinreb *et al.*, 2011), suggesting that it may contribute to the overall neuroprotective and antiapoptotic effects of its parental compound rasagiline (Bar-Am *et al.*, 2004, 2007, 2010a). This review presents fundamental studies concerning the pharmacology and mechanisms of neuroprotective action of rasagiline and its metabolites in PD and discusses their potential for disease modification.

II. Multiple Activities of Rasagiline

A. MAO-B Inhibition

MAO-A and -B inhibitory activities of rasagiline and the racemic compound, AGN1135, have been compared to those of selegiline *in vitro* and *in vivo*, in various tissues (e.g., brain, intestine, and liver) of mice, rats, cats, and monkeys (Kupsch *et al.*, 2001; Lamensdorf *et al.*, 1996; Riederer and Youdim, 1986; Youdim, 2006; Youdim and Bakhle, 2006; Youdim *et al.*, 2001a,b). Rasagiline and its racemic form were shown to possess highly potent selective irreversible inhibitory activity of MAO-B *in vitro*; the degree of selectivity of rasagiline for inhibition of MAO-B, compared to MAO-A was similar to that of selegiline (Youdim *et al.*, 2001a). However, *in vivo*, rasagiline is about 10 times more potent than selegiline and maintains its selective MAO-B inhibition in the liver and brain on chronic (21 days) oral dosage with ED_{50} values of 0.014 ± 0.002 and 0.013 ± 0.001 mg/kg, respectively (Youdim *et al.*, 2001a). This may explain the greater inhibitory potency of rasagiline in human platelet MAO-B (Chen *et al.*, 2007; Thebault *et al.*, 2004).

The *S*-isomer of rasagiline, TVP-1022, was demonstrated to be relatively inactive as a MAO-inhibitor *in vitro* and *in vivo* (Sterling *et al.*, 1998; Youdim *et al.*, 2001a). The full explanation for the markedly different degree of stereo selectivity for MAO inhibition between these closely allied chemical compounds has become apparent with the three-dimensional modeling of the MAO active site (Binda *et al.*, 2004; Youdim, 1978). Indeed, the crystallographic study by Binda *et al.* (2005) has clearly shown that, while rasagiline enters the MAO-B active site cleft and binds to it irreversibly, TVP-1022 needs to make a 180° turn to enter the cleft and slowly, but eventually, inhibits the enzyme (Youdim *et al.*, 2006).

N-propargyl-containing MAO inhibitors, such as clorgyline, pargyline, selegiline, and rasagiline, are known as suicide inhibitors, as they inactivate MAO-A and -B selectively and irreversibly by time- and concentration-mechanism via interacting covalently with the cysteinyl-FAD cofactor of the enzymes at their active center (Bach *et al.*, 1988; Kalir *et al.*, 1981; Kearney *et al.*, 1971; Maycock *et al.*, 1976; Salach *et al.*, 1979; Youdim, 1978). It has been shown that this interaction takes place between the *N*-propargyl group and N-5 of the FAD isoalloxazine component of liver and brain MAO and that 1 mol of the inhibitor inactivates 1 mol of the enzyme (Maycock *et al.*, 1976; Salach, 1978).

Results obtained from studies on rat tissues complemented findings in nonhuman primate (monkey) brains, where rasagiline was given chronically for 7 days at various doses and MAO-A and -B activities were measured in several brain regions, including caudate nucleus, globus pallidus, cerebral cortex, and hippocampus (Götz *et al.*, 1998). Rasagiline was shown to be a potent selective inhibitor of MAO-B in the caudate nucleus and globus pallidus, where similar to human brain, the activity of MAO-B is fourfold higher than that of MAO-A (Götz *et al.*, 1998). In human volunteers and Parkinsonian subjects, oral administration of rasagiline (0.5–1 mg/day) was associated with nearly complete inhibition of platelet MAO-B activity following 1-week drug administration (Chen *et al.*, 2007; Thebault *et al.*, 2004). It is most likely that this potency is reflected in the human brain, as it has been shown that the 10-mg daily dose of selegiline, as given to Parkinsonian subjects, can fully inhibit platelet MAO-B and has a similar MAO-B inhibitory potency in the brain, resulting in increased DA levels in the caudate nucleus, substantia nigra (SN), globus pallidus, and thalamus, with limited MAO-A inhibition (Riederer and Youdim, 1986; Riederer *et al.*, 1982). The recovery of MAO-A and -B activity after *in vivo* inhibition by rasagiline is related to the synthesis of the enzyme protein and significantly differs between tissues, as shown in rat liver, small intestine, and brain. The small intestine MAO-B activities had the fastest recovery (26% inhibition after 3 days), while brain MAO-B activity had the slowest recovery (80% inhibition after 3 days; Youdim *et al.*, 2001a). Similarly, the recovery of MAO-B activity in rat brain was lower than that in liver MAO-B (Youdim and Tipton, 2002).

1. *MAO-B-Inhibition Mediated Increase in DA via 2-Phenylethylamine*

Previous studies demonstrated that chronic treatment with a low MAO-B selective dose of selegiline or rasagiline increased DA in the rat striatum (Youdim *et al.*, 2001a). The mechanism underlying this effect is not fully understood, as inhibition of neither MAO-A nor MAO-B affects the steady state of brain dopamine: only when both forms are inhibited does DA activity increase (Green *et al.*, 1977). The elevation of central levels of the substrates DA and 2-phenylethylamine (2-PEA) resulting from inhibition of MAO-B has been proposed to play a part (Reynolds *et al.*, 1978a). Upon inhibition of degradation, 2-PEA, which has similar pharmacological activity to amphetamine, releases DA from the neuron terminals to the synapse. 2-PEA is not only a highly potent releaser of DA but also an inhibitor of neuronal uptake (Baker *et al.*, 1976).

B. Neuroprotection and Neurorescue in Cell Culture and Animal Models of Neurodegeneration

Rasagiline exerts its primary effect in PD presumably by MAO-B inhibition, resulting in a slower metabolism of endogenous and exogenous DA, thus providing symptomatic benefits (Finberg *et al.*, 1996, 1998). In addition, rasagiline has been shown in preclinical studies, to have broad neuroprotective/neurorestorative activities against a variety of neurotoxins in *in vivo* and neuronal cell cultures models of neurodegeneration, which may contribute to its possible disease-modifying activity. *In vivo* studies have described the neuroprotective effect of rasagiline in the MPTP model in mice and monkeys (Heikkila *et al.*, 1985; by preventing the conversion to 1-methyl-4-phenylpyridinium (MPP^+)); in the focal ischemia rat model (Speiser *et al.*, 1999); in stroke-prone spontaneously hypertensive rats (Eliash *et al.*, 2001); and in a neurotrauma mice model of head injury (Huang *et al.*, 1999). More prominently, rasagiline was found to possess an *in vivo* neurorestorative activity in SNpC neurons, when given posttreatment with MPTP (Sagi *et al.*, 2007) or the proteasome inhibitor, lactacystin (Zhu *et al.*, 2008).

In *in vitro* studies, the protective activity of rasagiline includes attenuation of cell death in partially differentiated rat pheochromocytoma PC-12 cells deprived of oxygen-glucose (Abu-Raya *et al.*, 1999), serum and nerve growth factor (NGF; Tatton *et al.*, 2002; Weinreb *et al.*, 2004), and neuroprotection against the endogenous neurotoxin *N*-methyl-(*R*)-salsolinol (*N*-M-(*R*)-Sal) (Akao *et al.*, 2002a,b; Maruyama *et al.*, 2001b, 2002b), 6-hydroxydopamine (6-OHDA; Maruyama *et al.*, 2000b, 2001a), 3-morpholinosydnonimine (SIN-1; a peroxynitrite donor,

NO; Maruyama *et al.*, 2001a, 2002a), ethanol (Ou *et al.*, 2009, 2010), and glutamate toxicity in human SH-SY5Y neuroblastoma cells (Finberg *et al.*, 1999).

Structure–activity studies provide evidence that the *N*-propargyl moiety of rasagiline and selegiline promotes neuronal survival via similar neuroprotective/ neurorescue pathways, thus enlightening the importance of this moiety for the novel activities of rasagiline (Bar-Am *et al.*, 2004, 2005). Indeed, we have shown that propargylamine significantly attenuated cell death induced by serum deprivation in neuronal cells (Bar-Am *et al.*, 2005; Weinreb *et al.*, 2004; Yogev-Falach *et al.*, 2003). The potent MAO-B inhibitory activity of rasagiline, which resides in the interaction of its *N*-propargyl moiety with FAD cofactor of the enzyme (Maycock *et al.*, 1976; Youdim *et al.*, 2001c), is not a prerequisite for its neuroprotective activity (Maruyama *et al.*, 2000c, 2002a; Sagi *et al.*, 2007). This is supported by the following observations: the putative neuroprotective mechanism of rasagiline and selegiline was established in cell lines that do not express MAO-B enzyme activity (PC-12 and neuroblastoma SH-SY5Y cells); CGP 3466 (a propargylamine derivative; Waldmeier *et al.*, 2000) and TVP-1022 (the (*S*)-enantiomer of rasagiline; Maruyama *et al.*, 2001a, 2002b; Yogev-Falach *et al.*, 2003), which are devoid of MAO-B inhibitory activity, exert neuroprotective effects similar to rasagiline and selegiline in cell cultures and animal models of neurodegeneration.

C. Antiapoptotic Action of Rasagiline

1. *Mitochondria and Related Cell Signaling Pathways*

Apoptosis and necrosis are modes of cell death that play an integral part in a variety of biological processes implicated in neurodegenerative diseases (e.g., PD, AD, and Huntington's disease), cardiovascular diseases, and diabetes (Jenner, 1998; Mizuno *et al.*, 1995; Smith *et al.*, 2000). The initial phase of neuronal apoptosis is triggered in response to an induction phase resulting from a variety of insults, including xenobiotics, Parkinsonism-causing endogenous and exogenous neurotoxins (*N*-M-(*R*)-Sal, 6-OHDA, and MPTP), and oxidative stress (Linert *et al.*, 1996; Loschmann *et al.*, 1994). This results in an accompanied change in mitochondrial membrane permeability (MMP), decline in mitochondrial membrane potential ($\Delta\psi m$), opening of mitochondrial permeability transition pore (MpTp) complex, inhibition of the ubiquitin–proteasome complex, release of mitochondrial cytochrome *c*, activation of caspases (in particular caspase 3), and induction of neuronal death (Bernardi *et al.*, 2001; Cheng *et al.*, 2001; Gomez-Lazaro *et al.*, 2008; Junn and Mouradian, 2001; Kroemer and Reed, 2000; Przedborski *et al.*, 2004; Vander Heiden and Thompson, 1999). The important role of MpTp in apoptotic cell death is supported by the findings that this complex, particularly the voltage dependent anion channel (VDAC) and adenine nucleotide translocase (ANT), is a direct functional target for the mitochondrial Bcl-2 family proteins. Antiapoptotic members (e.g., Bcl-2, Bcl-xL, Bcl-W) stabilize MpTp, while proapoptotic members (e.g., Bax, Bak, Bad, and Bid) promote and increase the changes in MMP (Cheng *et al.*, 2001; Vander Heiden and Thompson, 1999).

Previous studies demonstrated that maintenance of the mitochondria integrity and related proteins and enzymes is the main neuroprotective target of rasagiline, selegiline, and other related propargylamines (Akao *et al.*, 2002b; Ebadi *et al.*, 2002; Magyar *et al.*, 2006; Maruyama *et al.*, 2000b; Riederer and Laux, 2011; Sharma *et al.*, 2003). Indeed, it was shown that in serum- and NGF-deprived PC-12 cells, selegiline and its metabolite, desmethylselegiline, attenuated neuronal death by altering the levels or subcellular localization of proteins that affect MMP, scavenging oxidative radicals, or regulating specific signaling molecules (e.g., glyceraldehyde 3-phosphate dehydrogenase (GAPDH), superoxide dismutase (SOD), glutathione peroxidase *c*-JUN, Bcl-2, and Bax; (Tatton *et al.*, 2002)). *In vivo* studies demonstrated that rasagiline and selegiline increased the activity of the antioxidant enzymes, SOD, and catalase, in the rat brain after systemic drug administration (Carrillo *et al.*, 2000; Kitani *et al.*, 2000). Rasagiline possesses efficient neuroprotective activity against a range of insults associated with various cytotoxic processes, such as impaired mitochondrial function, ubiquitin–proteasome system inhibition, mitochondrial cytochrome *c* release, caspase 3 activation, and nucleosomal DNA fragmentation (Abu-Raya *et al.*, 1999, 2002; Akao *et al.*, 2002b; Bar-Am *et al.*, 2004; Maruyama *et al.*, 2000a,b; Ou *et al.*, 2009, 2010; Tatton *et al.*, 2002). This inhibitory activity of rasagiline was accompanied by an increase of antiapoptotic genes and proteins of the Bcl-2 family (e.g., Bcl-2 and Bcl-xL), as well as a decrease of proapoptotic members (e.g., Bax and Bad; Maruyama *et al.*, 2002a; Tatton *et al.*, 2002; Weinreb *et al.*, 2004; Zhu *et al.*, 2008; Fig. 2).

Previous studies provided evidence whereby the activation/regulation of Bcl-2 protein family, associated with protein kinase C (PKC) signaling pathway, mediated a range of neuroprotective effects induced by rasagiline, propargylamine, and 1-(*R*)-aminoindan (Bar-Am *et al.*, 2004, 2005, 2007; Weinreb *et al.*, 2004; Yogev-Falach *et al.*, 2003). The role for PKC activation was supported by *in vitro* studies demonstrated that rasagiline markedly induced the phosphorylation of PKC, specifically that of the prosurvival PKC isoforms, PKCα and PKCε (Bar-Am *et al.*, 2004; Weinreb *et al.*, 2004). In addition, the specific PKC inhibitor, GF109203X, markedly reversed rasagiline-suppressive effects on the cleavage and activity of caspase 3, indicating that PKC signaling cascade is essential for the drug-induced neuronal protective effect (Fig. 2; Bar-Am *et al.*, 2004, 2005, 2007; Weinreb *et al.*, 2004). Further, the mitogen-activated protein kinase (MAPK)/extracellular signal-regulated kinase (ERK) pathway, which has been shown to inhibit cell death in a number of systems, was found to be upregulated by rasagiline and propargylamine-related compounds (Fig. 2; Yogev-Falach *et al.*, 2002, 2003).

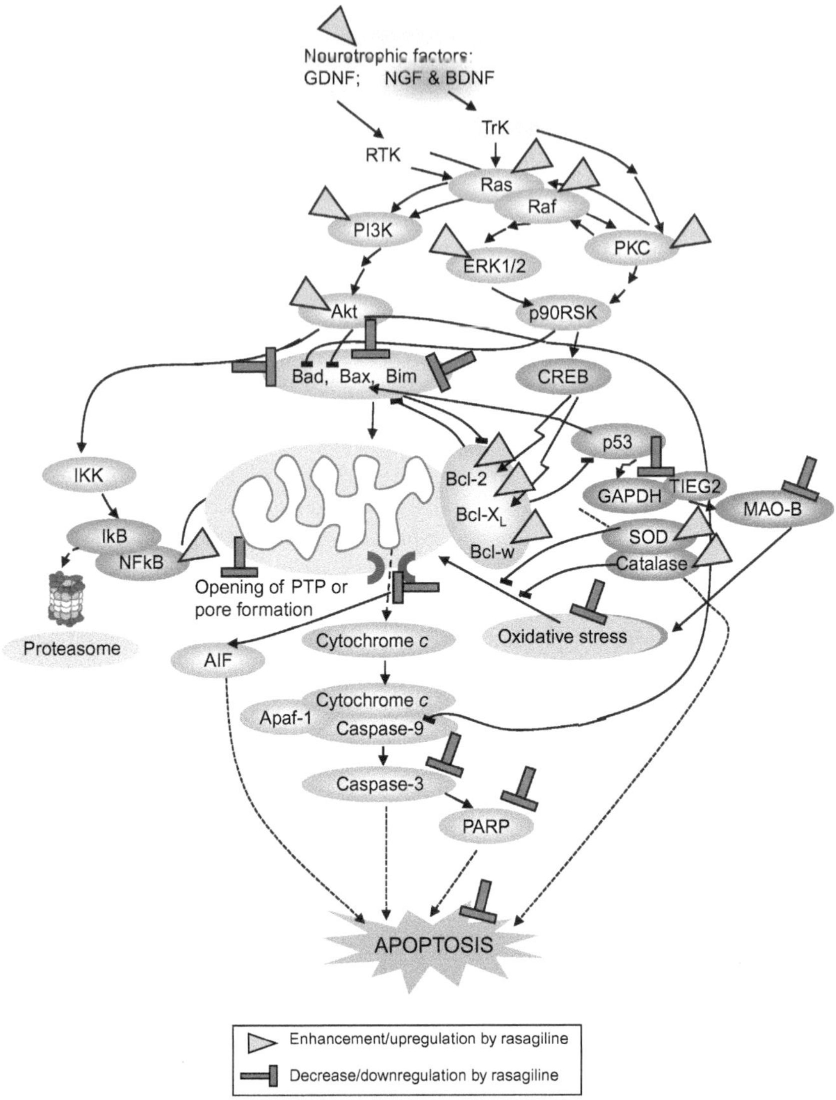

FIG. 2. Schematic illustration depicting potential points for intervention by rasagiline in apoptotic and survival pathways. The neuroprotective mechanism of action of the monoamine oxidase-B (MAO-B) inhibitor, anti-Parkinsonian drug, rasagiline, has been demonstrated to comprise protection of the mitochondrial viability and mitochondrial permeability transition pore by preventing the decline in membrane potential ($\Delta\psi m$), regulating cell survival Bcl-2 family proteins, and attenuating the cell death cascade initiated by proapoptotic Bcl-2 family proteins and caspase-3. In addition, rasagiline was found to block the nuclear translocation of glyceraldehyde-3-phosphate dehydrogenase (GAPDH) and ultimately inhibit MAO-B at the transcriptional level. Rasagiline was suggested to suppress oxidative

2. *Blockade of GAPDH Translocation to the Nucleus*

Other lines of research provided substantial evidence indicating that propargylamines exert antiapoptotic effects by means of an interaction with GAPDH (Tatton *et al.*, 2000). In addition, of being an intermediate enzyme in the glycolytic pathway, GAPDH has been reported to participate in other processes such as in initiation of cellular apoptosis or programmed cell death. Studies with antisense oligonucleotides showed that GAPDH is necessary for apoptosis to proceed in cerebrocortical neurons and PC-12 cells (Ishitani *et al.*, 1996; Sawa *et al.*, 1997). During apoptosis, GAPDH accumulates in the nucleus, and that increase has been proposed to constitute an initial event in the promotion of cell death (Tatton *et al.*, 2000). Carlile *et al.* (2000) demonstrated that propargylamines exert protection by binding to a central channel formed by the GAPDH tetramer and maintaining GAPDH as a dimer in the cytosol. In this form, GAPDH is stabilized and cannot translocate to the nucleus, thereby preventing upregulation of antiapoptotic proteins. In support, rasagiline and other propargylamine-containing compounds prevented the nuclear translocation of GAPDH and thus allowing the neurons to upregulate the levels of protective molecules such as Bcl-2, SOD, and GSH and thereby to prevent oxidative stress, maintain the mitochondrial membrane potential, and block apoptosis (Carlile *et al.*, 2000; Maruyama *et al.*, 2001b; Olanow, 2006; Youdim *et al.*, 2006).

The ability of propargylamines to interfere with apoptosis signaling pathways and increase neuronal survival appears to be independent of MAO-B inhibition. Indeed, the doses of selegiline required to protect rat facial motorneurons were lower than those reported necessary to inhibit the enzyme, and CGP 3466 (a propargylamine derivative) and TVP-1022 (the (*S*)-enantiomer of rasagiline), which do not inhibit MAO-B, were also protective both *in vivo* and *in vitro* (Waldmeier *et al.*, 2000). Further, selegiline protects against the toxicity of MPP^+, the product of the MAO-B reaction, even in cell lines that lack MAO B (for review, see Youdim *et al.*, 2006).

Studies to explore the molecular events that underlie the nuclear translocation of GAPDH in apoptosis revealed the occurrence of S-nitrosylation on GAPDH upon activation of nitric oxide synthase, which triggers its binding to Siah1 and translocation of the complex to the nucleus (Hara *et al.*, 2005). Here, GAPDH stabilizes the rapidly turning over Siah1, enhancing its E3 ubiquitin ligase activity, and causes

stress through increasing antioxidant enzyme activities, such as superoxide dismutase (SOD) and catalase. These neuroprotective properties have been also associated with increased generation of neurotrophic factors (nerve growth factor, NGF; brain-derived neurotrophic factor, BDNF; glial cell line-derived neurotrophic factor, GDNF) and activation of neurotrophic tyrosine kinase receptor (TRK) and Ret protein tyrosine kinase (RTK) pathways. This is turn causes stimulation of cell signal transduction cascades, the extracellular mitogen-activated protein kinases (MAPK), protein kinase C (PKC), Ras-phosphatidylinositol 3-kinase (PI3K)-Akt/NFkB. Adapted from Weinreb *et al.* (2010).

cell death. Recently, a novel role for GAPDH has been reported in ethanol-induced cellular damage. Ethanol elicited nuclear translocation of GAPDH, which induced the transcriptional activation of MAO-B via interaction of GAPDH with transforming growth factor-β-inducible early gene 2 (TIEG2; Fig. 2).

D. Induction of Neurotrophic Factors

Neurotrophic factors (NTFs) are secreted peptides playing important roles in proliferation, differentiation, and survival of neurons in the peripheral and CNS during development (McAllister, 2001), as well as in the synaptic activity and plasticity of mature neurons (Murer *et al.*, 2001). The family of NTFs comprises several structurally and functionally related molecules: (i) NGF-superfamily consists of two structurally related proteins, namely the brain-derived neurotrophic factor (BDNF) and NGF (Siegel and Chauhan, 2000); (ii) glial cell line-derived neurotrophic factor (GDNF) family; (iii) neurokine or neuropoietin superfamily; and (iv) nonneuronal growth factor-superfamily (Siegel and Chauhan, 2000). Individual NTFs may have different effects on the same population of neurons through differential activation of signaling mechanisms (Siegel and Chauhan, 2000).

It has been shown that rasagiline can upregulate gene expression of the NTFs, BDNF, and GDNF (Bar-Am *et al.*, 2005; Maruyama *et al.*, 2004; Weinreb *et al.*, 2004), suggesting a possible association between the neuroprotective mechanism of the drug and upregulation of NTFs. Indeed, BDNF and GDNF have been shown to play critical role in neuroprotective and neurorescue processes of several neurons, including dopaminergic and cholinergic, and promote survival of major neuronal types affected in neurodegenerative diseases (Murer *et al.*, 2001). Postmortem human studies demonstrated reduced striatal BDNF and GDNF expression levels in PD patients (Parain *et al.*, 1999; Hunot *et al.*, 1996), thus suggesting their involvement in the pathophysiology of PD.

Both rasagiline and selegiline were shown to induce mRNA expression levels of β-NGF and BDNF in rat midbrains (Weinreb *et al.*, 2009). However, the regulatory effect of rasagiline was more evident and significantly higher, compared to that of selegiline. In support, rasagiline was previously reported to enhance gene expression of β-NGF and its downstream transcription factors: NGF-1A binding protein (EGR1) and the early growth response protein-3 (EGR3) in response to MPTP *in vivo* (Sagi *et al.*, 2007). In addition, chronic low dose of rasagiline (0.05 mg/kg) administered to mice subsequently to MPTP treatment (Sagi *et al.*, 2007) rescued dopaminergic neurons in the SNpC and induced a number of cell signaling mediators associated with the TK receptor pathway, including ShcC, SOS, AF6, Rin1, and Ras, in parallel with a specific increase in the TK receptor downstream effecter phosphatidylinositol 3-kinase (PI3K)-proteins. The low dose of rasagiline was deliberately chosen not to inhibit brain MAO-B, while doses between 0.25–1 mg/kg that inhibit the enzyme did not produce neurorestoration in this model (Sagi *et al.*, 2007). This study demonstrated the essentiality of activation of Ras-PI3K-Akt survival pathway in the induction of β-NGF transcript in rasagiline-mediated neurorescue effect (Fig. 2).

E. Regulation of APP Proteolytic Processing

An additional neuroprotective effect, shared by the propargylamine-derivative compounds, is related to their ability to regulate the processing of APP by the nonamyloidogenic α-secretase pathway (Vetrivel and Thinakaran, 2006).

The generation of β-amyloid (Aβ), by enzymatic cleavage of APP, has been at the center of AD research. APP can be processed in at least two alternative, mutually exclusive posttranslational pathways: (i) In the amyloidogenic pathway, APP is sequentially cleaved by β and γ secretases, at the N- and C-terminal of the Aβ domain, respectively, to release Aβ. (ii) In the nonamyloidogenic pathway, APP is cleaved by α-secretase, within the Aβ sequence and generates a secreted form of soluble APP (sAPPα), thus precluding the formation of the amyloidogenic Aβ (Vetrivel and Thinakaran, 2006). Secreted sAPPα has been implicated to possess neurotrophic and neuroprotective properties (Chen and Tang, 2006; Chow *et al.*, 2010; Hampel *et al.*, 2010). Available evidence has demonstrated that α-secretase activity could be regulated via PKC, MAPK, and TK receptor signaling pathways, calcium ion (Ca^{2+}), and hormonal signaling (Mills and Reiner, 1999a,b; Zhang and Xu, 2007).

Previously, it was indicated that selegiline could increase the release of sAPPα via activation of the MAPK- and PKC-dependent pathways in neuronal cell cultures (Hu *et al.*, 2008; Yang *et al.*, 2007; Yogev-Falach *et al.*, 2003). Selegiline induced the phosphorylation of PKC and its substrate myristoylated alanine-rich C kinase substrate (MARCKS) and promoted the translocation of the two main isoforms involved in the pathogenesis of AD, PKCα, and PKCε (Lanni *et al.*, 2004), from the cytosolic to the membrane fraction (Yang *et al.*, 2007). In addition, rasagiline was found to have neuroprotective properties against Aβ-induced cytotoxicity and to induce sAPPα release (Yogev-Falach *et al.*, 2003). The increase of sAPPα release was blocked by a hydroxamic acid-based metalloprotease inhibitor, indicating that the effect was mediated via α-secretase activity (Yogev-Falach *et al.*, 2003). Rasagiline-induced sAPPα release was also significantly reduced by inhibitors of PKC and MAPK signaling cascades (Yogev-Falach *et al.*, 2002, 2003). This suggests the involvement of MAPK-and PKC-dependent pathways in rasagiline-stimulated sAPPα release. Animal studies demonstrated that treatment with rasagiline markedly decreased the levels of cell-associated, full-length APP in the mice hippocampus, which indicated that APP processing can also be regulated by rasagiline under *in vivo* conditions (Bar-Am *et al.*, 2004). These properties of rasagiline may be of clinical value toward accelerating nonamyloidogenic APP, thereby reducing Aβ levels. In addition, it was shown that chronic administration of rasagiline upregulated phospho-PKC levels and the expression of the PKCα and PKCε isoenzymes and elevated the levels of *p*-MARCKS and those of receptors for activated C kinase 1 (RACK1) in mice hippocampus, further indicating that the mechanism by which rasagiline affects APP processing may be related to PKC-associated signaling (Bar-Am *et al.*, 2004, 2010b; Hu *et al.*, 2008; Yang *et al.*, 2007; Yogev-Falach *et al.*, 2002, 2003).

Structure–activity relationship studies among rasagiline-related compounds suggest the crucial role of the N-propargyl moiety on these molecules for the regulatory effect on APP processing, because propargylamine itself induced the secretion of sAPPα and significantly increased MAPK phosphorylation with similar potency to that of rasagiline and its derivatives (Yogev-Falach *et al.*, 2003).

III. The Major Metabolite of Rasagiline, 1-(*R*)-Aminoindan

Rasagiline is primarily metabolized to 1-(*R*)-aminoindan (Fig. 1), by hepatic CYP1A2-mediated N-dealkylation followed by renal excretion of the parent conjugated compound and its metabolites (Chen and Swope, 2005). A kinetic and crystallographic analysis revealed that 1-(*R*)-aminoindan is not a substrate for MAO oxidation but, conversely, is a weak reversible inhibitor (Binda *et al.*, 2005).

Various neuroprotective studies demonstrated that the major metabolite of rasagiline, 1-(*R*)-aminoindan possesses beneficial pharmacological effects in animal and cell culture models (Bar-Am *et al.*, 2007; Speiser *et al.*, 1998). Indeed, 1-(*R*)-aminoindan exerts neuroprotective properties against the Parkinsonian neurotoxin, 6-OHDA in PC-12 cells, prevents cell death in a cytotoxic model of human neuroblastoma SK-*N*-SH cells in high density culture (Bar-Am *et al.*, 2007), and confers a significant neuroprotective effect against hydrogen peroxide (H_2O_2)-induced damage in SH-SY5Y neuroblastoma cells and rat primary cortical neurons (Bar-Am *et al.*, 2010a). Recent studies demonstrated that 1-(*R*)-aminoindan decreased ethanol-induced cell death through the inhibition of GAPDH-MAO-B-mediated apoptotic signaling pathway (Ou *et al.*, 2009, 2010). Moreover, *in vivo* chronic administration of 1-(*R*)-aminoindan prevented the reduction in the levels of striatal DA and its metabolites, dihydroxyphenylacetic acid (DOPAC), and homovanillic acid (HVA), and reduced the increase of DA turnover in the striatum in 6-OHDA and lactacystin rat models of PD (Weinreb *et al.*, 2011). In accordance, 1-(*R*)-aminoindan was also shown to retard impaired dopaminergic-related behavioral responses induced by 6-OHDA and lactacystin (Weinreb *et al.*, 2011). These findings suggest that 1-(*R*)-aminoindan may contribute to the overall neuroprotective activity of rasagiline.

IV. Summary and Conclusions

Compelling evidence points to the existence of an intrinsic antiapoptotic mechanism of neuroprotection operated by rasagiline toward stabilization of the MMP, induction of the antiapoptotic protein Bcl-2 and NTFs, attenuation of GAPDH translocation to the nucleus and regulation of antioxidant enzymes (Naoi and Maruyama, 2009). Structure–activity studies among rasagiline and selegiline-related compounds provide evidence that the *N*-propargyl moiety promotes neuronal survival via similar neuroprotective/neurorescue antiapoptotic pathways, thus enlightening the importance of this moiety for their prosurvival activities. In support, the neuroprotective action of rasagiline has been proven in various animal models of PD, as well as other neurodegenerative disorders, such as AD and amyotrophic lateral sclerosis. Although the mechanism by which rasagiline acts is not completely understood, it is likely that regulation of signaling pathways, as PKC and MAPK/ERK, is pivotal to its protective activity (Fig. 2). Compared to selegiline, the higher neuroprotective potency of rasagiline, which has been described in several preclinical studies (Chen and Ly, 2006; Chen *et al.*, 2007), may, in part, be associated with their different metabolites. Indeed, the major metabolite of rasagiline, 1-(*R*)-aminoindan has been demonstrated to stimulate beneficial neuroprotective effects by itself in animal and cell culture models (Bar-Am *et al.*, 2010a).

All together, the crucial neuroprotective role played by propargylamine and 1-(*R*)-aminoindan and the ability of rasagiline itself to restore and activate endogenous neurotrophins and mediators of classical survival pathways in the affected brain areas indicate that rasagiline may possess disease-modifying activity leading to incrementing clinical improvement.

Acknowledgment

The support of Technion Research and Development (Haifa, Israel), Teva Pharmaceutical Co. (Netanya, Israel), Stein Foundation, and Friedman Foundation (Philadelphia, USA) is gratefully acknowledged.

References

Abu-Raya, S., Blaugrund, E., Trembovler, V., Shilderman-Bloch, E., Shohami, E., and Lazarovici, P. (1999). Rasagiline, a monoamine oxidase-B inhibitor, protects NGF-different PC12 cells against oxygen-glucose deprivation. *J. Neurosci. Res.* **58**, 456–463.

Abu-Raya, S., Tabakman, R., Blaugrund, E., Trembovler, V., and Lazarovici, P. (2002). Neuroprotective and neurotoxic effects of monoamine oxidase-B inhibitors and derived metabolites under ischemia in PC12 cells. *Eur. J. Pharmacol.* **434**, 109–116.

Ahlskog, J.E., and Uitti, R.J. (2010). Rasagiline, Parkinson neuroprotection, and delayed-start trials: still no satisfaction? *Neurology* **74**, 1143–1148.

Akao, Y., Maruyama, W., Yi, H., Shamoto-Nagai, M., Youdim, M.B.H., and Naoi, M. (2002a). An anti-Parkinson's disease drug, N-propargyl-1(R)-aminoindan (rasagiline), enhances expression of anti-apoptotic Bcl-2 in human dopaminergic SH-SY5Y cells. *Neurosci. Lett.* **326**, 105–108.

Akao, Y., Maruyama, W., Shimizu, S., Yi, H., Nakagawa, Y., Shamoto-Nagai, M., Youdim, M.B., Tsujimoto, Y., and Naoi, M. (2002b). Mitochondrial permeability transition mediates apoptosis induced by N- methyl(R)salsolinol, an endogenous neurotoxin, and is inhibited by Bcl-2 and rasagiline, N-propargyl-1(R)-aminoindan. *J. Neurochem.* **82**, 913–923.

Bach, A.W., Lan, N.C., Johnson, D.L., Abell, C.W., Bembenek, M.E., Kwan, S.W., Seeburg, P.H., and Shih, J.C. (1988). cDNA cloning of human liver monoamine oxidase A and B: molecular basis of differences in enzymatic properties. *Proc. Natl. Acad. Sci. USA* **85**, 4934–4938.

Baker, G.B., Bertollini, A., del Carmine, R., Martin, I.L., and Raiteri, M. (1976). Effects of *p*-chloro-beta-phenylethylamine on the uptake and release of putative amine neurotransmitters in rat brain [proceedings]. *Br. J. Pharmacol.* **58**, 420 pp.

Bar-Am, O., Amit, T., Sagi, Y., and Y, M.B.H. (2004a). Contrasting neuroprotective and neurotoxic actions of respective metabolites of anti-Parkinson drugs rasagiline and selegiline. *Neuroscience Letters* **355**, 169–172.

Bar-Am, O., Yogev-Falach, M., Amit, T., Sagi, Y., and Y, M.B.H. (2004b). Regulation of protein kinase C by the anti-Parkinson drug, MAO-B inhibitor, rasagiline and its derivatives, in vivo. *J. Neurochem.* **89**, 1119–1125.

Bar-Am, O., Weinreb, O., Amit, T., and Youdim, M.B. (2005). Regulation of Bcl-2 family proteins, neurotrophic factors, and APP processing in the neurorescue activity of propargylamine. *FASEB J.* **19**, 1899–1901.

Bar-Am, O., Amit, T., and Youdim, M.B. (2007). Aminoindan and hydroxyaminoindan, metabolites of rasagiline and ladostigil, respectively, exert neuroprotective properties in vitro. *J. Neurochem.* **103**, 500–508.

Bar-Am, O., Weinreb, O., Amit, T., and Youdim, M.B. (2010a). The neuroprotective mechanism of 1-(R)-aminoindan, the major metabolite of the anti-parkinsonian drug rasagiline. *J. Neurochem.* **112**, 1131–1137.

Bar-Am, O., Amit, T., Weinreb, O., Youdim, M.B., and Mandel, S. (2010b). Propargylamine containing compounds as modulators of proteolytic cleavage of amyloid-beta protein precursor: involvement of mitogen activated protein kinase and protein kinase C activation. *J. Alzheimers Dis.* **21**, 361–371.

Bernardi, P., Petronilli, V., Di Lisa, F., and Forte, M. (2001). A mitochondrial perspective on cell death. *Trends Biochem. Sci.* **26**, 112–117.

Bernheimer, H., Birkmayer, W., and Hornykiewicz, O. (1962). Behavior of monoamine oxidase in the brain of man after therapy with monoamine oxidase inhibitors. *Wien. Klin. Wochenschr.* **74**, 558–559.

Binda, C., Hubalek, F., Li, M., Herzig, Y., Sterling, J., Edmondson, D.E., and Mattevi, A. (2004). Crystal structures of monoamine oxidase B in complex with four inhibitors of the N-propargylaminoindan class. *J. Med. Chem.* **47**, 1767–1774.

Binda, C., Hubalek, F., Li, M., Herzig, Y., Sterling, J., Edmondson, D.E., and Mattevi, A. (2005). Binding of rasagiline-related inhibitors to human monoamine oxidases: a kinetic and crystallographic analysis. *J. Med. Chem.* **48**, 8148–8154.

Birkmayer, W., and Hornykiewicz, O. (1961). The L-3,4-dioxyphenylalanine (DOPA)-effect in Parkinson-akinesia. *Wien. Klin. Wochenschr.* **73**, 787–788.

Birkmayer, W., and Hornykiewicz, O. (1962). The L-dihydroxyphenylalanine (L-DOPA) effect in Parkinson's syndrome in man: on the pathogenesis and treatment of Parkinson akinesis. *Archiv fur Psychiatrie und Nervenkrankheiten, vereinigt mit Zeitschrift fur die gesamte Neurologie und Psychiatrie* **203**, 560–574.

Birkmayer, W., Riederer, P., Youdim, M.B.H., and Linauer, W. (1975a). The potentiation of the anti akinetic effect after L-dopa treatment by an inhibitor of MAO-B, Deprenil. *J. Neural Transm.* **36**, 303–326.

Birkmayer, W., Danielczyk, W., Neumayer, E., and Riederer, P. (1975b). Dopaminergic supersensitivity in parkinsonism. *Adv. Neurol.* **9**, 121–129.

Birkmayer, W., Riederer, P., Ambrozi, L., and Youdim, M.B.H. (1977). Implications of combined treatment with 'Madopar' and L-deprenil in Parkinson's disease. A long-term study. *Lancet* **1**, 439–443.

Birkmayer, W., Knoll, J., Riederer, P., and Youdim, M.B.H. (1983). (−)-Deprenyl leads to prolongation of L-dopa efficacy in Parkinson's disease. *Mod. Probl. Pharmacopsychiatry* **19**, 170–176.

Birkmayer, W., Knoll, J., Riederer, P., Youdim, M.B.H., Hars, V., and Marton, J. (1985). Increased life expectancy resulting from addition of L-deprenyl to Madopar treatment in Parkinson's disease: a longterm study. *J. Neural Transm.* **64**, 113–127.

Carlile, G.W., Chalmers-Redman, R.M., Tatton, N.A., Pong, A., Borden, K.E., and Tatton, W.G. (2000). Reduced apoptosis after nerve growth factor and serum withdrawal: conversion of tetrameric glyceraldehyde-3-phosphate dehydrogenase to a dimer. *Mol. Pharmacol.* **57**, 2–12.

Carlsson, A., Lindqvist, M., Magnusson, T., and Waldeck, B. (1958). On the presence of 3-hydroxytyramine in brain. *Science* **127**, 471.

Carrillo, M.C., Minami, C., Kitani, K., Maruyama, W., Ohashi, K., Yamamoto, T., Naoi, M., Kanai, S., and Youdim, M.B.H. (2000). Enhancing effect of rasagiline on superoxide dismutase and catalase activities in the dopaminergic system in the rat. *Life Sci.* **67**, 577–585.

Chen, J.J., and Ly, A.V. (2006). Rasagiline: a second-generation monoamine oxidase type-B inhibitor for the treatment of Parkinson's disease. *Am. J. Health Syst. Pharm.* **63**, 915–928.

Chen, J.J., and Swope, D.M. (2005). Clinical pharmacology of rasagiline: a novel, second-generation propargylamine for the treatment of Parkinson disease. *J. Clin. Pharmacol.* **45**, 878–894.

Chen, Y., and Tang, B.L. (2006). The amyloid precursor protein and postnatal neurogenesis/neuroregeneration. *Biochem. Biophys. Res. Commun.* **341**, 1–5.

Chen, J.J., Swope, D.M., and Dashtipour, K. (2007). Comprehensive review of rasagiline, a second-generation monoamine oxidase inhibitor, for the treatment of Parkinson's disease. *Clin. Ther.* **29**, 1825–1849.

Cheng, E.H., Wei, M.C., Weiler, S., Flavell, R.A., Mak, T.W., Lindsten, T., and Korsmeyer, S.J. (2001). BCL-2, BCL-X(L) sequester BH3 domain-only molecules preventing BAX- and BAK-mediated mitochondrial apoptosis. *Mol. Cell* **8**, 705–711.

Chow, V.W., Mattson, M.P., Wong, P.C., and Gleichmann, M. (2010). An overview of APP processing enzymes and products. *Neuromol. Med.* **12**, 1–12.

Collins, G.G., Sandler, M., Williams, E.D., and Youdim, M.B.H. (1970). Multiple forms of human brain mitochondrial monoamine oxidase. *Nature* **225**, 817–820.

de la Fuente-Fernandez, R., Schulzer, M., Mak, E., and Sossi, V. (2010). Trials of neuroprotective therapies for Parkinson's disease: problems and limitations. *Parkinsonism Relat. Disord.* **16**, 365–369.

deMarcaida, J.A., Schwid, S.R., White, W.B., Blindauer, K., Fahn, S., Kieburtz, K., Stern, M., and Shoulson, I. (2006). Effects of tyramine administration in Parkinson's disease patients treated with selective MAO-B inhibitor rasagiline. *Mov. Disord.* **21**, 1716–1721.

Ebadi, M., Sharma, S., Shavali, S., and El, Refaey H. (2002). Neuroprotective actions of selegiline. *J. Neurosci. Res.* **67**, 285–289.

Eliash, S., Speiser, Z., and Cohen, S. (2001). Rasagiline and its (S) enantiomer increase survival and prevent stroke in salt-loaded stroke-prone spontaneously hypertensive rats. *J. Neural Transm.* **108**, 909–923.

Finberg, J.P., and Youdim, M.B. (1985). Modification of blood pressure and nictitating membrane response to sympathetic amines by selective monoamine oxidase inhibitors, types A and B, in the cat. *Br. J. Pharmacol.* **85**, 541–546.

Finberg, J.P.M., Sabbagh, A., and Youdim, M.B.H. (1980). Pharmacology of selective propargyl "suicide" inhibitors of monoamine oxidase. In: E. Usdin, T. Sourkes, and M.B.H. Youdim (Eds.), *Enzymes and Neurotransmitters in Mental Disease.* Wiley, Chichester, pp. 205–219.

Finberg, J.P., Tenne, M., and Youdim, M.B.H. (1981a). Tyramine antagonistic properties of AGN 1135, an irreversible inhibitor of monoamine oxidase type B. *Br. J. Pharmacol.* **73**, 65–74.

Finberg, J.P.M., Tenne, M., and Youdim, M.B.H. (1981b). Selective irreversible propargyl derivative inhibitors of monoamine oxidase (MAO) without the cheese effect. In: M.B.H. Youdim and E.S. Peykel (Eds.), *Monoamine Oxidase Inhibitors—The State of the Art.* Wiley, Chichester, pp. 31–41.

Finberg, J.P., Lamensdorf, I., Commissiong, J.W., and Youdim, M.B.H. (1996). Pharmacology and neuroprotective properties of rasagiline. *J. Neural Transm. Suppl.* **48**, 95–101.

Finberg, J.P., Wang, J., Bankiewicz, K., Harvey-White, J., Kopin, I.J., and Goldstein, D.S. (1998). Increased striatal dopamine production from L-DOPA following selective inhibition of monoamine oxidase B by R(+)-N-propargyl-1-aminoindan (rasagiline) in the monkey. *J. Neural Transm. Suppl.* **52**, 279–285.

Finberg, J.P., Lamensdorf, I., Weinstock, M., Schwartz, M., and Youdim, M.B.H. (1999). Pharmacology of rasagiline (N-propargyl-1R-aminoindan). *Adv. Neurol.* **80**, 495–499.

Gomez-Lazaro, M., Galindo, M.F., Concannon, C.G., Segura, M.F., Fernandez-Gomez, F.J., Llecha, N., Comella, J.X., Prehn, J.H., and Jordan, J. (2008). 6-Hydroxydopamine activates the mitochondrial apoptosis pathway through p38 MAPK-mediated, p53-independent activation of Bax and PUMA. *J. Neurochem.* **104**, 1599–1612.

Goren, T., Adar, L., Sasson, N., and Weiss, Y.M. (2010). Clinical pharmacology tyramine challenge study to determine the selectivity of the monoamine oxidase type B (MAO-B) inhibitor rasagiline. *J. Clin. Pharmacol.* **50**, 1420–1428.

Götz, M.E., Breithaupt, W., Sautter, J., Kupsch, A., Schwarz, J., Oertel, W.H., Youdim, M.B.H., Riederer, P., and Gerlach, M. (1998). Chronic TVP-1012 (rasagiline) dose–activity response of monoamine oxidases A and B in the brain of the common marmoset. *J. Neural Transm. Suppl.* **52**, 271–278.

Green, A.R., Mitchell, B.D., Tordoff, A.F., and Youdim, M.B.H. (1977). Evidence for dopamine deamination by both type A and type B monoamine oxidase in rat brain in vivo and for the degree of inhibition of enzyme necessary for increased functional activity of dopamine and 5-hydroxytryptamine. *Br. J. Pharmacol.* **60**, 343–349.

Hampel, H., Shen, Y., Walsh, D.M., Aisen, P., Shaw, L.M., Zetterberg, H., Trojanowski, J.Q., and Blennow, K. (2010). Biological markers of amyloid beta-related mechanisms in Alzheimer's disease. *Exp. Neurol.* **223**, 334–346.

Hara, M.R., Agrawal, N., Kim, S.F., Cascio, M.B., Fujimuro, M., Ozeki, Y., Takahashi, M., Cheah, J.H., Tankou, S.K., Hester, L.D., Ferris, C.D., Hayward, S.D., *et al.* (2005). S-nitrosylated GAPDH initiates apoptotic cell death by nuclear translocation following Siah1 binding. *Nat. Cell Biol.* **7**, 665–674.

Heikkila, R.E., Duvoisin, R.C., Finberg, J.P., and Youdim, M.B.H. (1985). Prevention of MPTP-induced neurotoxicity by AGN-1133 and AGN-1135, selective inhibitors of monoamine oxidase-B. *Eur. J. Pharmacol.* **116**, 313–317.

Hu, M.K., Liao, Y.F., Chen, J.F., Wang, B.J., Tung, Y.T., Lin, H.C., and Lee, K.P. (2008). New 1,2,3,4-tetrahydroisoquinoline derivatives as modulators of proteolytic cleavage of amyloid precursor proteins. *Bioorg. Med. Chem.* **16**, 1957–1965.

Huang, W., Chen, Y., Shohami, E., and Weinstock, M. (1999). Neuroprotective effect of rasagiline, a selective monoamine oxidase-B inhibitor, against closed head injury in the mouse. *Eur. J. Pharmacol.* **366**, 127–135.

Hunot, S., Bernard, V., Faucheux, B., Boissiere, F., Leguern, E., Brana, C., Gautris, P.P., Guerin, J., Bloch, B., Agid, Y., and Hirsch, E.C. (1996). Glial cell line-derived neurotrophic factor (GDNF) gene expression in the human brain: a post mortem in situ hybridization study with special reference to Parkinson's disease. *J. Neural Transm. (Budapest)* **103**, 1043–1052.

Ishitani, R., Sunaga, K., Hirano, A., Saunders, P., Katsube, N., and Chuang, D.M. (1996). Evidence that glyceraldehyde-3-phosphate dehydrogenase is involved in age-induced apoptosis in mature cerebellar neurons in culture. *J. Neurochem.* **66**, 928–935.

Jenner, P. (1998). Oxidative mechanisms in nigral cell death in Parkinson's disease. *Mov. Disord.* **13**, 24–34.

Johnson, J.P. (1968). Some observations upon a new inhibitor of monoamine oxidase in brain tissue. *Biochem. Pharmacol.* **17**, 1285–1297.

Junn, E., and Mouradian, M.M. (2001). Apoptotic signaling in dopamine-induced cell death: the role of oxidative stress, p38 mitogen-activated protein kinase, cytochrome c and caspases. *J. Neurochem.* **78**, 374–383.

Kalir, A., Sabbagh, A., and Youdim, M.B.H. (1981). Selective acetylenic 'suicide' and reversible inhibitors of monoamine oxidase types A and B. *Br. J. Pharmacol.* **73**, 55–64.

Kearney, E.B., Salach, J.I., Walker, W.H., Seng, R.L., Kenney, W., Zeszotek, E., and Singer, T.P. (1971). The covalently-bound flavin of hepatic monoamine oxidase. 1. Isolation and sequence of a flavin peptide and evidence for binding at the 8alpha position. *Eur. J. Biochem.* **24**, 321–327.

Kitani, K., Minami, C., Maruyama, W., Kanai, S., Ivy, G.O., and Carrillo, M.C. (2000). Common properties for propargylamines of enhancing superoxide dismutase and catalase activities in the dopaminergic system in the rat: implications for the life prolonging effect of (−)deprenyl. *J. Neural Transm. Suppl.* **60**, 139–156.

Knoll, J., and Magyar, K. (1972). Some puzzling pharmacological effects of monoamine oxidase inhibitors. *Adv. Biochem. Psychopharmacol.* **5**, 393–408.

Knoll, J., Ecseri, Z., Kelemen, K., Nievel, J., and Knoll, B. (1965). Phenylisopropylmethylpropinylamine (E-250), a new spectrum psychic energizer. *Arch. Int. Pharmacodyn. Ther.* **155**, 154–164.

Kroemer, G., and Reed, J.C. (2000). Mitochondrial control of cell death. *Nat. Med.* **6**, 513–519.

Kupsch, A., Sautter, J., Gotz, M., Breithaupt, W., Schwarz, J., Youdim, M.B.H., Riederer, P., Gerlach, M., and Oertel, W.H. (2001). Monoamine oxidase-inhibition and MPTP-induced neurotoxicity in the non-human primate: comparison of rasagiline (TVP 1012) with selegiline. *J. Neural Transm.* **108**, 985–1009.

Lamensdorf, I., Youdim, M.B., and Finberg, J.P. (1996). Effect of long-term treatment with selective monoamine oxidase A and B inhibitors on dopamine release from rat striatum in vivo. *J. Neurochem.* **67**, 1532–1539.

Lanni, C., Mazzucchelli, M., Porrello, E., Govoni, S., and Racchi, M. (2004). Differential involvement of protein kinase C alpha and epsilon in the regulated secretion of soluble amyloid precursor protein. *Eur. J. Biochem.* **271**, 3068–3075.

Lees, A.J., Shaw, K.M., Kohout, L.J., Stern, G.M., Elsworth, J.D., Sandler, M., and Youdim, M.B.H. (1977). Deprenyl in Parkinson's disease. *Lancet* **2**, 791–795.

Linert, W., Herlinger, E., Jameson, R.F., Kienzl, E., Jellinger, K., and Youdim, M.B.H. (1996). Dopamine, 6-hydroxydopamine, iron, and dioxygen—their mutual interactions and possible implication in the development of Parkinson's disease. *Biochim. Biophys. Acta* **1316**, 160–168.

Loschmann, P.A., Lange, K.W., Wachtel, H., and Turski, L. (1994). MPTP-induced degeneration: interference with glutamatergic toxicity. *J. Neural Transm. Suppl.* **43**, 133–143.

Magyar, K., Vizi, E.S., Ecseri, Z., and Knoll, J. (1967). Comparative pharmacological analysis of the optical isomers of phenyl-isopropyl-methyl-propinylamine (E-250). *Acta Physiol. Acad. Sci. Hung.* **32**, 377–387.

Magyar, K., Palfi, M., Jenei, V., and Szoko, E. (2006). Deprenyl: from chemical synthesis to neuroprotection. *J. Neural Transm.* **71**, 143–156.

Magyar, K., Szende, B., Jenei, V., Tabi, T., Palfi, M., and Szoko, E. (2010). R-deprenyl: pharmacological spectrum of its activity. *Neurochem. Res.* **35**, 1922–1932.

Maruyama, W., Akao, Y., Youdim, M.B.H., and Naoi, M. (2000a). Neurotoxins induce apoptosis in dopamine neurons: protection by N-propargylamine-1(R)- and (S)-aminoindan, rasagiline and TV1022. *J. Neural Transm. Suppl.* **60**, 171–186.

Maruyama, W., Yamamoto, T., Kitani, K., Carrillo, M.C., Youdim, M.B.H., and Naoi, M. (2000b). Mechanism underlying anti-apoptotic activity of a (−) deprenyl-related propargylamine, rasagiline. *Mech. Ageing Dev.* **116**, 181–191.

Maruyama, W., Youdim, M.B.H., and Naoi, M. (2001a). Antiapoptotic properties of rasagiline, N-propargylamine-1(R)-aminoindan, and its optical (S)-isomer, TV1022. *Ann. N. Y. Acad. Sci.* **939**, 320–329.

Maruyama, W., Akao, Y., Youdim, M.B., Davis, B.A., and Naoi, M. (2001b). Transfection-enforced Bcl-2 overexpression and an anti-Parkinson drug, rasagiline, prevent nuclear accumulation of glyceraldehyde-3-phosphate dehydrogenase induced by an endogenous dopaminergic neurotoxin, N-methyl(R)salsolinol. *J. Neurochem.* **78**, 727–735.

Maruyama, W., Takahashi, T., Youdim, M.B.H., and Naoi, M. (2002a). The anti-parkinson drug, rasagiline, prevents apoptotic DNA damage induced by peroxynitrite in human dopaminergic neuroblastoma SH-SY5Y cells. *J. Neural Transm.* **109**, 467–481.

Maruyama, W., Akao, Y., Carrillo, M.C., Kitani, K., Youdium, M.B., and Naoi, M. (2002b). Neuroprotection by propargylamines in Parkinson's disease: suppression of apoptosis and induction of prosurvival genes. *Neurotoxicol. Teratol.* **24**, 675–682.

Maruyama, W., Nitta, A., Shamoto-Nagai, M., Hirata, Y., Akao, Y., Youdim, M.B.H., Furukawa, S., Nabeshima, T., and Naoi, M. (2004). N-Propargyl-1 (R)-aminoindan, rasagiline, increases glial cell line-derived neurotrophic factor (GDNF) in neuroblastoma SH-SY5Y cells through activation of NF-kappaB transcription factor. *Neurochem. Int.* **44**, 393–400.

Maycock, A.L., Abeles, R.H., Salach, J.I., and Singer, T.P. (1976). The structure of the covalent adduct formed by the interaction of 3-dimethylamino-1-propyne and the flavine of mitochondrial amine oxidase. *Biochemistry* **15**, 114–125.

McAllister, A.K. (2001). Neurotrophins and neuronal differentiation in the central nervous system. *Cell. Mol. Life Sci.* **58**, 1054–1060.

Mills, J., and Reiner, P.B. (1999a). Mitogen-activated protein kinase is involved in N-methyl-D-aspartate receptor regulation of amyloid precursor protein cleavage. *Neuroscience* **94**, 1333–1338.

Mills, J., and Reiner, P.B. (1999b). Regulation of amyloid precursor protein cleavage. *J. Neurochem.* **72**, 443–460.

Mizuno, Y., Ikebe, S., Hattori, N., Nakagawa-Hattori, Y., Mochizuki, H., Tanaka, M., and Ozawa, T. (1995). Role of mitochondria in the etiology and pathogenesis of Parkinson's disease. *Biochim. Biophys. Acta* **1271**, 265–274.

Murer, M.G., Yan, Q., and Raisman-Vozari, R. (2001). Brain-derived neurotrophic factor in the control human brain, and in Alzheimer's disease and Parkinson's disease. *Prog. Neurobiol.* **63**, 71–124.

Naoi, M., and Maruyama, W. (2009). Functional mechanism of neuroprotection by inhibitors of type B monoamine oxidase in Parkinson's disease. *Expert Rev. Neurother.* **9**, 1233–1250.

Naoi, M., and Maruyama, W. (2010). Monoamine oxidase inhibitors as neuroprotective agents in age-dependent neurodegenerative disorders. *Curr. Pharm. Des.* **16**, 2799–2817.

O'Carroll, A.M., Fowler, C.J., Phillips, J.P., Tobbia, I., and Tipton, K.F. (1983). The deamination of dopamine by human brain monoamine oxidase. Specificity for the two enzyme forms in seven brain regions. *Naunyn Schmiedebergs Arch. Pharmacol.* **322**, 198–202.

Olanow, C.W. (2006). Rationale for considering that propargylamines might be neuroprotective in Parkinson's disease. *Neurology* **66**, S69–S79.

Olanow, C.W., Hauser, R.A., Jankovic, J., Langston, W., Lang, A., Poewe, W., Tolosa, E., Stocchi, F., Melamed, E., Eyal, E., and Rascol, O. (2008). A randomized, double-blind, placebo-controlled, delayed start study to assess rasagiline as a disease modifying therapy in Parkinson's disease (the ADAGIO study): rationale, design, and baseline characteristics. *Mov. Disord.* **23**, 2194–2201.

Olanow, C.W., Rascol, O., Hauser, R., Feigin, P.D., Jankovic, J., Lang, A., Langston, W., Melamed, E., Poewe, W., Stocchi, F., and Tolosa, E. (2009). A double-blind, delayed-start trial of rasagiline in Parkinson's disease. *N. Engl. J. Med.* **361**, 1268–1278.

Ou, X.M., Lu, D., Johnson, C., Chen, K., Youdim, M.B., Rajkowska, G., and Shih, J.C. (2009). Glyceraldehyde-3-phosphate dehydrogenase-monoamine oxidase B-mediated cell death-induced by ethanol is prevented by rasagiline and 1-R-aminoindan. *Neurotox. Res.* **16**, 148–159.

Ou, X.M., Stockmeier, C.A., Meltzer, H.Y., Overholser, J.C., Jurjus, G.J., Dieter, L., Chen, K., Lu, D., Johnson, C., Youdim, M.B., Austin, M.C., Luo, J., *et al.* (2010). A novel role for glyceraldehyde-3-phosphate dehydrogenase and monoamine oxidase B cascade in ethanol-induced cellular damage. *Biol. Psychiatry* **67**, 855–863.

Parain, K., Murer, M.G., Yan, Q., Faucheux, B., Agid, Y., Hirsch, E., and Raisman-Vozari, R. (1999). Reduced expression of brain-derived neurotrophic factor protein in Parkinson's disease substantia nigra. *Neuroreport* **10**, 557–561.

Parkinson Study Group, (1989). Effect of deprenyl on the progression of disability in early Parkinson's disease. *N. Engl. J. Med.* **321**, 1364–1371.

Parkinson Study Group (1993). Effects of tocopherol and deprenyl on the progression of disability in early Parkinson's disease. *N. Engl. J. Med.* **328**, 176–183.

Parkinson Study Group (1996). Impact of deprenyl and tocopherol treatment on Parkinson's disease in DATATOP subjects not requiring levodopa. *Ann. Neurol.* **39**, 29–36.

Parkinson Study Group (1998). Mortality in DATATOP: a multicenter trial in early Parkinson's disease. *Ann. Neurol.* **43**, 318–325.

Parkinson Study Group (2002). A controlled trial of rasagiline in early Parkinson disease: the TEMPO Study. *Arch. Neurol.* **59**, 1937–1943.

Parkinson Study Group (2004). A controlled, randomized, delayed-start study of rasagiline in early Parkinson disease. *Arch. Neurol.* **61**, 561–566.

Parkinson Study Group (2005). A randomized placebo-controlled trial of rasagiline in levodopa-treated patients with Parkinson disease and motor fluctuations: the PRESTO study. *Arch. Neurol.* **62**, 241–248.

Przedborski, S., Tieu, K., Perier, C., and Vila, M. (2004). MPTP as a mitochondrial neurotoxic model of Parkinson's disease. *J. Bioenerg. Biomembr.* **36**, 375–379.

Reynolds, G.P., Riederer, P., Sandler, M., Jellinger, K., and Seemann, D. (1978a). Amphetamine and 2-phenylethylamine in post-mortem Parkinsonian brain after (−) deprenyl administration. *J. Neural Transm.* **43**, 271–277.

Reynolds, G.P., Elsworth, J.D., Blau, K., Sandler, M., Lees, A.J., and Stern, G.M. (1978b). Deprenyl is metabolized to methamphetamine and amphetamine in man. *Br. J. Clin. Pharmacol.* **6**, 542–544.

Riederer, P., and Laux, G. (2011). MAO-inhibitors in Parkinson's disease. *Exp. Neurobiol.* **20**, 1–17.

Riederer, P., and Youdim, M.B.H. (1986). Monoamine oxidase activity and monoamine metabolism in brains of parkinsonian patients treated with l-deprenyl. *J. Neurochem.* **46**, 1359–1365.

Riederer, P., Reynolds, G.P., Youdim, M.B.H., and Jellinger, K. (1982). Chemical structure and pharmacological action. In: K.U.E. Kamijo and T. Nagatsu (Eds.), *Monoamine Oxidase—Basic and Clinical Frontiers*, Vol. 564. pp. 345–350.

Sabbagh, A., and Youdim, M.B.H. (1978). Selective inhibition of monoamine oxidase type B by propargyl-containing drugs. *Isr. J. Med. Sci.* **14**, 1097.

Sagi, Y., Mandel, S., Amit, T., and Youdim, M.B. (2007). Activation of tyrosine kinase receptor signaling pathway by rasagiline facilitates neurorescue and restoration of nigrostriatal dopamine neurons in post-MPTP-induced parkinsonism. *Neurobiol. Dis.* **25**, 35–44.

Salach, J.I. Jr. (1978). Preparation of monoamine oxidase from beef liver mitochondria. *Methods Enzymol.* **53**, 495–501.

Salach, J.I., Detmer, K., and Youdim, M.B.H. (1979). The reaction of bovine and rat liver monoamine oxidase with [14 C]-clorgyline and [14 C]-deprenyl. *Mol. Pharmacol.* **16**, 234–241.

Sandler, M., Glover, V., Ashford, A., and Stern, G.M. (1978). Absence of "cheese effect" during deprenyl therapy: some recent studies. *J. Neural Transm.* **43**, 209–215.

Sano, H. (1960). Biochemistry of the extrapyramidal system Shinkei Kennkyu No Shinpo. *Parkinsonism Relat. Disord.* **5**, 42–48.

Sawa, A., Khan, A.A., Hester, L.D., and Snyder, S.H. (1997). Glyceraldehyde-3-phosphate dehydrogenase: nuclear translocation participates in neuronal and nonneuronal cell death. *Proc. Natl. Acad. Sci. USA* **94**, 11669–11674.

Sharma, S.K., Carlson, E.C., and Ebadi, M. (2003). Neuroprotective actions of Selegiline in inhibiting 1-methyl, 4-phenyl, pyridinium ion (MPP+)-induced apoptosis in SK-N-SH neurons. *J. Neurocytol.* **32**, 329–343.

Siegel, G.J., and Chauhan, N.B. (2000). Neurotrophic factors in Alzheimer's and Parkinson's disease brain. *Brain Res. Brain Res. Rev.* **33**, 199–227.

Simpson, L.L. (1978). Evidence that deprenyl, A type B monoamine oxidase inhibitor, is an indirectly acting sympathomimetic amine. *Biochem. Pharmacol.* **27**, 1591–1595.

Smith, M.A., Rottkamp, C.A., Nunomura, A., Raina, A.K., and Perry, G. (2000). Oxidative stress in Alzheimer's disease. *Biochim. Biophys. Acta* **1502**, 139–144.

Speiser, Z., Levy, R., and Cohen, S. (1998). Effects of N-propargyl-1-(R)aminoindan (rasagiline) in models of motor and cognition disorders. *J. Neural Transm.* **52**, 287–300.

Speiser, Z., Mayk, A., Eliash, S., and Cohen, S. (1999). Studies with rasagiline, a MAO-B inhibitor, in experimental focal ischemia in the rat. *J. Neural Transm.* **106**, 593–606.

Sterling, J., Veinberg, A., Lerner, D., Goldenberg, W., Levy, R., Youdim, M., and Finberg, J. (1998). (R)(+)-N-propargyl-1-aminoindan (rasagiline) and derivatives: highly selective and potent inhibitors of monoamine oxidase B. *J. Neural Transm. Suppl.* **52**, 301–305.

Szende, B., Barna, G., and Magyar, K. (2010). Cytoprotective effect of (−)-deprenyl, (−)desmethyl-deprenyl and (−)deprenyl-N-oxide on glutathione depleted A-2058 melanoma cells. *J. Neural Transm.* **117**, 695–698.

Tatton, W.G., Chalmers-Redman, R.M., Elstner, M., Leesch, W., Jagodzinski, F.B., Stupak, D.P., Sugrue, M.M., and Tatton, N.A. (2000). Glyceraldehyde-3-phosphate dehydrogenase in neurodegeneration and apoptosis signaling. *J. Neural Transm. Suppl.* **60**, 77–100.

Tatton, W.G., Chalmers-Redman, R.M., Ju, W.J., Mammen, M., Carlile, G.W., Pong, A.W., and Tatton, N.A. (2002). Propargylamines induce antiapoptotic new protein synthesis in serum- and nerve growth factor (NGF)-withdrawn, NGF-differentiated PC-12 cells. *J. Pharmacol. Exp. Ther.* **301**, 753–764.

Thebault, J.J., Guillaume, M., and Levy, R. (2004). Tolerability, safety, pharmacodynamics, and pharmacokinetics of rasagiline: a potent, selective, and irreversible monoamine oxidase type B inhibitor. *Pharmacotherapy* **24**, 1295–1305.

Vander Heiden, M.G., and Thompson, C.B. (1999). Bcl-2 proteins: regulators of apoptosis or of mitochondrial homeostasis? *Nat. Cell Biol.* **1**, E209–E216.

Vetrivel, K.S., and Thinakaran, G. (2006). Amyloidogenic processing of beta-amyloid precursor protein in intracellular compartments. *Neurology* **66**, S69–S73.

Waldmeier, P.C., Spooren, W.P., and Hengerer, B. (2000). CGP 3466 protects dopaminergic neurons in lesion models of Parkinson's disease. *Naunyn Schmiedebergs Arch. Pharmacol.* **362**, 526–537.

Weinreb, O., Bar-Am, O., Amit, T., Chillag-Talmor, O., and Youdim, M.B.H. (2004). Neuroprotection via pro-survival protein kinase C isoforms associated with Bcl-2 family members. *FASEB J.* **18**, 1471–1473.

Weinreb, O., Amit, T., Sagi, Y., Drigues, N., and Youdim, M.B. (2009). Genomic and proteomic study to survey the mechanism of action of the anti-Parkinson's disease drug, rasagiline compared with selegiline, in the rat midbrain. *J. Neural Transm.* **116**, 1457–1472.

Weinreb, O., Amit, T., Bar-Am, O., and Youdim, M.B. (2010). Rasagiline: a novel anti-Parkinsonian monoamine oxidase-B inhibitor with neuroprotective activity. *Prog. Neurobiol.* **92**, 330–344.

Weinreb, O., Bar-Am, O., Prosolovich, K., Amit, T., and Youdim, M.B. (2011). Does 1-(R)-aminoindan possess neuroprotective properties against experimental Parkinson's disease? *Antioxid. Redox Signal.* **14**, 767–775.

Yang, H.Q., Ba, M.W., Ren, R.J., Zhang, Y.H., Ma, J.F., Pan, J., Lu, G.Q., and Chen, S.D. (2007). Mitogen activated protein kinase and protein kinase C activation mediate promotion of sAPPalpha secretion by deprenyl. *Neurochem. Int.* **50**, 74–82.

Yogev-Falach, M., Amit, T., Bar-Am, O., Sagi, Y., Weinstock, M., and Youdim, M.B.H. (2002). The involvement of mitogen-activated protein (MAP) kinase in the regulation of amyloid precursor protein processing by novel cholinesterase inhibitors derived from rasagiline. *FASEB J.* **16**, 1674–1676.

Yogev-Falach, M., Amit, T., Bar-AM, O., and Youdim, M.B.H. (2003). The importance of propargylamine moiety in the anti-Parkinson drug rasagiline and its derivatives for MAPK-dependent amyloid precursor protein processing. *FASEB J.* **17**, 2325–2327.

Youdim, M.B.H. (1978). The active centers of monoamine oxidase types "A" and "B": binding with (14 C)-clorgyline and (14 C)-deprenyl. *J. Neural Transm.* **43**, 199–208.

Youdim, M.B.H. (2003). Rasagiline: an anti-parkinson drug with neuroprotective activity. *Expert Rev. Neurother.* **3**, 737–749.

Youdim, M.B. (2006). The path from anti Parkinson drug selegiline and rasagiline to multifunctional neuroprotective anti Alzheimer drugs ladostigil and m30. *Curr. Alzheimer Res.* **3**, 541–550.

Youdim, M.B., and Bakhle, Y.S. (2006). Monoamine oxidase: isoforms and inhibitors in Parkinson's disease and depressive illness. *Br. J. Pharmacol.* **147**(Suppl. 1), S287–S296.

Youdim, M.B., and Tipton, K.F. (2002). Rat striatal monoamine oxidase-B inhibition by l-deprenyl and rasagiline: its relationship to 2-phenylethylamine-induced stereotypy and Parkinson's disease. *Parkinsonism Relat. Disord.* **8**, 247–253.

Youdim, M.B., Gross, A., and Finberg, J.P. (2001a). Rasagiline [N-propargyl-1R(+)-aminoindan], a selective and potent inhibitor of mitochondrial monoamine oxidase B. *Br. J. Pharmacol.* **132**, 500–506.

Youdim, M.B., Wadia, A., Tatton, W., and Weinstock, M. (2001b). The anti-Parkinson drug rasagiline and its cholinesterase inhibitor derivatives exert neuroprotection unrelated to MAO inhibition in cell culture and in vivo. *Ann. N. Y. Acad. Sci.* **939**, 450–458.

Youdim, M.B.H., Gross, A., and Finberg, J.P.M. (2001c). Rasagiline [N-Propargyl-1R(+)-aminoindant], A selective and potent inhibitor of mitochondrial monoamine oxidase B. *Br. J. Pharmacol.* **132**, 500–506.

Youdim, M.B., Edmondson, D., and Tipton, K.F. (2006). The therapeutic potential of monoamine oxidase inhibitors. *Nat. Rev.* **7**, 295–309.

Zhang, Y.W., and Xu, H. (2007). Molecular and cellular mechanisms for Alzheimer's disease: understanding APP metabolism. *Curr. Mol. Med.* **7**, 687–696.

Zhu, W., Xie, W., Pan, T., Jankovic, J., Li, J., Youdim, M.B., and Le, W. (2008). Comparison of neuroprotective and neurorestorative capabilities of rasagiline and selegiline against lactacystin-induced nigrostriatal dopaminergic degeneration. *J. Neurochem.* **105**, 1970–1978.

RASAGILINE IN PARKINSON'S DISEASE

L.M. Chahine and M.B. Stern

Penn Comprehensive Neuroscience Center, University of Pennsylvania, Philadelphia, Pennsylvania, USA

Abstract

It has long been recognized that monoamine oxidase (MAO) inhibitors have a role in the management of Parkinson's disease (PD). The MAO-B inhibitor rasagiline has neuroprotective effects in animal models, mediated partly by its antiapoptotic activity. Rasagiline has been shown to be effective as monotherapy for early PD and as an adjunct to dopaminergic therapy. Clinical trials have also shown putative disease-modifying effects, though rasagiline's potential to alter the long-term course of PD remains controversial. Given the demonstrated benefits of rasagiline, along with its safety and tolerability profile, it has an important role to play in PD therapy.

I. Introduction

For over six decades, it has been recognized that monoamine oxidase (MAO) inhibitors have a role to play in Parkinson's disease (PD) pharmacotherapy (Birkmayer and Hornykiewicz, 1962; Langston and Koller, 1991; Sano *et al.*,

INTERNATIONAL REVIEW OF NEUROBIOLOGY, VOL. 100
DOI: 10.1016/B978-0-12-386467-3.00008-X

0074-7742/11 $35.00

1960; Sano, 2000; Bernheimer *et al.*, 1962). MAO is an enzyme located in the outer mitochondrial membrane that catalyzes the oxidation of monoamines, including dopamine, norepineprhine, and serotonin. It exists in two isoforms: MAO-A and MAO-B; dopamine is oxidatively deaminated to 2,4-dihydroxyphenylacetic acid and homovanillic acid (HVA) by either MAO isoform. In the brain, MAO is found in glial cells (Konradi *et al.*, 1987), and MAO-B accounts for approximately 80% of MAO activity in the human basal ganglia (Chen and Swope, 2005; Kalaria *et al.*, 1988). Inhibition of MAO-B increases striatal dopamine, hence the rationale for MAO-B inhibitor treatment in PD.

MAO inhibitors were first recognized to have psychotropic effects in the 1950s. In 1952, Fox introduced iproniazid for the treatment of tuberculosis and serendipitously noted its beneficial psychotropic effects (Lopez-Munoz and Alamo, 2009; Youdim, 1995; Fox 1952). That same year, inhibition of liver and brain mitochondrial MAO activity by iproniazid was demonstrated by Zeller and Barsky (1952). Later, Crane (1956a,b) and Kline (1958) reported further on the psychotropic effects in tuberculosis patients being treated with iproniazid, prompting the pharmaceutical company Hoffman La Roche to synthesize various MAO inhibitors in the early 1960s. However, due to the side effects of nonselective MAO inhibition, selective MAO-B inhibitors were pursued. Selegiline (or L-deprenyl), the first widely available MAO-B inhibitor, was introduced as a symptomatic and possibly neuroprotective agent in PD and was extensively studied in clinical trials, though its amphetamine-like metabolite L-methamphetamine prompted the search for alternative MAO-B inhibitors (see also Chapter 'Selective inhibitors of monoamine oxidase type B and the "cheese effect"' by Finberg and Gillman). Rasagiline, or *N*-propargyl-1-(*R*)aminoindan, is an optical isomer of selegiline and, like selegiline, is a selective and irreversible MAO-B inhibitor, but without amphetamine-like metabolites. This review discusses the putative mechanism of action of neuroprotection by rasagiline, results from clinical trials investigating the symptomatic and disease-modifying effects of rasagiline, and rasagiline's safety profile.

II. *In Vivo* and *In Vitro* Neuroprotective Effects of Rasagiline

Rasagiline has been shown *in vivo* or *in vitro* to protect against neuronal damage due to various insults including serum deprivation (Finberg *et al.*, 1998), oxygen deprivation (Abu-Raya *et al.*, 1999), ischemia (Speiser *et al.*, 1999.), anoxia (Speiser *et al.*, 1998), and head trauma (Finberg *et al.*, 1999). Rasagiline is protective to dopaminergic cells, though it may not be protective to GABAergic cells (Finberg *et al.*, 1998; Goggi *et al.*, 2000).

The *in vivo* effects of rasagiline have been studied in both rats and primates. In primates, rasagiline attenuates conversion of the toxin 1-methyl-4-phenyl-1,2,3,6-tetrahydropyridine (MPTP) into the neurotoxin 1-methyl-4-phenylpyridinium (MPP^+) (Kupsch *et al.*, 2001). As compared to primates, rats have a greater amount of brain MAO-A. In rats, rasagiline increases striatal dopamine (as measured through striatal microdialysis) when administered chronically (daily over a 3-week period), but not in a single dose. The metabolites of dopamine 3,4-dihydrophenylacetic acid (DOPAC) and HVA were not, however, correspondingly reduced (Finberg *et al.*, 1996). It has been hypothesized that increased striatal dopamine upon chronic exposure to rasagiline may result from increased endogenous β-penylethylamine (PEA) levels, an amine normally metabolized by MAO-B, that normally increases release and reduces reuptake of dopamine by neurons. Unlike in rats, in primates, immediate (as opposed to chronic) MAO-B inhibition is sufficient to cause reductions in dopamine metabolism, likely resulting from the greater levels of MAO-B in primate as compared to rat brains (Finberg *et al.*, 1999).

The neuroprotective effects of rasagiline *in vitro* are mediated by the propargyl moiety; its effects are not, however, fully explained by its inhibition of MAO-B and may result from the metabolite of rasagiline, aminoindan (Youdim *et al.*, 2001a,b). While it has been suggested based on *in vitro* studies that some metabolites of selegiline are neurotoxic (Bar Am *et al.*, 2004), other data have shown that at least one metabolite of the *R*-enantiomer of selegiline, deprenyl-*N*-oxide, may increase mitotic activity (Magyar *et al.*, 2010).

Addition of rasagiline to rat fetal mesencephalic cells caused increased neurite outgrowth and an increase in tyrosine hydroxylase positive neurons (Finberg *et al.*, 1996). In addition, rasagiline reduced cell death among cultured rat hippocampal neurons exposed to glutamate (Finberg *et al.*, 1999). A neuronal trophic effect has been suggested as well (Finberg *et al.*, 1998). In MPTP-treated mice, a 10-day treatment with rasagiline was associated with reduced dopaminergic cell loss, possibly mediated by activation of the tyrosine kinase receptor pathway, ligands of which include nerve growth factor, glial cell line-derived neurotrophic factor (GDNF), and brain-derived neurotrophic factor (BDNF; Sagi *et al.*, 2007).

The putative neuroprotective effects of rasagiline may result from inhibition of apoptosis. In neuroblastoma cell lines, rasagiline reduced apoptosis via stabilization of the mitochondrial membrane potential (Maruyama *et al.*, 2002). With mitochondrial stress, translocation of glyceraldehyde-3-phosphate dehydrogenase (GAPDH) into the cell nucleus reduces transcription of antiapoptotic genes and rasagiline binds to GAPDH, preventing its nuclear translocation (Maruyama *et al.*, 2001). Rasagiline thus exerts antiapoptotic effects in animal models through an effect on gene regulation, including upregulation of antiapoptotic genes of the Bcl family and downregulation of proapoptotic genes such as *Bad* and *Bax* (Jenner, 2004; Maruyama *et al.*, 2000a,b; Youdim *et al.*, 2005), as well as through upregulation of protein kinase C, deficiencies of which have been implicated in PD (Youdim *et al.*, 2005). Rasagiline also increases activity of antioxidant enzymes such as superoxide dismutase and catalase (Youdim *et al.*, 2005; Maruyama *et al.*, 2000a,b).

Obviously, interpretation of animal data, and extrapolation to humans, should be done with caution, given the differences in brain structure, enzyme expression, and drug metabolism between different species, but animal data does show culpable putative mechanisms of neuroprotection mediated by rasagiline.

III. Pharmacokinetics and Pharmacodynamics

Rasagiline mesylate is the salt form of rasagiline that is commercially available. It has a similar molecular weight to selegiline, allowing for comparisons of their activities based on dosages or molar concentrations (Chen and Swope, 2005). While rasagiline has equal MAO-B inhibitory potency to selegiline *in vitro*, *in vivo* (in both rats and humans) it has at least five times the potency of selegiline (Youdim *et al.*, 2001a,b). This greater potency is unlikely of clinical significance *in vivo*. Rasagiline is dosed once daily, whereas selegiline is dosed either once or twice per day. Rasagiline is absorbed in the gastrointestinal tract, and food does not affect its absorption; it can be administered with or without food. It crosses the blood–brain barrier readily. Rasagiline is metabolized by the hepatic CYP1A2 system to 1-*R*-aminoindan, which does not have MAO inhibitory activity. Because rasagiline is exclusively hepatically cleared, its use should be avoided in those with significant hepatic impairment (Chen and Swope, 2005; Thebault *et al.*, 2004).

While the plasma half-life of rasagline is 1.34 h, this does not correlate with the duration of action of rasagiline, as it is an irreversible MAO-B inhibitor, such that recovery of MAO-B activity is dependent on *de novo* enzyme synthesis. Studies of rasagiline that have included washout phases indicate that clinically apparent symptomatic benefits from MAO-B inhibition may occur for at least 6 weeks after cessation of the drug (Rabey *et al.*, 2000). However, controversy exists as to the length of symptomatic benefit that persists and the half-time of MAO-B in the brain (e.g., time required for 50% of the baseline amount to be resynthesized) after withdrawal of MAO-B inhibitors. Positron emission tomography (PET) studies using ^{11}C-L-deprenyl as a radiotracer to examine the recovery of MAO-B in healthy volunteers (Freedman *et al.*, 2005) and PD patients (Fowler *et al.*, 1994) suggested long half-times of MAO-B (40 days), but this technique likely overestimates the time required for MAO-B resynthesis (Riederer *et al.*, 2004). The half-time of MAO-B in the brain has been estimated at 10 days (Riederer *et al.*, 2004) based on animal (Green *et al.*, 1977; Youdim and Tipton, 2002) and human (PD and healthy control) studies (Clarke *et al.*, 2003a,b) using selegiline.

Serum platelet MAO-B activity has been suggested as a bioassay that may reflect brain MAO-B inhibition (Thebault *et al.*, 2004), though the correlation between platelet and brain MAO-B is not known, and MAO-B turnover is likely more rapid in brain compared to serum (Clarke *et al.*, 2003a).

A. Drug Interaction Considerations

Serotonin syndrome is a potentially fatal reaction due to serotoninergic overstimulation that presents with, among other features, mental status changes, myoclonus, and autonomic instability. Because MAO catalyzes the oxidation of serotonin, concern has been raised about the risk of serotonin syndrome in patients taking rasagiline along with other serotoninergic agents, and the rasagiline package insert lists antidepressants in this context. No instances of serotonin syndrome have been reported in trials of rasagiline. In the PRESTO trial (Parkinson Study Group, 2005; Schwid and Parkinson Study Group, 2005), adverse effects were not more common in patients being treated with the combination of a selective serotonin reuptake inhibitor and rasagiline as compared with placebo. Of 1361 PD patients treated with rasagiline in clinical trials, 323 were taking an antidepressant as well; no events associated with serotonin toxicity were noted, and PD patients taking antidepressants were not more likely to discontinue rasagiline (Panisset *et al.*, 2007). The concomitant use of rasagiline with selective serotonin reuptake inhibitors is therefore likely safe (see also Chapter 'Selective inhibitors of monoamine oxidase type B and the "cheese effect"' by Finberg and Gillman). On the other hand, based on a severe reaction characterized by delirium, muscle rigidity, and hyperthermia reported with concomitant administration of meperidine and selegiline, and concerns that such an interaction may apply to all MAO inhibitors, it has been suggested that the combination of rasagiline and meperidine be avoided (Zornberg *et al.*, 1991). Inhibitors of CYP1A2, such as ciprofloxacin and fluvoxamine, should be used with caution in patients being treated with rasagiline, as they may increase rasagiline levels significantly (Chen and Swope, 2005).

IV. Evidence for the Symptomatic Benefits of Rasagiline in PD

Several randomized trials have demonstrated the efficacy of rasagiline as a symptomatic therapy for PD. Early pilot studies showed rasagiline to be well tolerated (Rabey *et al.*, 2000; Stern *et al.*, 2004) and suggested efficacy of rasagiline as a symptomatic therapy (Stern *et al.*, 2004). Subsequent trials demonstrated efficacy of rasagiline as monotherapy for early PD and as an adjunct to dopaminergic therapy (Table I). Of note, these trials have been sponsored by the manufacturers of the commercially available form of rasagiline.

Table I

EVIDENCE FOR THE SYMPTOMATIC BENEFITS OF RASAGILINE IN PARKINSON'S DISEASE FROM RANDOMIZED CONTROLLED TRIALS[a].

Authors	Study name	Design	Main eligibility criteria	Treatment groups: number enrolled	Primary outcome measures	Selected secondary outcome measures	Results	Significant differences in adverse events between treatment groups
Parkinson Study Group (2002)	TEMPO: (TVP-1012) in Early Monotherapy for Parkinson's disease Outpatients	Multicenter, 26-week, parallel-group, randomized, double-blind, placebo-controlled (with 26-week extension, delayed-start phase)	*De novo*, idiopathic PD	Rasagiline 1 mg/day: 134 Rasagiline 2 mg/day: 132 Placebo: 138	Change in total UPDRS score between baseline and week 26	Need for levodopa therapy UPDRS motor and ADL subscales	Better overall UPDRS performance in 1 mg and 2 mg groups compared to placebo No difference in time to need levodopa therapy among the three groups Greater improvements in UPDRS motor and ADL subscales and quality of life measures in 1- and 2-mg groups compared to placebo	None; 4 mmHg increase in SBP in 2 mg rasagiline group ($p = 0.02$)
Parkinson Study Group (2005)	PRESTO: Parkinson's Disease Study Group Rasagiline	Multicenter, 26-week, parallel-group, randomized,	Levodopa-treated idiopathic PD with $\geq$ 2.5 hours	Rasagiline 0.5 mg/day: 164 Rasagiline 1 mg/day:	Change from baseline in mean total	CGI, change in baseline in UPDRS,	Significant reduction in mean total daily off time in rasagiline groups compared to placebo (1.85, 1.41, and 0.91 h in 1-mg group, 0.5-	Adverse events more common in rasagiline groups (% patients in 1-mg, 0.5-mg, and

	Efficacy & Safety in the Treatment of "OFF"	double-blind, placebo-controlled	of daily off time	149 Placebo: 159 (76% of those enrolled completed study without protocol deviation)	daily off time	UPDRS ADL subscale	mg group, and placebo groups, respectively) Significant improvements CGI and UPDRS ADL in both rasagiline groups compared to placebo	placebo groups respectively): weight loss (9.4%, 2.4%, 2.5%), vomiting (6.7%, 3.7%, 1.3%), anorexia (5.4% , 1.8%, 0.6%), and balance difficulties (3.4%, 5.5%, 0.6%). Dyskinesias occurred 18% of rasagiline-treated patients compared to 10% of placebo-treated patients
Rascol *et al.* (2005)	LARGO: Lasting Effect in Adjunct Therapy with Rasagiline Given Once Daily	Multicenter, 18-week, parallel-group, randomized, double-blind, placebo-controlled	Levodopa-treated idiopathic PD patients with motor fluctuations for at least 1 h daily in off-state	Rasagiline 1 mg/day: 231 Entacapone 200 mg with each dose of levodopa: 227 Placebo: 229	Change from baseline in mean total daily off time	CGI, UPDRS ADL and motor subscores	Significant reduction in mean total daily off time in both rasagiline and entacapone groups (1.18 and 1.2 h, respectively, three times more than reduction in placebo group) Significant improvements in CGI and UPDRS ADL and motor subscales	None; postural hypotension occurred in 2% of patients in rasagiline and entacapone groups

[a] ADL, activities of daily living; CGI, clinical global impression (of investigator); PDQUALIF, Parkinson's Disease Quality-of-Life Scale; SBP, systolic blood pressure; UPDRS, Unified Parkinson's Disease Rating Scale.

The TEMPO trial (Parkinson Study Group, 2002) assessed the efficacy of rasagiline as monotherapy in *de novo* PD patients but had an extension phase to assess longer term effects (see below). The symptomatic benefits of rasagiline as monotherapy, while demonstrated statistically, are modest; the Unified Parkinson's Disease Rating Scale (UPDRS) score was four points less in treatment arms compared to placebo. For reference, untreated PD patients have an average annual rate of decline in total UPDRS scores of 10 points (Kieburtz, 2003; Parkinson Study Group, 1993). Rasagiline has not been compared to dopamine agonists or levodopa monotherapy in clinical trials, but in clinical trials of the latter agents, effects on UPDRS scores have been greater early on (Holloway *et al.*, 2004; Fahn *et al.*, 2004; Rascol, 2003).

As mentioned, rasagiline is also efficacious as adjunctive therapy in PD to reduce off time/increase on time. In the PRESTO study (Parkinson Study Group, 2005), levodopa-treated patients treated with rasagiline had a reduction in off time by approximately 84–108 min compared with placebo (with nearly equal increases in on time), though with a mild increase in troublesome dyskinesias in the 1-mg rasagiline group. While reductions in off time have not been compared directly for rasagiline and dopamine agonists, in trials of pramipexole, reductions in off time were comparable (Clarke *et al.*, 2000). In the LARGO study (Rascol *et al.*, 2005), in which patients were randomized to rasagiline, entacapone, or placebo, rasagiline 1 mg daily reduced off time in PD patients on optimum levodopa treatment, with an accompanying increase in on time without troublesome dyskinesias, and without necessitating significant reductions in levodopa. The reductions in off time, approximately 70 min, were comparable for rasagiline and entacapone (though the study was not powered to compare rasagiline with entacapone directly) and were almost three times greater than reductions in the placebo group. There was some suggestion that rasagiline improved postural instability and freezing, whereas entacapone did not.

Rasagiline has also been shown to have a positive impact on activities of daily living (ADLs) (Parkinson Study Group, 2004, 2005; Rascol *et al.*, 2005; Hauser *et al.*, 2009) and on quality of life. In the TEMPO trial (Parkinson Study Group, 2002; Biglan *et al.*, 2006), quality of life, as assessed by the Parkinson Disease Quality of Life Questionnaire (PDQUALIF), improved from baseline to week 26 in both groups receiving rasagiline (2.91 and 2.74 unit differences as compared to placebo in the 2 and 1-mg rasagiline groups, respectively) but worsened in the placebo group. While baseline PDQUALIF scores correlated strongly with UPDRS scores, there was only a weak correlation between UPDRS and PDQUALIF at 26 and 52 weeks, suggesting that the beneficial effects of rasagiline on quality of life extended beyond those captured by the UPDRS. Differences in self-image/sexuality and social role domains accounted for a large proportion of the differences in PDQUALIF scores from baseline to last measurement among the treatment groups, postulated to be possibly due to an antidepressant effect of rasagiline, but also likely due to the impact on motor symptoms. PDQUALIF scores had slightly worsened by the end of the extension phase of the TEMPO study (at 52 weeks from baseline). The PRESTO study showed a trend toward improvement in quality of life in the 0.5 mg rasagiline group (Parkinson Study Group, 2005).

While in the United States, rasagiline is prohibitively costly for some patients (depending on their insurance coverage), it has been demonstrated that when taking into account quality-adjusted life-years and other measures, concomitant use of rasagiline and levodopa may not only be cost-effective but also cost-saving from a societal perspective (Hudry *et al.*, 2006; Haycox *et al.*, 2009).

V. Evidence for Disease Modification with Rasagiline in PD

The fundamental problem in demonstrating disease-modifying effects of pharmacologic agents in PD stems from the lack of well-defined objective measures of PD progression. Time to need levodopa as well as changes in UPDRS scores have been used as measures of neuroprotection; UPDRS scores do not, however, necessarily allow for a distinction between symptomatic and neuroprotective effects (Parkinson Study Group, 1993). A novel proparyglamine that does not have MAO-B inhibitory activity and therefore no symptomatic benefits (TCH346) did not show evidence for efficacy as a neuroprotectant (with the primary endpoint being time to need levodopa), raising concern that benefits obtained with selegiline and rasagiline may be symptomatic rather than neuroprotective (Olanow *et al.*, 2006). Washout periods have been utilized as a means of eliminating the symptomatic effect, but the theoretically prolonged MAO-B inhibition and other pharmacodynamic effects of MAO-B inhibitors make this method unreliable (Rabey *et al.*, 2000; Parkinson Study Group, 1993, 2004). Delayed-start trials, in which the group initially treated with placebo is later given active treatment, have also been utilized as a means of distinguishing symptomatic effects from neuroprotective ones, because by the end of the delayed-start phase, the symptomatic effects of the drug are expected to be balanced between treatment groups. In other words, if a drug has purely symptomatic effects, patients who receive the drug after a delayed period of time would "catch up" with those who had received the drug from study onset, whereas if a neuroprotective effect is present, those who received the drug for longer periods of time would show sustained improvements over those who did not.

Two delayed-start trials have examined the efficacy of rasagiline as a neuroprotectant in PD, both funded by manufacturers of the drug. In the first, an extension of the TEMPO trial (Parkinson Study Group, 2002), patients were initially randomized to one of three treatment arms: rasagiline 1 mg, rasagiline

2 mg, or placebo for 6 months. At 6 months, those in the rasagiline 1 and 2-mg arms continued their assigned treatments, but those in the placebo arm were assigned to rasagiline 2 mg (unless patients were felt to require initial dopaminergic medications in the first 6 months of the trial, in which case they entered the active treatment phase earlier, at the time when additional dopaminergic therapy was required). Blinding was maintained for the entire study period. The primary measure of efficacy was change in total UPDRS score from baseline until the end of the study, at 52 weeks. The intention-to-treat cohort of the extension phase of the TEMPO trial included 371 patients; of those, 46 did not complete the entire 26-week placebo-controlled phase because of need for additional dopaminergic therapy, and therefore entered the active treatment phase with rasagiline at earlier than 6 months. Subjects treated with 1 or 2 mg of rasagiline had less progression in total UPDRS scores than subjects who delayed rasagiline therapy for 6 months. Those treated with rasagiline 1 or 2 mg for the entire 52-week period showed a 1.82 and 2.29 point adjusted difference in UPDRS score, respectively, compared to those who received placebo in the first 6 months followed by rasagiline 2 mg for the subsequent 6 months.

A further long-term open-label extension of the TEMPO trial followed PD patients treated with rasagiline for up to 6.5 years (Hauser *et al.*, 2009). A dose of 1 mg was used for all patients in the open-label extension trial, given that the 2-mg dose had not been shown to provide greater symptomatic benefit. Three hundred and six PD patients participated in the open-label extension; mean follow-up time was 3.5 years, and 177 individuals were followed for a total of 6.5 years (from the time rasagiline was started). The adjusted mean difference in change from baseline total UPDRS score in the early- versus the delayed-start group was 2.5 units (a 16% mean relative difference) for the entire 6.5-year observation period. There was no difference in mean time to additional dopaminergic treatment in the early- versus the delayed-start groups (1.5 and 1.8 years, respectively), and levodopa equivalents did not differ between the groups. In addition, there was no difference in onset of dyskinesias. While the results of the TEMPO extension phase were interpreted as implying possible neuroprotection, delayed-start trials cannot account for the possibility that early treatment may sensitize striatal neurons or somehow enhance the efficiency of dopaminergic release (Parkinson Study Group, 2004).

The need for longer-duration trials to confirm the findings of the extension of the TEMPO trial prompted the second large delayed-start trial of rasagiline in PD, the "Assess Rasagiline as a Disease Modifying Therapy in Parkinson's Disease," or ADAGIO study (Olanow *et al.*, 2008, 2009). The ADAGIO study was a multicenter, 72-week, parallel-group, randomized, double-blind, placebo-controlled trial. The main inclusion criterion was untreated PD with disease duration of less than 18 months. Early-start PD patients received a total of either 1 or 2 mg of rasagiline daily for 72 weeks, whereas delayed-start participants received 36 weeks of placebo followed by rasagiline 1 or 2 mg daily for 36 weeks. There were three primary endpoints: rate (slope estimate) of UPDRS progression per week from weeks 12–36 in placebo versus rasagiline groups, change from baseline to week 72 total UPDRS, and noninferiority of slope estimates for the rate of change from baseline UPDRS score between weeks 48 and 72 in the early- compared to the delayed-start groups. The secondary endpoint was change in total UPDRS score between baseline and last observation at week 36. A total of 1176 PD patients were enrolled in the study, 581 in the early- and 595 in the delayed-start groups.

The rasagiline 1-mg daily dose met all three primary endpoints. The change in slope of UPDRS scores per week was slower for rasagiline compared to placebo, and there was a 2.82 point increase in total UPDRS score in the early-start group compared to a 4.5 point increase in the delayed-start rasagiline 1-mg group (a 1.7 point difference). The estimates of the change in UPDRS scores between 48 and 72 weeks showed noninferiority of the response between the early- and delayed-start groups. Rasagiline 2-mg daily did not meet all three primary endpoints: the change in total UPDRS score between baseline and 72 weeks did not differ between early- and delayed-start groups. In a *post hoc* analysis, the difference in change in UPDRS scores from baseline to week 72 between the 2 mg early- and delayed-start groups was significantly greater among the subgroup that had baseline UPDRS scores in the highest quartile as compared to subjects in the other three quartiles. Both the rasagiline 1 and 2-mg doses met the secondary endpoint. While there is no clear explanation for why the rasagiline 1-mg dose met all endpoints whereas the 2-mg dose did not, it has been suggested that the symptomatic effect of the 2-mg dose may have masked a benefit associated with early-start treatment.

Whether or not the delayed-start trial design is optimal for assessing disease modification in PD is in itself controversial (Ahlskog and Uitti, 2010). While the TEMPO and ADAGIO studies provide compelling evidence that rasagiline is disease modifying, the results are by no means conclusive.

VI. Adverse Effects

While rasagiline and selegiline have not been compared in clinical trials, their symptomatic benefits are likely similar (Rascol, 2003). An advantage of rasagiline over selegiline largely relates to its improved side effect profile. As mentioned, selegiline is metabolized to L-methamphetamine, whereas rasagiline is metabolized to aminoindan, which lacks amphetamine-like properties. The side effects caused by selegiline, exerted through L-methemphetamine, include a possible sympathomimetic effect, though the clinical significance of this effect is not well established (see also Chapter 'Selective inhibitors of monoamine oxidase type B and the "cheese effect"' by Finberg and Gillman). Selegiline may also possibly have neurotoxic effects and psychiatric side effects (Chen and Swope, 2005; Bar Am *et al.*,, 2004; Churchyard *et al.*, 1997). A similar significant sympathomimetic effect has not been observed with rasagiline despite extensive blood pressure monitoring in clinical trials (Stern *et al.*, 2004), though a 4 mmHg increase in systolic blood pressure was observed in patients in the 2-mg rasagiline group in the TEMPO study (Parkinson Study Group, 2002), but not in the 1-mg group. As mentioned above, in the PRESTO study, 18% of rasagiline-treated patients experienced dyskinesias compared to 10% of placebo-treated patients; of note, the study protocol limited adjustments in levodopa beyond the first 6 weeks of the trial (Parkinson Study Group, 2005).

In randomized controlled trials of rasagiline, there have been no significant differences between rasagiline and placebo groups in terms of serious adverse events. In clinical trials of rasagiline, adverse events commonly reported in the treatment arm have included infection (Parkinson Study Group, 2002, 2004) and headache (Stern *et al.*, 2004; Parkinson Study Group, 2002, 2004; Olanow *et al.*, 2009) (Table I). In the PRESTO trial, weight loss, vomiting, anorexia, and balance difficulties were more common in the rasagiline groups compared to placebo (Parkinson Study Group, 2005). Other adverse events reported, occurring in approximately 10% or less of patients in treatment or placebo groups, have included dizziness, nausea, postural hypotension, back pain, fatigue, syncope, sleep disturbance, hallucinations, falls, and accidental injury (Parkinson Study Group, 2005, 2002; Stern *et al.*, 2004; Rascol *et al.*, 2005; Lew *et al.*, 2010). A postmarketing observational study found dizziness, headache, nausea, and vomiting to be the most common side effects seen with rasagiline (Reichmann and Jost, 2010).

In the various trials of rasagiline, concomitant administration of dopamine agonists and amantadine was allowed, and based on the results, rasagiline is safely administered with the latter drugs (Parkinson Study Group, 2005; Rascol *et al.*, 2005), even in PD patients older than 70 (Rascol *et al.*, 2005). Age-related increases in adverse events was not reported in clinical trials of rasagiline; *post hoc* analyses of the TEMPO and PRESTO trials have revealed a higher risk of serious adverse in older as compared to younger patients in both the rasagiline and placebo groups, including confusion, symptomatic orthostatic hypotension, and hallucinations (Goetz *et al.*, 2006). There was not, however, a statistical interaction between age and rasagiline exposure, suggesting that while older adults are more vulnerable to adverse effects, in general, rasagiline is generally well tolerated alone or in combination with levodopa, regardless of age, and that there are no specific tolerability or monitoring precautions required in older PD patients treated with rasagiline.

Because of the detection of two cases of melanoma during the TEMPO study (Parkinson Study Group, 2004; Lew *et al.*, 2010), in the PRESTO trial

(Parkinson Study Group, 2005), melanomas were actively monitored for. Melanoma was detected in one patient prior to starting study drug and in three patients in the rasagiline arms, but since only one-third as many patients in the study received placebo as compared to rasagiline, the implications of these findings are unclear (Parkinson Study Group, 2005). Until further information is available, surveillance for melanoma in PD patients, in general, is warranted.

A. "Cheese Reaction"

As mentioned, MAO exits in two isoforms, MAO-A and MAO-B. MAO-A is the isoform most prevalent in the intestines and sympathetic nerves, whereas more than 80% of human striatal MAO is the MAO-B isoform. Tyramine, an amine present in various fermented foods such as aged cheeses and wine, is metabolized predominantly by MAO-A, and inhibition of intestinal MAO-A leads to increased absorption of tyramine. Tyramine leads to sympathetic stimulation by causing release of noradrenaline from sympathetic neurons, and increased circulating tyramine can therefore lead to a severe hypertensive reaction (Youdim, 1995; Youdim and Weinstock, 2004). At doses administered in routine clinical practice, both selegiline and rasagiline lead to selective MAO-B inhibition: rasagiline is at least 60 times more potent at inhibiting MAO-B as compared to MAO-A (Youdim *et al.*, 2001a,b; though at supratherapeutic doses, MAO-A inhibition does occur). A tyramine-rich diet has been estimated to contain around 50 mg of tyramine (over a 24-hour period). A subset of patients in the TEMPO (Parkinson Study Group, 2005, 2002) and PRESTO (Parkinson Study Group, 2005) trials were given an oral tyramine challenge, 75 and 50 mg, respectively, at the end of the study and underwent cardiovascular monitoring. No significant hemodynamic changes were seen in the TEMPO trial participants. In the PRESTO study, there were no significant group differences in blood pressure mean changes in the treatment groups. Three participants did have nonsustained blood pressure elevations suggestive of a possible pressor response, and four patients (three in the 0.5-mg rasagiline group and one in the placebo group) had systolic blood pressure elevations of 30 mm Hg or more (Parkinson study Group, 2001; deMarcaida *et al.*, 2006). The blood pressure elevations occurred, however, after more than 110 min of tyramine ingestion, and these blood pressure changes were felt unlikely to be due to concomitant administration of rasagiline and tyramine but rather perhaps due to PD-related autonomic instability, motor fluctuations, or other factors. Based on the latter, dietary restrictions are considered unnecessary in patients being treated with the rasagiline 0.5–2 mg daily.

VII. Conclusion

Since its introduction over a decade ago, rasagiline has been extensively studied. Rasagiline has a good tolerability and safety profile and has symptomatic benefits both as monotherapy in early PD and as an adjunct to dopaminergic therapy. Improvements not only in motor control but also quality of life and ADL further support the use of rasagiline in PD. Whether or not rasagiline's neuroprotective effects in the laboratory translate into a disease-modifying effect in patients, as suggested by the TEMPO and ADAGIO trails, requires further study with novel, more objective measures of PD progression.

References

Abu-Raya, S., Blaugrund, E., Trembovler, V., Shilderman-Bloch, E., Shohami, E., and Lazarovici, P. (1999). Rasagiline, a monoamine oxidase-B inhibitor, protects NGF-differentiated PC12 cells against oxygen-glucose deprivation. *J. Neurosci. Res.* **58**(3), 456–463.

Ahlskog, J.E., and Uitti, R.J. (2010). Rasagiline, Parkinson neuroprotection, and delayed-start trials: still no satisfaction? *Neurology* **74**(14), 1143–1148.

Bar Am, O., Amit, T., and Youdim, M.B. (2004). Contrasting neuroprotective and neurotoxic actions of respective metabolites of anti-Parkinson drugs rasagiline and selegiline. *Neurosci. Lett.* **355**(3), 169–172.

Bernheimer, H., Birkmayer, W., and Hornykiewicz, O. (1962). Behavior of monoamine oxidase in the brain of man after therapy with monoamine oxidase inhibitors. *Wien. Klin. Wochenschr.* **74**, 558–559.

Biglan, K.M., Schwid, S., Eberly, S., Blindauer, K., Fahn, S., Goren, T., Kieburtz, K., Oakes, D., Plumb, S., Siderowf, A., Stern, M., Shoulson, I., *et al.* (2006). Rasagiline improves quality of life in patients with early Parkinson's disease. *Mov. Disord.* **21**(5), 616–623.

Birkmayer, W., and Hornykiewicz, O. (1962). The L-dihydroxyphenylalanine (L-DOPA) effect in Parkinson's syndrome in man: on the pathogenesis and treatment of Parkinson akinesis. *Arch. Psychiatr. Nervenkr. Z. gesamte Neurol. Psychiatr.* **203**, 560–574.

Chen, J.J., and Swope, D.M. (2005). Clinical pharmacology of rasagiline: a novel, second-generation propargylamine for the treatment of Parkinson disease. *J. Clin. Pharmacol.* **45**(8), 878–894.

Churchyard, A., Mathias, C.J., Boonkongchuen, P., and Lees, A.J. (1997). Autonomic effects of selegiline: possible cardiovascular toxicity in Parkinson's disease. *J. Neurol. Neurosurg. Psychiatry* **63**(2), 228–234.

Clarke, C.E., Speller, J.M., and Clarke, J.A. (2000). Pramipexole for levodopa-induced complications in Parkinson's disease. *Cochrane Database Syst. Rev.* **3**(3), CD002261.

Clarke, A., Brewer, F., Johnson, E.S., Mallard, N., Hartig, F., Taylor, S., and Corn, T.H. (2003a). A new formulation of selegiline: improved bioavailability and selectivity for MAO-B inhibition. *J. Neural Transm.* **110**(11), 1241–1255.

Clarke, A., Johnson, E.S., Mallard, N., Corn, T.H., Johnston, A., Boyce, M., Warrington, S., and MacMahon, D.G. (2003b). A new low-dose formulation of selegiline: clinical efficacy, patient preference and selectivity for MAO-B inhibition. *J. Neural Transm.* **110**(11), 1257–1271.

Crane, G.E. (1956a). Further studies on iproniazid phosphate; isonicotinil-isopropyl-hydrazine phosphate marsilid. *J. Nerv. Ment. Dis.* **124**(3), 322–331.

Crane, G.E. (1956b). The psychiatric side-effects of iproniazid. *Am. J. Psychiatry* **112**(7), 494–501.

deMarcaida, J.A., Schwid, S.R., White, W.B., Blindauer, K., Fahn, S., Kieburtz, K., Stern, M., and Shoulson, I.Parkinson Study Group TEMPO & PRESTO Tyramine Substudy Investigators and Coordinators (2006). Effects of tyramine administration in Parkinson's disease patients treated with selective MAO-B inhibitor rasagiline. *Mov. Disord.* **21**(10), 1716–1721.

Fahn, S., Oakes, D., Shoulson, I., Kieburtz, K., Rudolph, A., Lang, A., Olanow, C.W., Tanner, C., and Marek, K.Parkinson Study Group (2004). Levodopa and the progression of Parkinson's disease. *N. Engl. J. Med.* **351**(24), 2498–2508.

Finberg, J.P., Lamensdorf, I., Commissiong, J.W., and Youdim, M.B. (1996). Pharmacology and neuroprotective properties of rasagiline. *J. Neural Transm. Suppl.* **48**, 95–101.

Finberg, J.P., Takeshima, T., Johnston, J.M., and Commissiong, J.W. (1998). Increased survival of dopaminergic neurons by rasagiline, a monoamine oxidase B inhibitor. *Neuroreport* **9**(4), 703–707.

Finberg, J.P., Lamensdorf, I., Weinstock, M., Schwartz, M., and Youdim, M.B. (1999). Pharmacology of rasagiline (N-propargyl-1R-aminoindan). *Adv. Neurol.* **80**, 495–499.

Fowler, J.S., Volkow, N.D., Logan, J., Wang, G.J., MacGregor, R.R., Schyler, D., Wolf, A.P., Pappas, N., Alexoff, D., and Shea, C. (1994). Slow recovery of human brain MAO B after Ldeprenyl Selegeline withdrawal. *Synapse* **18**(2), 86–93.

Fox, H.H. (1952). The chemical approach to the control of tuberculosis. *Science* **116**(3006), 129–134.

Freedman, N.M., Mishani, E., Krausz, Y., Weininger, J., Lester, H., Blaugrund, E., Ehrlich, D., and Chisin, R. (2005). In vivo measurement of brain monoamine oxidase B occupancy by rasagiline, using (11)C-l-deprenyl and PET. *J. Nucl Med.* **46**(10), 1618–1624.

Goetz, C.G., Schwid, S.R., Eberly, S.W., Oakes, D., and Shoulson, I.Parkinson Study Group TEMPO and PRESTO Investigators (2006). Safety of rasagiline in elderly patients with Parkinson disease. *Neurology* **66**(9), 1427–1429.

Goggi, J., Theofilopoulos, S., Riaz, S.S., Jauniaux, E., Stern, G.M., and Bradford, H.F. (2000). The neuronal survival effects of rasagiline and deprenyl on fetal human and rat ventral mesencephalic neurones in culture. *Neuroreport* **11**(18), 3937–3941.

Green, A.R., Mitchell, B.D., Tordoff, A.F., and Youdim, M.B. (1977). Evidence for dopamine deamination by both type A and type B monoamine oxidase in rat brain in vivo and for the degree of inhibition of enzyme necessary for increased functional activity of dopamine and 5-hydroxytryptamine. *Br. J. Pharmacol.* **60**(3), 343–349.

Hauser, R.A., Lew, M.F., Hurtig, H.I., Ondo, W.G., Wojcieszek, J., and Fitzer-Attas, C.J.TEMPO Open-label Study Group (2009). Long-term outcome of early versus delayed rasagiline treatment in early Parkinson's disease. *Mov. Disord.* **24**(4), 564–573.

Haycox, A., Armand, C., Murteira, S., Cochran, J., and Francois, C. (2009). Cost effectiveness of rasagiline and pramipexole as treatment strategies in early Parkinson's disease in the UK setting: an economic Markov model evaluation. *Drugs Aging* **26**(9), 791–801.

Holloway, R.G., Shoulson, I., Fahn, S., Kieburtz, K., Lang, A., Marek, K., McDermott, M., Seibyl, J., Weiner, W., Musch, B., Kamp, C., Welsh, M., *et al.* (2004). Pramipexole vs levodopa as initial treatment for Parkinson disease: a 4-year randomized controlled trial. *Arch. Neurol.* **61**(7), 1044–1053.

Hudry, J., Rinne, J.O., Keranen, T., Eckert, L., and Cochran, J.M. (2006). Cost-utility model of rasagiline in the treatment of advanced Parkinson's disease in Finland. *Ann. Pharmacother.* **40**(4), 651–657.

Jenner, P. (2004). Preclinical evidence for neuroprotection with monoamine oxidase-B inhibitors in Parkinson's disease. *Neurology* **63**(no. 7 Suppl. 2), S13–S22.

Kalaria, R.N., Mitchell, M.J., and Harik, S.I. (1988). Monoamine oxidases of the human brain and liver. *Brain* **111**(6), 1441–1451.

Kieburtz, K. (2003). Designing neuroprotection trials in Parkinson's disease. *Ann. Neurol.* **53**(Suppl. 3), S100–S107discussion S107–S109.

Kline, N.S. (1958). Clinical experience with iproniazid (marsilid). *J. Clin. Exp. Psychopathol.* **19**(no. 2 Suppl. 1), 72–78, discussion 78–79.

Konradi, C., Riederer, P., Jellinger, K., and Denney, R. (1987). Cellular action of MAO inhibitors. *J. Neural Transm. Suppl.* **25**, 15–25.

Kupsch, A., Sautter, J., Gotz, M.E., Breithaupt, W., Schwarz, J., Youdim, M.B., Riederer, P., Gerlach, M., and Oertel, W.H. (2001). Monoamine oxidase-inhibition and MPTP-induced neurotoxicity in the non-human primate: comparison of rasagiline (TVP 1012) with selegiline. *J.Neural. transm.* **108**(8–9), 985–1009.

Langston, J.W., and Koller, W.C. (1991). The next frontier in Parkinson's disease: presymptomatic detection. *Neurology* **41**(no. 5 Suppl. 2), 5–7.

Lew, M.F., Hauser, R.A., Hurtig, H.I., Ondo, W.G., Wojcieszek, J., Goren, T., and Fitzer-Attas, C.J. (2010). Long-term efficacy of rasagiline in early Parkinson's disease. *Int. J. Neurosci.* **120**(6), 404–408.

Lopez-Munoz, F., and Alamo, C. (2009). Monoaminergic neurotransmission: the history of the discovery of antidepressants from 1950s until today. *Curr. Pharm. Des.* **15**(14), 1563–1586.

Magyar, K., Szende, B., Jenei, V., Tabi, T., Palfi, M., and Szoko, E. (2010). R-deprenyl: pharmacological spectrum of its activity. *Neurochem. Res.* **35**(12), 1922–1932.

Maruyama, W., Akao, Y., Youdim, M.B., and Naoi, M. (2000a). Neurotoxins induce apoptosis in dopamine neurons: protection by N-propargylamine-1(R)- and (S)-aminoindan, rasagiline and TV1022". *J. Neural Transm. Suppl.* **60**(60), 171–186.

Maruyama, W., Yamamoto, T., Kitani, K., Carrillo, M.C., Youdim, M., and Naoi, M. (2000b). Mechanism underlying anti-apoptotic activity of a (−)deprenyl-related propargylamine, rasagiline. *Mech. Ageing Dev.* **116**(2–3), 181–191.

Maruyama, W., Akao, Y., Youdim, M.B., Davis, B.A., and Naoi, M. (2001). Transfection-enforced Bcl-2 overexpression and an anti-Parkinson drug, rasagiline, prevent nuclear accumulation of glyceraldehyde-3-phosphate dehydrogenase induced by an endogenous dopaminergic neurotoxin, N-methyl(R)salsolinol. *J. Neurochem.* **78**(4), 727–735.

Maruyama, W., Takahashi, T., Youdim, M., and Naoi, M. (2002). The anti-Parkinson drug, rasagiline, prevents apoptotic DNA damage induced by peroxynitrite in human dopaminergic neuroblastoma SH-SY5Y cells. *J. Neural Transm.* **109**(4), 467–481.

Olanow, C.W., Schapira, A.H., LeWitt, P.A., Kieburtz, K., Sauer, D., Olivieri, G., Pohlmann, H., and Hubble, J. (2006). TCH346 as a neuroprotective drug in Parkinson's disease: a double-blind, randomised, controlled trial. *Lancet Neurol.* **5**(12), 1013–1020.

Olanow, C.W., Hauser, R.A., Jankovic, J., Langston, W., Lang, A., Poewe, W., Tolosa, E., Stocchi, F., Melamed, E., Eyal, E., and Rascol, O. (2008). A randomized, double-blind, placebo-controlled, delayed start study to assess rasagiline as a disease modifying therapy in Parkinson's disease (the ADAGIO study): rationale, design, and baseline characteristics. *Mov. Disord.* **23**(15), 2194–2201.

Olanow, C.W., Rascol, O., Hauser, R., Feigin, P.D., Jankovic, J., Lang, A., Langston, W., Melamed, E., Poewe, W., Stocchi, F., and Tolosa, E.ADAGIO Study Investigators (2009). A double-blind, delayed-start trial of rasagiline in Parkinson's disease. *N. Engl. J. Med.* **361**(13), 1268–1278.

Panisset, M., Schwid, S., Ondo, W., Fitzer-Attas, C., and Chen, J.J. (2007). Safety of concomitant therapy with rasagiline and antidepressants in Parkinson's disease. *Mov. Disord.* **22**(Suppl. 16), S104–S105.

Parkinson Study Group, (1993). Effects of tocopherol and deprenyl on the progression of disability in early Parkinson's disease. The Parkinson Study Group. *N. Engl. J. Med.* **328**(3), 176–183.

Parkinson study Group (2001). Tyramine challenge to assess the safety of rasagiline monotherapy in a placebo-controlled multicenter trial for early Parkinson's disease (the TEMPO study) *Neurology* **56**, A345.

Parkinson Study Group, (2002). A controlled trial of rasagiline in early Parkinson disease: the TEMPO Study. *Arch. Neurol.* **59**(12), 1937–1943.

Parkinson Study Group, (2004). A controlled, randomized, delayed-start study of rasagiline in early Parkinson disease. *Arch. Neurol.* **61**(4), 561–566.

Parkinson Study Group, (2005). A randomized placebo-controlled trial of rasagiline in levodopa-treated patients with Parkinson disease and motor fluctuations: the PRESTO study. *Arch. Neurol.* **62** (2), 241–248.

Rabey, J.M., Sagi, I., Huberman, M., Melamed, E., Korczyn, A., Giladi, N., Inzelberg, R., Djaldetti, R., Klein, C., and Berecz, G.Rasagiline Study Group (2000). Rasagiline mesylate, a new MAO-B inhibitor for the treatment of Parkinson's disease: a double-blind study as adjunctive therapy to levodopa. *Clin. Neuropharmacol.* **23**(6), 324–330.

Rascol, O. (2003). Monoamine oxidase inhibitors–is it time to up the TEMPO? *Lancet Neurol.* **2**(3), 142–143.

Rascol, O., Brooks, D.J., Melamed, E., Oertel, W., Poewe, W., Stocchi, F., and Tolosa, E.LARGO study group (2005). Rasagiline as an adjunct to levodopa in patients with Parkinson's disease and motor fluctuations (LARGO, Lasting effect in Adjunct therapy with Rasagiline Given Once daily, study): a randomised, double-blind, parallel-group trial. *Lancet* **365**(9463), 947–954.

Reichmann, H., and Jost, W.H. (2010). Efficacy and tolerability of rasagiline in daily clinical use—a post-marketing observational study in patients with Parkinson's disease. *Eur. J. Neurol.* **17**(9), 1164–1171.

Riederer, P., Lachenmayer, L., and Laux, G. (2004). Clinical applications of MAO-inhibitors. *Curr. Med. Chem.* **11**(15), 2033–2043.

Sagi, Y., Mandel, S., Amit, T., and Youdim, M.B. (2007). Activation of tyrosine kinase receptor signaling pathway by rasagiline facilitates neurorescue and restoration of nigrostriatal dopamine neurons in post-MPTP-induced parkinsonism. *Neurobiol. Dis.* **25**(1), 35–44.

Sano, H. (2000). Biochemistry of the extrapyramidal system Shinkei Kennkyu No Shinpo, Advances in Neurological Sciences. (ISSN 0001–8724) Tokyo, October (1960);5:42–48. *Parkinsonism Relat. Disord.* **6**(1), 3–6.

Sano, I., Taniguchi, K., Gamo, T., Takesada, M., and Kakimoto, Y. (1960). The catecholamines in the central nervous system. *Klin. Wochenschr.* **38**, 57–62.

Schwid, S.R.Parkinson Study Group (2005). Safety of rasagiline in combination with serotonion reuptake inhibitors. *Ann. Neurol.* **58**(Suppl. 9), S56.

Speiser, Z., Katzir, O., Rehavi, M., Zabarski, T., and Cohen, S. (1998). Sparing by rasagiline (TVP-1012) of cholinergic functions and behavior in the postnatal anoxia rat. *Pharmacol. Biochem. Behav.* **60** (2), 387–393.

Speiser, Z., Mayk, A., Eliash, S., and Cohen, S. (1999). Studies with rasagiline, a MAO-B inhibitor, in experimental focal ischemia in the rat. *J. Neural Transm.* **106**(7–8), 593–606.

Stern, M.B., Marek, K.L., Friedman, J., Hauser, R.A., LeWitt, P.A., Tarsy, D., and Olanow, C.W. (2004). Double-blind, randomized, controlled trial of rasagiline as monotherapy in early Parkinson's disease patients. *Mov. Disord* **19**(8), 916–923.

Thebault, J.J., Guillaume, M., and Levy, R. (2004). Tolerability, safety, pharmacodynamics, and pharmacokinetics of rasagiline: a potent, selective, and irreversible monoamine oxidase type B inhibitor. *Pharmacotherapy* **24**(10), 1295–1305.

Youdim, M.B. (1995). The advent of selective monoamine oxidase A inhibitor antidepressants devoid of the cheese reaction. *Acta Psychiatr. Scand. Suppl.* **386**, 5–7.

Youdim, M.B., and Tipton, K.F. (2002). Rat striatal monoamine oxidase-B inhibition by l-deprenyl and rasagiline: its relationship to 2-phenylethylamine-induced stereotypy and Parkinson's disease. *Parkinsonism Relat. Disord.* **8**(4), 247–253.

Youdim, M.B., and Weinstock, M. (2004). Therapeutic applications of selective and non-selective inhibitors of monoamine oxidase A and B that do not cause significant tyramine potentiation. *Neurotoxicology* **25**(1–2), 243–250.

Youdim, M.B., Gross, A., and Finberg, J.P. (2001a). Rasagiline [N-propargyl-1R(+)-aminoindan], a selective and potent inhibitor of mitochondrial monoamine oxidase B. *Br. J. Pharmacol.* **132**(2), 500–506.

Youdim, M.B., Wadia, A., Tatton, W., and Weinstock, M. (2001b). The anti-Parkinson drug rasagiline and its cholinesterase inhibitor derivatives exert neuroprotection unrelated to MAO inhibition in cell culture and in vivo. *Ann. N. Y. Acad. Sci.* **939**, 450–458.

Youdim, M.B., Bar Am, O., Yogev-Falach, M., Weinreb, O., Maruyama, W., Naoi, M., and Amit, T. (2005). Rasagiline: neurodegeneration, neuroprotection, and mitochondrial permeability transition. *J. Neurosci. Res.* **79**(1–2), 172–179.

Zeller, E.A., and Barsky, J. (1952). In vivo inhibition of liver and brain monoamine oxidase by 1-Isonicotinyl-2-isopropyl hydrazine. *Proc. Soc. Exp. Biol. Med.* **81**(2), 459–461, Society for Experimental Biology and Medicine (New York, N.Y.).

Zornberg, G.L., Bodkin, J.A., and Cohen, B.M. (1991). Severe adverse interaction between pethidine and selegiline. *Lancet* **337**(8735), 246.

SELECTIVE INHIBITORS OF MONOAMINE OXIDASE TYPE B AND THE "CHEESE EFFECT"

John P.M. Finberg[1] and Ken Gillman[2]

[1]Department of Molecular Pharmacology, Rappaport Faculty of Medicine, Technion, Haifa, Israel
[2]Psychotropical Research, Bucasia, Queensland, Australia

Abstract

Potentiation of the cardiovascular and other effects of dietary tyramine by monoamine oxidase (MAO) inhibitors (cheese effect) has been a major limitation to clinical use of these drugs. The discovery that MAO exists in two distinct isoforms, MAO-A and MAO-B, together with the development of selective inhibitors of each isoform, enabled the understanding that selective inhibition of MAO-A, or inhibition of both isoforms, will cause cheese effect, but selective inhibition of MAO-B can be elicited without dangerous pressor reaction. This development has permitted the introduction of selective MAO-B inhibitors to clinical

DOI: 10.1016/B978-0-12-386467-3.00009-1

medicine for treatment of Parkinson's disease. This review describes the basic mechanisms involved in cheese effect, as well as providing information on tyramine levels in a variety of foodstuff, and surveys clinical information from tyramine pressor testing with the selective MAO-B inhibitors, selegiline and rasagiline.

I. Introduction

Inhibitors of the enzyme monoamine oxidase (MAO) were initially used clinically for the treatment of affective disorders, but their use fell to a low level after reports of a serious hypertensive reaction when patients treated with these drugs consumed foods such as yellow cheese which have a high content of the catecholamine releaser substance, tyramine (Blackwell and Mabbitt, 1965). The enzyme MAO exists in two isoforms, known as MAO-A and MAO-B, which have different affinities for substrates and inhibitors. The isoform MAO-A shows selectivity for noradrenaline, adrenaline, and serotonin (5-HT) as substrates and is inhibited *inter alia* by clorgyline, whereas MAO-B shows higher affinity for β-phenylethylamine (PEA) and benzylamine and is inhibited by selegiline (l-deprenyl) and rasagiline. Tyramine and dopamine show similar affinity for both isoforms. The tyramine interaction ("cheese effect") was first seen in patients treated with an MAOI which inhibits both enzyme forms, particularly tranylcypromine (TCP) or phenelzine. Subsequently, drugs were developed which show marked selectivity for one isoform or the other by virtue of their selective binding affinity. Preclinical research has established that selective inhibition of MAO-A, generally using clorgyline, which is a very selective and irreversible inhibitor of this enzyme, causes cheese effect, whereas selective inhibition of MAO-B by selegiline (L-deprenyl) or rasagiline does not (Finberg and Tenne, 1982; Finberg and Youdim, 1983; Youdim and Finberg, 1987). Similar results were found with the irreversible MAO-B selective inhibitor MDL-72145 (Fozard *et al.*, 1985). The metabolism of endogenous dopamine is reduced by both enzyme isoforms (O'Carroll *et al.*, 1983). This important property of the MAO-B inhibitors, that is, enhancement of dopamine's *in vivo* action without potentiation of tyramine effect, has enabled their extensive use in the treatment of Parkinson's disease. This brief review will expand on the preclinical and clinical aspects of this development. The pharmacological consequences of selective inhibition of MAO-A or MAO-B is a complex topic, which involves not only the biochemical aspects of the enzyme inhibition itself but also the consequences of selective localization of the isoforms in the different neuronal and other cell types in the body. For further information on this topic, the reader is referred to more detailed reviews (Youdim and Finberg, 1983; Youdim *et al.*, 1988, 2006).

II. Pharmacology of Tyramine

Tyramine (4-hydroxyphenethylamine, *para*-tyramine) is a substance widely found in nature, as it can be formed from the amino acid tyrosine by the action of the ubiquitous enzyme aromatic amino acid decarboxylase (AAADC). Because of its high sensitivity to oxidation by MAO, tyramine's endogenous levels are very low, and tyramine is one of a group of biogenic amines (BAs) which are referred to as "trace amines" (for review, see Berry, 2004). Other members of this group include amines such as octopamine and PEA, which themselves possess important physiological effects, and it is important to appreciate that several of these amines have a high turnover rate, although their steady-state levels are very low, for example, levels of *p*-tyramine in rat brain are about 1/1000 those of dopamine (Berry, 2004; Philips *et al.*, 1974). Indeed, the high affinity of tyramine for MAO enabled the original detection of this enzyme by Hare (1928).

The pharmacology of tyramine was originally studied by Dale and Dixon in 1909 (quoted from Hare, 1928), who showed that its physiological effects were similar to those of adrenaline. Subsequently, Burn and Rand (1958) described the essential difference between tyramine and agents such as adrenaline and noradrenaline, in that whereas reserpinization potentiated the effects of noradrenaline and adrenaline, it eliminated those of tyramine, leading eventually to the conclusion that tyramine caused its effects by releasing noradrenaline from neuronal storage sites. The basis for tyramine's sympathomimetic action is that it is a substrate for the noradrenaline transporter molecule (NET) situated in the plasma membrane of sympathetic nerves and adrenal medulla, as well as the vesicular transporter (VMAT) situated in the membrane of storage vesicles. The initial transport of tyramine molecules into the neuron activates reverse transport of noradrenaline molecules situated in the cytoplasmic compartment just inside the plasma membrane, and tyramine also releases monoamines stored in vesicular storage granules (Trendelenburg *et al.*, 1987). In fact, tyramine has been shown to release noradrenaline from isolated storage vesicles on a stoichiometric basis, but the precise basis for this stoichiometry is still unclear.

As MAO is an intraneuronal enzyme, which is bound to the mitochondrial outer membrane, tyramine molecules within the axoplasm are susceptible to enzymatic destruction by MAO. Tyramine is also a substrate for the dopamine β-hydroxylase enzyme found in storage vesicles, so following uptake into the storage granules, it will be converted to octopamine (β,4-dihydroxyphenylethylamine). Accumulation of octopamine, which has low affinity for α-adrenoceptors, is thought to be one of the mechanisms by which chronic administration of MAO inhibitors causes reduced systemic blood pressure (BP) and orthostatic hypotension (Fischer *et al.*, 1965). Another possible mechanism is increased effects of monoamine neurotransmitters at receptor sites in brain stem cardiovascular control areas. Tyramine administered directly to the striatum by microdialysis also causes release of dopamine from dopaminergic varicosities in the brain (Schmidt and Ferger, 2004), but systemically administered tyramine will have little or no central nervous system (CNS) effect, as it does not pass the blood–brain barrier. Recent work, however, has shown the existence of specific trace amine-associated receptors (TAAR), and so the physiological properties of this amine (and other trace amines) may be considerably more important than had been thought for many years, including a physiological role in neuromodulation (Berry, 2004; Pirri *et al.*, 2009).

III. Mechanism of "Cheese Effect"

The mechanism of tyramine potentiation by MAOI occurs at several levels. Tyramine is a substrate for both isoforms A and B of MAO, and cellular expression of each isoform is an important consideration in the action of enzyme inhibitors. Tyramine contained in food stuff will be released into the intestinal lumen, and a large proportion will be metabolized by MAO-A in the intestinal wall (Hasan *et al.*, 1988). Following absorption into the circulation, most of the remaining tyramine molecules will be metabolized in the liver (which contains a high concentration of MAO-A and MAO-B, see Finberg and Youdim, 1983; Youdim *et al.*, 1988). Thus first-pass clearance of tyramine is very high, and following a tyramine-containing meal, tyramine blood levels will normally be very low. If sufficient tyramine reaches the circulation, it will be available for uptake by postganglionic sympathetic neurons, inducing release of noradrenaline; however, if MAO is intact, then a considerable portion of the amine will be deaminated within the neuron. Also, as axoplasmatic noradrenaline levels are kept low by the action of intraneuronal MAO, little noradrenaline will be available for reverse transport into the synapse.

Inhibition of MAO can therefore potentiate tyramine effects by several mechanisms: (a) increased absorption of tyramine across the intestinal barrier, (b) reduced hepatic metabolism, these two leading to elevated blood tyramine concentration, (c) enhanced life of tyramine within sympathetic neurons, (d) increased soluble pool of noradrenaline available for release, and (e) enhanced release of noradrenaline and adrenaline from adrenal medullary cells.

Although tyramine is a substrate for both subtypes of MAO (A and B), because of the selective expression of the MAO subtypes in different cells and tissues, selective inhibition of MAO-A causes a much larger potentiation of tyramine than selective inhibition of MAO-B. Following oral tyramine administration, systemic absorption of the amine is increased by selective inhibition of MAO-A (Bieck *et al.*,

1993; Hasan *et al.*, 1988). At the neuronal level, selective inhibition of MAO-A causes tyramine potentiation, as it is the A subtype which is selectively expressed in sympathetic neurons (Jarrott and Iversen, 1971). The importance of neuronal MAO inhibition in the tyramine potentiation phenomenon can be seen in isolated tissue experiments, as selective inhibition of MAO-A in such pharmacological experiments produces a marked tyramine potentiation, whereas selective MAO-B inhibition has little effect, and tyramine concentration in the solution bathing the tissue is not changed by inhibition of MAO (Finberg and Tenne, 1982). An important consequence of the neuronal level potentiation of the action of tyramine is that the effects of amines which are not substrates for MAO (e.g., amphetamine, ephedrine) will also be potentiated by inhibition of MAO-A.

IV. Cardiovascular Actions of Tyramine

Pharmacological effects of tyramine on the circulation are complex. By virtue of releasing noradrenaline from postganglionic sympathetic varicosities, tyramine will activate α and β_1 adrenoceptors in blood vessels and myocardium, and release of adrenaline will in addition activate β_2-adrenoceptors (mainly producing vasodilatation, but there is also a small population of these receptors in myocardium). In addition, however, it has become apparent that TAAR are also expressed in heart and vasculature, so that some of the actions of tyramine may be mediated by direct activation of these receptors, as some vascular responses to tyramine are not inhibited by cocaine or reserpinization (Broadley, 2010; Frascarelli *et al.*, 2008a). Tyramine also releases other vasoactive substances, including dopamine, ATP, and other autacoids (see Broadley, 2010 for review). However, initial reports that tyramine releases dopamine from forearm vasculature were later shown to be due to contamination of dopamine in the tyramine preparation, and a dopamine-free tyramine preparation infused intraarterially produced only vasoconstriction without vasodilatation in forearm blood vessels (Jacob *et al.*, 2005). One of the first responses to intravenously administered tyramine is bradycardia (Lader *et al.*, 1970), which was originally thought to be due to cardiovascular reflexes, but in view of the plethora of tyramine's potential effects, may also be mediated by other mechanisms. Positive inotropic effects of tyramine also participate in the pressor response to tyramine and may be responsible for the initial pressor response to low doses of tyramine (Meck *et al.*, 2003a).

The preclinical data showing that selective inhibition of MAO-A caused cheese effect was found to be true also in human subjects. Clorgyline was shown to be a potent potentiator of tyramine effects in human subjects (Lader *et al.*, 1970), and at doses which cause marked reduction in plasma MHPG (a metabolite of noradrenaline by MAO and catechol *O*-methyltransferase), by 80%, clorgyline causes a similar degree of tyramine potentiation as that produced by phenelzine (Bieck *et al.*, 1993). Pharmacokinetic studies showed that apparent bioavailability of tyramine sulfate conjugate was increased fourfold by selegiline and eight- to ninefold by clorgyline or phenelzine, indicative of inhibition of MAO in gastrointestinal tract and liver, which increases plasma free tyramine available for conjugation (Bieck *et al.*, 1993).

V. Selegiline

In order to produce antidepressant effect using an MAO inhibitor, it is necessary to inhibit MAO-A, which is the major subtype involved in deamination of noradrenaline and 5-HT; however, inhibition of this isoform alone will potentiate tyramine effects. An important development in therapeutic use of MAO inhibitors was the discovery of selegiline (deprenyl), the first selective inhibitor of MAO-B, by Knoll and coworkers (Knoll and Magyar, 1972; Knoll *et al.*, 1978). In a variety of pharmacological preparations, both *in vitro* and *in vivo*, administration of selegiline at pharmacological doses was shown to inhibit, rather than potentiate, tyramine effects. This tyramine–inhibitory effect was attributed to the amphetamine-like property of the selegiline molecule, which is composed of (−)-methamphetamine together with a propargyl group, the latter being responsible for the irreversible enzyme inhibition (Knoll and Magyar, 1972). Selegiline, however, has a very low potency for inhibition of amine uptake. Its major metabolite (−)-methamphetamine is an effective inhibitor of amine uptake, but with IC_{50} values in the micromolar range, which is much higher than the concentrations which would be obtained after *in vivo* administration of selegiline (Glezer and Finberg, 2003; Lamensdorf *et al.*, 1999). By contrast to its inhibition of tyramine effects, Knoll and colleagues demonstrated that selegiline caused a powerful potentiation of the pharmacological effects of PEA, another indirectly acting sympathomimetic amine, in the cat (Knoll and Magyar, 1972). This amine, however, is a selective substrate for the MAO-B enzyme, so its potentiation is mainly the result of reduced metabolic breakdown, a fact which was shown using animals with acute hepatic ablation. The fact that selegiline does not potentiate tyramine effects at MAO-B selective doses permitted its introduction to the clinical treatment of Parkinson's disease (Birkmayer *et al.*, 1982), as dopamine behaves as a partial substrate of MAO-B in primate brain.

Selegiline is a propargyl-type suicide inhibitor, which, by virtue of its chemical similarity to dopamine, initially binds to the enzyme active site but there becomes converted to an adduct which combines covalently and irreversibly with the enzyme, inactivating it. Administration of a single submaximal dose of such an inhibitor will cause inactivation of a portion of the enzyme, but subsequent doses administered before the enzyme has time to reconstitute will progressively inactivate a greater proportion of the enzyme activity, the final activity level attained being a resultant of new enzyme synthesis and enzyme inactivation rates. Selegiline is metabolized extensively by first-pass hepatic metabolism via cytochrome P450 enzymes and has an oral bioavailability of about 10%. Selegiline metabolites include (−)-methamphetamine and (−)-amphetamine (Laine *et al.*, 2000).

VI. Clinical Use of Selegiline and Cheese Effect

Although selective inhibition of MAO-B does not produce tyramine potentiation, the selective MAO-B inhibitors available for clinical use, that is, selegiline and rasagiline, will inhibit MAO-A at higher doses (Youdim *et al.*, 2001), and these doses can cause cheese effect. In patient studies, it is currently impossible to determine the degree of MAO inhibition in the CNS; however, platelet MAO is only type B, and determination of platelet MAO-B therefore provides an indication of the extent of MAO-B inhibition in peripheral tissues. As these propargyl inhibitors efficiently pass the blood–brain barrier, it may be assumed that the extent of MAO-B inhibition for a given dose in the brain is very close to that in the periphery, and animal experiments show this to be true (see Waldmeier *et al.*, 1981). In human patients, the standard oral dose of 10 mg selegiline daily causes a nearly complete inhibition of platelet MAO-B, and on the average, a twofold potentiation of tyramine's pressor effect, as assessed by determination of the dose of tyramine necessary to increase systolic blood pressure (SBP) by 30 mmHg (Elsworth *et al.*, 1978), henceforth referred to as T30. This dose of selegiline is ineffective as an antidepressant, but an increase in the dose to 30 or more milligrams daily causes significant antidepressant effect (Mann *et al.*, 1989; Mendlewicz and Youdim, 1983). The higher dose of 60 mg daily (Sunderland *et al.*, 1994) produced biochemical changes indicative of extensive inhibition of MAO-A (Murphy *et al.*, 1998) and so could only be administered with a tyramine-restricted diet.

In recent years, two new dose forms of selegiline have been developed aimed to selectively increase inhibition of MAO in the brain as opposed to periphery, one is the selegiline transdermal system (STS), and the second is a liquid dose form aimed to increase absorption from the buccal cavity (Zydis selegiline). The transdermal application of the drug enables initial bypass of the hepatic metabolizing system, so that selegiline molecules will reach the circulation and distribute to brain and periphery. The peripheral compartment, however, will be more rapidly cleared of the drug by hepatic metabolism, and with continuous absorption, the degree of MAO inhibition in brain increases far more than in the periphery, resulting in significant inhibition of MAO-A as well as MAO-B in the brain, but mainly inhibition of MAO-B in the periphery (Mawhinney *et al.*, 2003). The effectiveness of STS as an antidepressant treatment, together with lack of significant tyramine potentiation, has now been shown in several clinical trials (Blob *et al.*, 2007; Pae *et al.*, 2007; Robinson and Amsterdam, 2008). Selegiline in this form had a modest antidepressant effect at the dose of 6 mg/24 h (Amsterdam, 2003). In a trial with healthy subjects, with a 6-mg/24-h patch, tyramine was administered in capsules. Following 3 days treatment with selegiline, the T30 ratio (i.e., T30 in unmedicated subjects divided by T30 in drug-treated subjects) was 1.85, as opposed to approximately 40 for TCP (Azzaro *et al.*, 2006). In another trial in healthy subjects using the 6-mg/24-h patch, the tyramine was given in the form of yellow cheese, and the amount administered was calculated based on known data. Amounts of cheese were adjusted to administer about 100 mg tyramine at breakfast and 400 mg at dinner. The designated pressor end point of a 30 mmHg increase in SBP over three consecutive measurements was not achieved in any of the 16 subjects (Blob *et al.*, 2007). Selegiline is available also in the STS form at calculated delivery doses of 9 and 12 mg/24 h, but tyramine interaction has not been assessed directly using these larger doses (Jessen *et al.*, 2008).

Buccally administered Zydis selegiline at a dose of 1.25 mg was found to inhibit platelet MAO-B to a similar extent as produced by 10 mg selegiline given orally. Following 14 days treatment, however, the T30 was reduced by half with orally administered selegiline but was unchanged by Zydis selegiline (Clarke *et al.*, 2003).

VII. Clinical Use of Rasagiline and Cheese Effect

The second selective irreversible inhibitor of MAO-B available for clinical use, rasagiline (*N*-propargyl-1-aminoindan), was authorized by the FDA in 2006. In this drug, the propargyl group is combined with 1-aminoindan, which is devoid of amphetamine-like activity (Finberg and Youdim, 2002; Glezer and Finberg, 2003). As with selegiline, selectivity for MAO-B inhibition is relative not absolute, with a ratio of IC_{50} for MAO-A to MAO-B of 51 in human brain homogenate (Youdim *et al.*, 2001). Rasagiline, however, is subject to a lower degree of hepatic first-pass metabolism than is selegiline and is about 10 times more potent on oral administration, both in rats and in man, with an oral daily dose for Parkinson's disease of 0.5–1.0 mg as opposed to 5–10 mg for selegiline.

Following initial demonstration of efficacy for improvement of Parkinsonian symptoms in monotherapy and together with L-dopa without dietary restrictions in TEMPO, PRESTO, and LARGO clinical trials (Parkinson, 2002; PRESTO,

2005; Rascol *et al.*, 2005), oral tyramine sensitivity was studied in a subset of patients who had completed the main PRESTO or TEMPO trials and suffered from fluctuating response to L-dopa (deMarcaida *et al.*, 2006). The doses of rasagiline used in these trials were 1 and 2 mg daily (TEMPO) and 0.5 and 1 mg daily (PRESTO). Tyramine (75 or 50 mg) was administered mixed with applesauce, yogurt, or ice cream following a low tyramine content meal, and specific endpoints were fixed (elevation in SBP of 30 mmHg for three consecutive readings, or reduction in heart rate of 40 beats/min over a period of at least 10 min). BP and heart rate were also determined during the study. In patients of the TEMPO study, none of the 55 patients treated with rasagiline had pressor responses achieving the end point values. In the PRESTO study, 4 of the 55 patients (3 on rasagiline 0.5 mg and 1 on placebo) had SBP elevations in excess of 30 mmHg, but none had reflex bradycardia or ECG changes. Since these pressor responses commenced more than 110 min after tyramine administration and did not subside within 2 h, they were considered not drug related, but the possibility of delayed gastrointestinal function in these Parkinsonian patients must also be considered (Isaacson, 2008). None of the patients on 1 mg rasagiline showed target hypertensive responses.

To obtain extra safety data, and with a stated aim of confirming selectivity of clinically used doses of rasagiline for inhibition of MAO-B, a new double-blind tyramine potentiation trial in healthy volunteers was constructed (Goren *et al.*, 2010). Several dose regimens of rasagiline were tested; the nonselective inhibitor phenelzine was included as a positive control and selegiline as a comparator drug. The T30 response was determined in the fasting state by administering tyramine in capsules at escalating doses up to a maximum of 800 mg. It should be noted that bioavailability of tyramine administered orally in capsules on an empty stomach is twice as high as that given in cheese (VanDenBerg *et al.*, 2003).

The 149 participants in this trial who completed the study were aged 40–70. Volunteers were excluded if they had a history of symptomatic reaction to tyramine-containing food, as was a single volunteer whose SBP reached 30 mmHg after 25 mg tyramine, and nine others who did not attain the 30 mmHg target rise in SBP during the baseline period even after 800 mg tyramine. The ratios of T30 control/T30 on drug were placebo 1.5, phenelzine (45 mg daily for 14 days) 17.3, selegiline (10 mg daily for 14 days) 2.5, rasagiline (1 mg daily for 14 days) 2.0, rasagiline (2 mg daily for 14 days) 3.3, rasagiline (2 mg daily for 30 days) 2.45, rasagiline (4 mg daily for 14 days) 4.5, and rasagiline (6 mg daily for 14 days) 5.1. An additional determination included in this study was measurement of plasma level of 3,4-dihydroxyphenylglycol (DHPG), as an index of MAO-A inhibition (circulating DHPG is produced mainly from the intraneuronal deamination of noradrenaline). The data presented with DHPG are difficult to interpret, as the variability of values was high, and statistical significance of differences were not supplied. Plasma levels of DHPG were reduced by about 20% in patients given phenelzine, and by about the same degree in those treated with 4 and 6 mg rasagiline daily; however, the tyramine sensitivity was increased much more in the phenelzine group. The data from this study therefore support that from others, in that selective inhibition of MAO-B increases tyramine sensitivity by a factor of about 2 (as opposed to 1.5 in the placebo group), and that serious reactions to large doses of tyramine (much more than would be ingested in food) are not anticipated at doses of MAO-B inhibitors selective for nearly complete inhibition of the enzyme in human patients (one patient at the highest dose of rasagiline experienced acute coronary syndrome, but little can be deduced from such an isolated observation). The results from this trial caused the FDA to reduce several food and medication restrictions on patients treated with rasagiline.

In an additional trial on patients participating in the PRESTO study, ambulatory BP was monitored during normal living conditions at home using a telephonic device (White *et al.*, 2008). Patients monitored their own BPs before and after meals, as well as supine and standing. A similar number (about 10–15%) of patients had postprandial elevation in SBP greater than 30 mmHg, but these cases were distributed equally between placebo and rasagiline groups. Additionally, the system detected no significant alteration in supine or standing BP in all patient groups. This data reinforce results from the other studies on tyramine which show great variability in individual BP responses to this amine, with a fraction of unmedicated subjects showing pressor responses to small amounts of tyramine as well as to low tyramine content meals.

VIII. Reversible Inhibitors of MAO-B

In an attempt to neutralize the cheese effect with potential antidepressant inhibitors of MAO-A, competitive reversible inhibitors have been developed (brofaromine, moclobemide). In the presence of elevated tyramine levels, these drugs will be displaced from the enzyme binding site, and so MAO will be able to metabolize the tyramine. Reversible inhibitors with selectivity for MAO-B have also been described (Fowler *et al.*, 1993), one of which (safinamide) is currently under clinical testing for antiparkinsonian efficacy, but no data are available on tyramine testing.

IX. Modification of L-Dopa Cardiovascular Response by Selective Inhibitors of MAO-B

Another aspect of MAO-B inhibitors and potentiation of cardiovascular effects of BAs are their possible interaction with L-dopa. When given together with a peripheral AAADC inhibitor, most of the pharmacological effects of

L-dopa are those exerted by dopamine produced in the CNS, including hypotension (Calne *et al.*, 1970). Without a peripheral AAADC inhibitor, and together with clorgyline or TCP, L-dopa caused pressor and tachycardic effects in rats, whereas selegiline and rasagiline did not cause these vascular and cardiac effects (Finberg *et al.*, 2006). This finding indicates that dopamine may cause its vasoconstrictor and cardiac effects by releasing noradrenaline from sympathetic varicosities (it causes vasodilator effects by direct activation of dopamine D1 receptors in renal and splanchnic vessels).

A survey of parkinsonian patients receiving selegiline and L-dopa (Churchyard *et al.*, 1997) found that the combination resulted in excess mortality, which was attributed to acute cardiovascular effects. These findings, however, are controversial (Pursiainen *et al.*, 2007). Dopamine generated from L-dopa in the CNS could cause activation of dopamine receptors or adrenoreceptors located in medullary cardiovascular centers, including those in the rostral ventrolateral medullary area, as well as modification of cardiovascular reflexes. These actions could cause hypotension by reduction of central sympathetic tone, in a similar way to the action of tricyclic antidepressant drugs (Lavian *et al.*, 1991). In a clinical trial of rasagiline in adjunct therapy with L-dopa, cardiovascular adverse effects were no greater than those with the comparator drug (entacapone) (Rascol *et al.*, 2005).

X. Tyramine Content in Foods

The medical literature historically has not reflected the research and published data that exist in the food chemistry field concerning BAs in food. Herein is an introductory explication followed by some detail on particular foods and beverages, with a few selected references from the vast food chemistry literature.

Tyramine and other BAs form in significant amount in comestibles usually only as a result of the decarboxylating activity of certain types of microorganisms that are contaminants in the food preparation process. The accumulation of tyramine depends on the physical medium (e.g., minced meat has a greater surface area), time, and temperature. Various other amines are also formed, histamine, putrescine, cadaverine, tyramine, tryptamine, 2-PEA, spermine, and spermidine are the most important, and all tend to accumulate in concert (Al Bulushi *et al.*, 2009; Shalaby, 1996). That is why smell is a helpful guide for what to avoid (however, tyramine can accumulate without food smelling "off"). Such amines have increasingly been monitored as part of the food hygiene audit process. In foods like yogurt and cheese, starter cultures that are devoid of decarboxylating organisms are now widespread practice. So far limits have been set only for histamine (varying between 50 and 200 mg/kg), but attention is being paid to all BAs. That is reflected in international standards of production of everything from fermented sausages to cheeses and beer. A "BA index" for Belgian lambic beer has been developed as part of a project to minimize BAs. European Union (EU) regulations have proposed a limit of 10 mg/L for tyramine.

As a result, tyramine levels in all foods are now generally lower than they have been in the past. So, for those who already follow healthy eating amounts and patterns, a low tyramine diet involves very few changes, because healthy amounts of foods, especially cheese, are around what is safe tyramine-wise: that is, 25 g of cheese is a healthy portion, and even matured cheeses infrequently contain more than 250 mg/kg. So a 25-g portion of such a cheese has a negligible amount of only 6 mg tyramine. Indeed, few foods have the degree of excess tyramine that might make the BP go *dangerously* high when eaten in healthy portion sizes in MAO-inhibited patients. The pressor response is a dose-related effect, which is why it is safe to "test" small quantities of some foods, for example, a favorite local cheese.

The oral T30 in *fasted* and *unmedicated* subjects given tyramine *capsules* is approximately 500 mg (range 200–800). It is a simple but time-consuming procedure so the number of tests done is relatively small, usually they are normal volunteers not patients with depressive illness and subgroups are even smaller; therefore, all these data must be interpreted with that in mind. Intersubject variation, based on these modest numbers of normal subjects ($n = 55$), is approximately threefold (Bieck and Antonin, 1989). For tyramine in a normal meal, the T30 range was found to be three to four times higher than fasting, that is, 800–2000 mg (Berlin *et al.*, 1989). The maximum plasma concentration of tyramine taken with a meal was reduced ($n = 7$) by 72% (VanDenBerg *et al.*, 2003); the time taken to maximal plasma concentration (T_{max}) fasting was 20 min, and with a meal was 75 min. After administration of TCP, the mean ED50 (dose required for 50% of subjects to experience an SBP elevation of *at least* 30 mmHg), *fasting with capsules*, was 8 mg ($n = 12$); however, the mean SBP increases when the same dose of tyramine was given as a piece of cheese was only 11 mmHg versus 46 mmHg with capsules (Berlin *et al.*, 1989; Bieck and Antonin, 1988, 1994). The T30 (TCP) would be expected to be even higher with a normal meal, which was indeed what Zimmer *et al.* found; the average T30 (*fasting/capsules*) was 16 mg, whereas, in the same subjects with tyramine in a meal, it was 35 mg (range, 20–50 mg, $n = 8$) (Zimmer *et al.*, 1990).

This information indicates that the most tyramine sensitive of subjects on an irreversible inhibitor of MAO-A and -B like TCP would be able to ingest 10 mg on an empty stomach (e.g., a tyramine-rich liquid) to get a 30–45 mmHg SPB increase, but about 30 mg as part of a meal. Average subjects would need to ingest somewhat more, about 50–60 mg. Such degrees (30–45 mmHg) of BP elevation are not dangerous. Dangerous elevations of BP almost certainly need to be considerably greater. This approximation is in close agreement with the original estimations made by Blackwell and Da Prada (Blackwell and Mabbitt, 1965; Da Prada and Zurcher, 1992).

Thus, for a majority of patients on TCP, the amount of tyramine needed in a meal to provoke a degree of BP elevation likely to be dangerous is probably in excess of 50 mg. As can be seen from the data below, modern food hygiene practices make that rather unlikely unless people are both careless and consume unhealthily large portions. For other drugs like rasagiline, that have much lower tyramine potentiation, the amount of tyramine needed to elevate BP dangerously would be commensurately higher, indeed heroic gluttony would be required to ingest excess tyramine.

The tyramine reaction consists of a thumping heartbeat and a progressive increase in BP. The pulse rate usually becomes slower, in response to the increase in BP (Bieck and Antonin, 1988). Tightness in the chest and pallor often occur. Symptoms usually start within 30 min of ingestion, and the duration of the reaction is about 2 h. Headache is not a reliable indicator of high BP. Death or morbidity from cheese reaction is extremely rare, possibly rarer than serious reactions to many modern drugs, for example, bleeding secondary to selective serotonin reuptake inhibitor drugs (SSRIs) (Gillman, 2011a; Opatrny *et al.*, 2008).

Low doses of tyramine increase SBP by increasing the force of cardiac contraction (positive inotropic effect) and by increasing the ejection fraction (Meck *et al.*, 2003b), that is, probably mediated via TAAR1. Octopamine, β-PEA, and tryptamine produce a negative inotropic effect (in rat ventricle *in vitro*) at micromolar concentrations (Frascarelli *et al.*, 2008b). Because these other amines are usually coingested with tyramine in foods, this may partly explain the lower pressor effect of tyramine in foods versus tyramine capsules given as comparators, in addition to their lower bioavailability. A similar mechanism resulting from the stimulation of multiple receptors may also be responsible for the variable presentation for histamine poisoning (scombroidosis). Elevated tyramine intake further increases BP by vasoconstriction, which is sensitive to α-adrenoceptor antagonists.

XI. Treatment of Cheese Reaction

Short-duration BP elevation well in excess of 200 mmHg occurs with various common activities. Modest levels of exercise with weights can raise the SBP even more, in excess of 400 mmHg. (Haykowsky *et al.*, 1996). Also, rapid reduction of acute hypertension (within 1–2 h) may of itself result in adverse effects (Flanigan and Vitberg, 2006). It is probably best to initiate treatment of acutely elevated BP *only* if there is evidence of end organ damage and *only* in a critical care setting. The most appropriate hypotensive agent will depend on the particular system affected (brain, heart, lungs, kidneys). Pain (from the associated headache) and anxiety both exacerbate hypertension, so doctors should remain calm and consider administration of a benzodiazepine, whilst instituting measures to assess end organ damage. Sublingual nifedipine is usually contra-indicated because it has an unpredictable effect (Gillman, 2011a).

XII. Specific Foods

It is first appropriate to observe that after more than 50 years of nonselective MAOI use, deaths have been rare (NB but lesser usage post 1968) and the vast majority of them occurred before the cheese reaction was recognized (21 deaths in 500 reports to the FDA before 1967). No deaths have been documented with foods other than cheese. Many of the deaths involved large quantities of cheese, prolonged symptoms which had not been acted on, and an absence of any medical treatment. Clinical experience indicates that dogmatic dictates from doctors, and rigid lists of prohibited foods, many of which are incorrect, tend to lead to cynicism and noncompliance. An explanatory and educational approach is more effective. Unfortunately, many of the MAOI diets given to patients, and those available on the Internet, even from educational establishments, still contain errors and omissions, over generalizations and unwarranted prohibitions based on poor evidence and understanding of the processes at work. A comprehensive review of recent references about tyramine and food with explanation and analysis is posted on the net (Gillman, 2011b).

As cheese is the only documented substance causing fatalities, and by far the most important for dietary indiscretion, it will be dealt with in more detail.

First, it can be stated with a great deal of confidence that only mature/"vintage" style cheeses (>3, usually >9, months aging) have a significantly elevated level of tyramine (i.e., >250 mg/kg). Fresh nonmatured, that is, unripened/unaged cheese styles and yogurt are always safe. Likewise, commercial "supermarket" cheeses are likely to be low in tyramine (<100 mg/kg) because budget prices do not pay for long warehouse aging.

Matured gourmet and "artisanal" cheeses can develop high levels (as much as 1000 mg/kg of tyramine), for example, matured Stilton, Cheddar, Parmigiano, manchego, compté (USS 25 g L). A large number of papers have measured levels in cheeses from many countries all over the world. After each food, there is a note in brackets indicating the estimated degree of risk of hypertension for those on potent MAO-AB inhibitors like TCP. The abbreviations are USS, usual serving size; degree of risk N, none; L, low; M, moderate; H, high. So, very high tyramine foods like mature artisan cheddar is: USS 25 g, risk M (i.e., has 25 mg of tyramine, so some sensitive people will get moderate BP elevation if it is eaten on an empty stomach, but most people eating it as part of a meal will not get problematic BP elevation). Examples from more recent findings: Portuguese traditional cheeses, Terrincho (Pintadoa *et al.*, 2008), all were < 100 mg/kg; blue cheese, Czech (Komprda *et al.*, 2008c; Novella-Rodriguez *et al.*, 2003), the mean was 380 mg/kg, different cheeses (vats) varied widely, from 10 mg/kg to 875 mg/kg; blue cheeses obtained from Spanish retail stores (20 samples) averaged 14 mg/kg (range from 0 to 1585 mg/kg); Dutch-type semi-hard cheeses mostly < 50 mg/kg, maximum 250 (Komprda *et al.*, 2008a,b). Even with the strongest of these cheeses, it would only be with large portion sizes of 50–100 g that the risk could be classified as moderate or high and with most of them it would be nonexistent or low even with large portion sizes.

XIII. Fish Sauces

Fish sauces (garum in classical Roman cooking) now are ubiquitous and are made from seafood, often anchovy, allowed to ferment approximately 140–200 days (for a recent review, see Cho *et al.*, 2006; Stute *et al.*, 2002; Zaman *et al.*, 2009). Examples: Korean fermented fish products average less than 50 mg/kg; liquid fish sauces made from a variety of sea foods, scallop, squid, etc., average 350, maximum (anchovy) 600 mg/kg (Cho *et al.*, 2006); commercial fish sauces from the Far East (45 samples) less than 200 mg/kg, maximum of 588 mg/kg (Stute *et al.*, 2002) (USS 10 ml, risk N).

XIV. Meat

For a review of polyamines in meat (and vegetables), see Kalac and Glória (2009). Fresh liver has no tyramine (Krausová *et al.*, 2006), but if stored badly or past its "use by" date when purchased and then kept in a domestic fridge that is not cold enough, it may soon become risky, especially when macerated into a bacterial growth medium called "liver pate," which is safe only if freshly made and properly refrigerated (i.e., *below* 4 °C). All dry cured meat products have only low levels of tyramine, as expected, generally < 5 mg/kg (Lorenzo *et al.*, 2007; Ruiz-Capillas and Jimenez-Colmenero, 2004). So "Parma ham," prosciutto, copa, etc., will all be safe (USS < 100 g risk N). With fermented "salami" type, sausages hygiene and low temperature processing are improving steadily, and more recent surveys have all found lower tyramine levels: Suzzi reviewed 20 studies from all over Europe (Suzzi and Gardini, 2003) and found tyramine was usually below 200 mg/kg (USS < 100 g risk N). Fresh and smoked fish is safe, but fermented fish, a speciality of the Baltic countries, for example, Strömming, rakfisk, etc., is high risk (Gillman, 2011b).

Marmite is made from residual brewer's yeast and has relatively high amounts of BAs, average 320 mg/kg of tyramine (Populin *et al.*, 2007), but portion size is small, so a large teaspoon full will contain only a few mg (USS 10 g risk N). Soy sauce, "tofu," "miso," etc., are also made by fermentation: most soy sauce samples measured have ranged between 10 and 200 mg/L (Ibe *et al.*, 2003), typical levels averaged 150 mg/L, with a rare maximum of 940 mg; Chinese soy (40 samples) were mostly less than 200 mg/L (Yongmeia *et al.*, 2009) (USS 10 ml risk N).

Kalac reported on more than 100 samples of Sauerkraut from seven countries, almost all <200 mg/kg, but a couple from Czech Rep. were 400–900 mg/kg (Kalac and Glória, 2009). Korean "kimchi" cabbage averaged 50 mg/kg, max. 120 mg/kg (Cho *et al.*, 2006) (USS 100 g risk N/L).

XV. Beer and Wine

Tyramine in liquids, which may be taken on an empty stomach, should be regarded as a special case because tyramine will be absorbed much more rapidly. For those who are tyramine sensitive, even 10 mg might provoke a reaction, even if the BP is unlikely to be sufficiently elevated for long enough to cause morbidity. If an MAOI-treated patient were to drink a liter of beer at 50 mg/L of tyramine, then morbidity might result (e.g., see Ottervanger *et al.*, 1993). However generally, wine and beer in moderation (two standard drinks in 2 h) are definitely safe. Modern commercial wines do not contain significant tyramine. Standards, and awareness of beer brewing hygiene issues, have increased since some of the older results, but some caution is still warranted: some rare "artisan/boutique" ones are risky on occasion (see Kalac and Glória, 2009). Some Belgian lambic beers have high tyramine content; of 21 samples, the mean tyramine was 28 mg/L and the maximum was nearly 70 mg/L (Loret *et al.*, 2005) (USS 500 ml risk M/H).

It would seem very likely that all standard commercial and modern beers all over the world will be safe (USS 500 ml risk N). A great majority have low tyramine (2–8 mg/L); a very few are up to 30–50 mg/L (for a review, see Kalac and Krízek, 2003). Tang looked at 18 beers all brewed in China, some European under license, and values were mostly 3–5 (max 7) mg/L (Tang *et al.*, 2009). Spanish beers had <2 mg/L (Cortacero-Ramırez *et al.*, 2007). About 17 domestic Turkish and 13 imported beers were evaluated (Anli *et al.*, 2006) and all were <2 mg/L. Shulman's group (Canada) looked at a total of 98 beer samples (79 different brands of beer) 15 years ago, only 4 of 98 beer samples were >10 mg/L (Tailor *et al.*, 1994) (USS 500 ml risk N).

XVI. Conclusions

A variety of preclinical as well as clinical studies with various forms of tyramine administration and using the two selective MAO-B inhibitors currently available for clinical use agree that tyramine sensitivity is increased by a factor of 2 following selective inhibition of MAO-B in human subjects. Data on individual sensitivity to tyramine is more difficult to interpret because of the occasional reports of large increases in systemic BP to quite small amounts of tyramine, both in normal subjects and in volunteers; however, even these responses are not greater than increases in BP from normal activities such as sport, and similar responses have been described in placebo-treated subjects. The special case of the Parkinsonian patient must also be considered, as they may suffer from delayed intestinal absorption due to autonomic nervous system disturbances, and also their cardiovascular reflexes may be impaired.

Acknowledgment

J. F. is a codeveloper of rasagiline and receives income from sales of the drug.

References

Al Bulushi, I., Poole, S., Deeth, H.C., and Dykes, G.A. (2009). Biogenic amines in fish: roles in intoxication, spoilage, and nitrosamine formation—a review. *Crit. Rev. Food Sci. Nutr.* **49**, 369–377.

Amsterdam, J.D. (2003). A double-blind, placebo-controlled trial of the safety and efficacy of selegiline transdermal system without dietary restrictions in patients with major depressive disorder. *J. Clin. Psychiatry* **64**, 208–214.

Anli, R.E., Vura, N., Demiray, S., and Mert, B. (2006). Biogenic amine content of beers consumed in Turkey and influence of storage conditions on biogenic amine formation. *J. Inst. Brewing* **112**, 267–274.

Azzaro, A.J., Vandenberg, C.M., Blob, L.F., Kemper, E.M., Sharoky, M., Oren, D.A., and Campbell, B.J. (2006). Tyramine pressor sensitivity during treatment with the selegiline transdermal system 6 mg/24 h in healthy subjects. *J. Clin. Pharmacol.* **46**, 933–944.

Berlin, I., Zimmer, R., Cournot, A., Payan, C., Pedarriosse, A.M., and Puech, A.J. (1989). Determination and comparison of the pressor effect of tyramine during long-term moclobemide and tranylcypromine treatment in healthy volunteers. *Clin. Pharmacol. Ther.* **46**, 344–351.

Berry, M.D. (2004). Mammalian central nervous system trace amines. Pharmacologic amphetamines, physiologic neuromodulators. *J. Neurochem.* **90**, 257–271.

Bieck, P.R., and Antonin, K.H. (1988). Oral tyramine pressor test and the safety of monoamine oxidase inhibitor drugs: comparison of brofaromine and tranylcypromine in healthy subjects. *J. Clin. Psychopharmacol.* **8**, 237–245.

Bieck, P.R., and Antonin, K.H. (1989). Tyramine potentiation during treatment with MAO inhibitors: brofaromine and moclobemide vs irreversible inhibitors. *J. Neural Transm. Suppl.* **28**, 21–31.

Bieck, P.B., and Antonin, K.-H. (1994). Tyramine potentiation during treatment with MAOIs. In: S. H. Kennedy (Ed.), *Clinical Advances in Monoamine Oxidase Inhibitor Therapies*. American Psychiatric Press, Washington & London, pp. 83–110.

Bieck, P.R., Antonin, K.H., and Schulz, R. (1993). Clinical pharmacology of monoamine oxidase inhibitors. In: H. Yasuhara, S.H. Parvez, K. Oguchi, M. Sandler, and T. Nagatsu (Eds.), *Monoamine Oxidase: Basic and Clinical Aspects*. VSP, Utrecht, pp. 177–196.

Birkmayer, W., Riederer, P., and Youdim, M.B. (1982). (−)Deprenyl in the treatment of Parkinson's disease. *Clin. Neuropharmacol.* **5**, 195–230.

Blackwell, B., and Mabbitt, L.A. (1965). Tyramine in cheese related to hypertensive crises after monoamine-oxidase inhibition. *Lancet* **1**, 938–940.

Blob, L.F., Sharoky, M., Campbell, B.J., Kemper, E.M., Gilmor, M.G., VanDenberg, C.M., and Azzaro, A.J. (2007). Effects of a tyramine-enriched meal on blood pressure response in healthy male volunteers treated with selegiline transdermal system 6 mg/24 hour. *CNS Spectr.* **12**, 25–34.

Broadley, K.J. (2010). The vascular effects of trace amines and amphetamines. *Pharmacol. Ther.* **125**, 363–375.

Burn, J.H., and Rand, M.J. (1958). The action of sympathomimetic amines in animals treated with reserpine. *J. Physiol.* **144**, 314–336.

Calne, D.B., Brennan, J., Spiers, A.S., and Stern, G.M. (1970). Hypotension caused by L-dopa. *Br. Med. J.* **1**, 474–475.

Cho, T.Y., Han, G.H., Bahn, K.N., Son, Y.W., Jang, M.R., Lee, C.H., Kim, S.H., B, K.D., and Kim, S. B. (2006). Evaluation of biogenic amines in Korean commercial fermented foods. *Korean J. Food Sci. Technol.* **38**, 730–737.

Churchyard, A., Mathias, C.J., Boonkongchuen, P., and Lees, A.J. (1997). Autonomic effects of selegiline: possible cardiovascular toxicity in Parkinson's disease. *J. Neurol. Neurosurg. Psychiatry* **63**, 228–234.

Clarke, A., Johnson, E.S., Mallard, N., Corn, T.H., Johnston, A., Boyce, M., Warrington, S., and MacMahon, D.G. (2003). A new low-dose formulation of selegiline: clinical efficacy, patient preference and selectivity for MAO-B inhibition. *J. Neural Transm.* **110**, 1257–1271.

Cortacero-Ramırez, S., Arraez-Roman, D., Antonio Segura-Carretero, A., and Fernandez-Gutierrez, A. (2007). Determination of biogenic amines in beers and brewing-process samples by capillary electrophoresis coupled to laser-induced fluorescence detection. *Food Chem.* **100**, 383–389.

Da Prada, M., and Zurcher, G. (1992). Tyramine content of preserved and fermented foods or condiments of Far Eastern cuisine. *Psychopharmacology (Berlin)* **106**(Suppl.), S32–S34.

deMarcaida, J.A., Schwid, S.R., White, W.B., Blindauer, K., Fahn, S., Kieburtz, K., Stern, M., and Shoulson, I. (2006). Effects of tyramine administration in Parkinson's disease patients treated with selective MAO-B inhibitor rasagiline. *Mov. Disord.* **21**, 1716–1721.

Elsworth, J.D., Glover, V., Reynolds, G.P., Sandler, M., Lees, A.J., Phuapradit, P., Shaw, K.M., Stern, G.M., and Kumar, P. (1978). Deprenyl administration in man: a selective monoamine oxidase B inhibitor without the 'cheese effect'. *Psychopharmacology (Berl)* **57**, 33–38.

Finberg, J.P., and Tenne, M. (1982). Relationship between tyramine potentiation and selective inhibition of monoamine oxidase types A and B in the rat vas deferens. *Br. J. Pharmacol.* **77**, 13–21.

Finberg, J.P., and Youdim, M.B. (1983). Selective MAO A and B inhibitors: their mechanism of action and pharmacology. *Neuropharmacology* **22**, 441–446.

Finberg, J.P., and Youdim, M.B. (2002). Pharmacological properties of the anti-Parkinson drug rasagiline; modification of endogenous brain amines, reserpine reversal, serotonergic and dopaminergic behaviours. *Neuropharmacology* **43**, 1110–1118.

Finberg, J.P., Gross, A., Bar-Am, O., Friedman, R., Loboda, Y., and Youdim, M.B. (2006). Cardiovascular responses to combined treatment with selective monoamine oxidase type B inhibitors and L-DOPA in the rat. *Br. J. Pharmacol.* **149**, 647–656.

Fischer, J.E., Horst, W.D., and Kopin, I.J. (1965). Beta-hydroxylated sympathomimetic amines as false neurotransmitters. *Br. J. Pharmacol. Chemother.* **24**, 477–484.

Flanigan, J.S., and Vitberg, D. (2006). Hypertensive emergency and severe hypertension: what to treat, who to treat, and how to treat. *Med. Clin. North Am.* **90**, 439–451.

Fowler, J.S., Volkow, N.D., Logan, J., Schlyer, D.J., MacGregor, R.R., Wang, G.J., Wolf, A.P., Pappas, N., Alexoff, D., Shea, C., *et al.* (1993). Monoamine oxidase B (MAO B) inhibitor therapy in Parkinson's disease: the degree and reversibility of human brain MAO B inhibition by Ro 19 6327. *Neurology* **43**, 1984–1992.

Fozard, J.R., Zreika, M., Robin, M., and Palfreyman, M.G. (1985). The functional consequences of inhibition of monoamine oxidase type B: comparison of the pharmacological properties of L-deprenyl and MDL 72145. *Naunyn Schmiedebergs Arch. Pharmacol.* **331**, 186–193.

Frascarelli, S., Ghelardoni, S., Chiellini, G., Vargiu, R., Ronca-Testoni, S., Scanlan, T.S., Grandy, D. K., and Zucchi, R. (2008). Cardiac effects of trace amines: pharmacological characterization of trace amine-associated receptors. *Eur. J. Pharmacol.* **587**, 231–236.

Gillman, P.K. (2011a). Advances pertaining to the pharmacology and interactions of irreversible nonselective monoamine oxidase inhibitors. *J. Clin. Psychopharmacol.* **31**, 66–74.

Gillman, P.K. (2011b). Monoamine oxidase inhibitors, dietary tyramine and drug interactions. http://psychotropical.com/pdfs/maois_diet_full.pdf[accessed Dec 2010].

Glezer, S., and Finberg, J.P. (2003). Pharmacological comparison between the actions of methamphetamine and 1-aminoindan stereoisomers on sympathetic nervous function in rat vas deferens. *Eur. J. Pharmacol.* **472**, 173–177.

Goren, T., Adar, L., Sasson, N., and Weiss, Y.M. (2010). Clinical pharmacology tyramine challenge study to determine the selectivity of the monoamine oxidase type B (MAO-B) inhibitor rasagiline. *J. Clin. Pharmacol.* **50**, 1420–1428.

Hare, M.L. (1928). Tyramine oxidase: a new enzyme system in liver. *Biochem. J.* **22**, 968–979.

Hasan, F., McCrodden, J.M., Kennedy, N.P., and Tipton, K.F. (1988). The involvement of intestinal monoamine oxidase in the transport and metabolism of tyramine. *J. Neural Transm. Suppl.* **26**, 1–9.

Haykowsky, M.J., Findlay, J.M., and Ignaszewski, A.P. (1996). Aneurysmal subarachnoid hemorrhage associated with weight training: three case reports. *Clin. J. Sport Med.* **6**, 52–55.

Ibe, A., Tabata, S., Sadamasu, Y., Yasui, A., Shimoi, T., Endoh, M., and Saito, K. (2003). Production of tyramine in "moromi" mash during soy sauce fermentation. *Shokuhin Eiseigaku Zasshi* **44**, 220–226.

Isaacson, S.H. (2008). Selective MAO-B inhibitors have low potential for the tyramine effect. *Mov. Disord.* **25**, 123–124.

Jacob, G., Gamboa, A., Diedrich, A., Shibao, C., Robertson, D., and Biaggioni, I. (2005). Tyramine-induced vasodilation mediated by dopamine contamination: a paradox resolved. *Hypertension* **46**, 355–359.

Jarrott, B., and Iversen, L.L. (1971). Noradrenaline metabolizing enzymes in normal and sympathetically denervated vas deferens. *J. Neurochem.* **18**, 1–6.

Jessen, L., Kovalick, L.J., and Azzaro, A.J. (2008). The selegiline transdermal system (emsam): a therapeutic option for the treatment of major depressive disorder. *Pharmacy Ther.* **33**, 212–246.

Kalac, P., and Glória, M.B.A. (2009). Biogenic amines in cheeses, wines, beers and sauerkraut. In: G. Dandrifosse (Ed.), *Biological Aspects of Biogenic Amines, Polyamines and Conjugates.* Transworld Research Network, Kerala, India, pp. 267–309.

Kalac, P., and Krízek, M. (2003). A review of biogenic amines and polyamines in beer. *J. Inst. Brewing* **109**, 123–128.

Knoll, J., and Magyar, K. (1972). Some puzzling pharmacological effects of monoamine oxidase inhibitors. *Adv. Biochem. Psychopharmacol.* **5**, 393–408.

Knoll, J., Ecsery, Z., Magyar, K., and Satory, E. (1978). Novel (−)deprenyl-derived selective inhibitors of B-type monoamine oxidase. The relation of structure to their action. *Biochem. Pharmacol.* **27**, 1739–1747.

Komprda, T., Burdychova, R., Dohnal, V., Cwikova, O., Sladkova, P., and Dvorackova, H. (2008a). Some factors influencing biogenic amines and polyamines content in Dutch-type semi-hard cheese. *Eur. Food Res. Technol.* **227**, 29–36.

Komprda, T., Burdychova, R., Dohnal, V., Cwikova, O., Sladkova, P., and Dvorackova, H. (2008b). Tyramine production in Dutch-type semi-hard cheese from two different producers. *Food Microbiol.* **25**, 219–227.

Komprda, T., Dohnal, V., and Závodníková, R. (2008c). Contents of some biologically active amines in a Czech blue-vein cheese. *Czech J. Food Sci.* **26**, 428–440.

Krausová, P., Kalač, P., Křížek, M., and Pelikánová, T. (2006). Content of biologically active polyamines in livers of cattle, pigs and chickens after animal slaughter. *Meat Sci.* **73**, 640–644.

Lader, M.H., Sakalis, G., and Tansella, M. (1970). Interactions between sympathomimetic amines and a new monoamine oxidase inhibitor. *Psychopharmacologia* **18**, 118–123.

Laine, K., Anttila, M., Huupponen, R., Maki-Ikola, O., and Heinonen, E. (2000). Multiple-dose pharmacokinetics of selegiline and desmethylselegiline suggest saturable tissue binding. *Clin. Neuropharmacol.* **23**, 22–27.

Lamensdorf, I., Porat, S., Simantov, R., and Finberg, J.P. (1999). Effect of low-dose treatment with selegiline on dopamine transporter (DAT) expression and amphetamine-induced dopamine release in vivo. *Br. J. Pharmacol.* **126**, 997–1002.

Lavian, G., Di Bona, G., and Finberg, J.P. (1991). Inhibition of sympathetic nerve activity by acute administration of the tricyclic antidepressant desipramine. *Eur. J. Pharmacol.* **194**, 153–159.

Lorenzo, J.M., Martínez, S., Francoa, I., and Carballo, J. (2007). Biogenic amine content during the manufacture of dry-cured lacón, a Spanish traditional meat product: effect of some additives. *Meat Sci.* **77**, 287–293.

Loret, S., Deloyer, P., and Dandrifosse, G. (2005). Levels of biogenic amines as a measure of the quality of the beer fermentation process: data from Belgian samples. *Food Chem.* **89**, 519–525.

Mann, J.J., Aarons, S.F., Wilner, P.J., Keilp, J.G., Sweeney, J.A., Pearlstein, T., Frances, A.J., Kocsis, J.H., and Brown, R.P. (1989). A controlled study of the antidepressant efficacy and side effects of (−)-deprenyl. A selective monoamine oxidase inhibitor. *Arch. Gen. Psychiatry* **46**, 45–50.

Mawhinney, M., Cole, D., and Azzaro, A.J. (2003). Daily transdermal administration of selegiline to guinea-pigs preferentially inhibits monoamine oxidase activity in brain when compared with intestinal and hepatic tissues. *J. Pharm. Pharmacol.* **55**, 27–34.

Meck, J.V., Martin, D.S., D'Aunno, D.S., and Waters, W.W. (2003). Pressor response to intravenous tyramine is a marker of cardiac, but not vascular, adrenergic function. *J. Cardiovasc. Pharmacol.* **41**, 126–131.

Mendlewicz, J., and Youdim, M.B. (1983). L-Deprenil, a selective monoamine oxidase type B inhibitor, in the treatment of depression: a double blind evaluation. *Br. J. Psychiatry* **142**, 508–511.

Murphy, D.L., Karoum, F., Pickar, D., Cohen, R.M., Lipper, S., Mellow, A.M., Tariot, P.N., and Sunderland, T. (1998). Differential trace amine alterations in individuals receiving acetylenic inhibitors of MAO-A (clorgyline) or MAO-B (selegiline and pargyline). *J. Neural Transm. Suppl.* **52**, 39–48.

Novella-Rodriguez, S., Veciana-Nogues, M.T., Izquerdo-Pulido, M., Vidal-Carou, M.C., Roig-Sagues, A.X., and Trujillo-Mesa, A.J. (2003). Distribution of biogenic amines and polyamines in cheese. *J. Food Sci.* **68**, 750–755.

O'Carroll, A.M., Fowler, C.J., Phillips, J.P., Tobbia, I., and Tipton, K.F. (1983). The deamination of dopamine by human brain monoamine oxidase. Specificity for the two enzyme forms in seven brain regions. *Naunyn Schmiedebergs Arch. Pharmacol.* **322**, 198–202.

Opatrny, L., Delaney, J.A., and Suissa, S. (2008). Gastro-intestinal haemorrhage risks of selective serotonin receptor antagonist therapy: a new look. *Br. J. Clin. Pharmacol.* **66**, 76–81.

Ottervanger, J.P., Klaver, M.M., Kolling, P., and Stricker, B.H. (1993). Intracranial hemorrhage following use of MAO inhibitor tranylcypromine and beer. *Nederlands Tijdschrift Voor Geneeskunde* **137**, 921–922.

Pae, C.U., Lim, H.K., Han, C., Neena, A., Lee, C., and Patkar, A.A. (2007). Selegiline transdermal system: current awareness and promise. *Prog. Neuropsychopharmacol. Biol. Psychiatry* **31**, 1153–1163.

Parkinson (2002). A controlled trial of rasagiline in early Parkinson disease: the TEMPO Study. *Arch. Neurol.* **59**, 1937–1943.

Philips, S.R., Durden, D.A., and Boulton, A.A. (1974). Identification and distribution of p-tyramine in the rat. *Can. J. Biochem.* **52**, 366–373.

Pintadoa, A.I.E., Pinho, O., Ferreira, I.M.P.L.V.O., Pintadoa, M.M.E., Gomesa, A.M.P., and Malcataa, F.X. (2008). Microbiological, biochemical and biogenic amine profiles of Terrincho cheese manufactured in several dairy farms. *Int. Dairy J.* **18**, 631–640.

Pirri, J.K., McPherson, A.D., Donnelly, J.L., Francis, M.M., and Alkema, M.J. (2009). A tyramine-gated chloride channel coordinates distinct motor programs of a Caenorhabditis elegans escape response. *Neuron* **62**, 526–538.

Populin, T., Moret, S., Truant, S., Lanfranco, S., and Conte, L.S. (2007). A survey on the presence of free glutamic acid in foodstuffs, with and without added monosodium glutamate. *Food Chem.* **104**, 1712–1717.

PRESTO (2005). A randomized placebo-controlled trial of rasagiline in levodopa-treated patients with Parkinson disease and motor fluctuations: the PRESTO study. *Arch. Neurol.* **62**, 241–248.

Pursiainen, V., Korpelainen, T.J., Haapaniemi, H.T., Sotaniemi, A.K., and Myllyla, V.V. (2007). Selegiline and blood pressure in patients with Parkinson's disease. *Acta Neurol. Scand.* **115**, 104–108.

Rascol, O., Brooks, D.J., Melamed, E., Oertel, W., Poewe, W., Stocchi, F., and Tolosa, E. (2005). Rasagiline as an adjunct to levodopa in patients with Parkinson's disease and motor fluctuations (LARGO, Lasting effect in Adjunct therapy with Rasagiline Given Once daily, study): a randomised, double-blind, parallel-group trial. *Lancet* **365**, 947–954.

Robinson, D.S., and Amsterdam, J.D. (2008). The selegiline transdermal system in major depressive disorder: a systematic review of safety and tolerability. *J. Affect. Disord.* **105**, 15–23.

Ruiz-Capillas, C., and Jimenez-Colmenero, F. (2004). Biogenic amine content in Spanish retail market meat products treated with protective atmosphere and high pressure. *Eur. Food Res. Technol.* **218**, 237–241.

Schmidt, N., and Ferger, B. (2004). The biogenic trace amine tyramine induces a pronounced hydroxyl radical production via a monoamine oxidase dependent mechanism: an in vivo microdialysis study in mouse striatum. *Brain Res.* **1012**, 101–107.

Shalaby, A.R. (1996). Significance of biogenic amines to food safety and human health. *Food Res. Int.* **29**, 675–690.

Stute, R.K., Petridis, H., Steinhart, A., and Biernoth, G. (2002). Biogenic amines in fish and soy sauce. *Eur. Food Res. Technol.* **215**, 101–107.

Sunderland, T., Cohen, R.M., Molchan, S., Lawlor, B.A., Mellow, A.M., Newhouse, P.A., Tariot, P.N., Mueller, E.A., and Murphy, D.L. (1994). High-dose selegiline in treatment-resistant older depressive patients. *Arch. Gen. Psychiatry* **51**, 607–615.

Suzzi, G., and Gardini, F. (2003). Biogenic amines in dry fermented sausages: a review. *Int. J. Food Microbiol.* **88**, 41–54.

Tailor, S.A., Shulman, K.I., Walker, S.E., Moss, J., and Gardner, D. (1994). Hypertensive episode associated with phenelzine and tap beer—a reanalysis of the role of pressor amines in beer. *J. Clin. Psychopharmacol.* **14**, 5–14.

Tang, T., Shi, T., Qian, K., Li, P., Li, J., and Cao, Y. (2009). Determination of biogenic amines in beer with pre-column derivatization by high performance liquid chromatography. *J. Chromatogr. B Analyt. Technol. Biomed. Life Sci.* **877**, 507–512.

Trendelenburg, U., Langeloh, A., and Bonisch, H. (1987). Mechanism of action of indirectly acting sympathomimetic amines. *Blood Vessels* **24**, 261–270.

VanDenBerg, C.M., Blob, L.F., Kemper, E.M., and Azzaro, A.J. (2003). Tyramine pharmacokinetics and reduced bioavailability with food. *J. Clin. Pharmacol.* **43**, 604–609.

Waldmeier, P.C., Felner, A.E. and Maitre, L. (1981). Long-term effects of selective MAO inhibitors on MAO activity and amine metabolism. In: M.B.H. Youdim and E.S. Paykel (Eds.), *Monoamine Oxidase Inhibitors, the State of the Art.* Wiley, New York, pp. 87–102.

White, W.B., Salzman, P., and Schwid, S.R. (2008). Transtelephonic home blood pressure to assess the monoamine oxidase-B inhibitor rasagiline in Parkinson disease. *Hypertension* **52**, 587–593.

Yongmeia, L., Xiaohonga, C., Meia, J., Xina, L., Rahmanb, N., Mingshenga, D., and Yanc, G. (2009). Biogenic amines in Chinese soy sauce. *Food Control* **20**, 593–597.

Youdim, M.B.H., and Finberg, J.P.M. (1983). Monoamine oxidase inhibitor antidepressants. In: D. G. Grahame-Smith and P.J. Cowen (Eds.), *Psychopharmacology.* Excerpta Medica, Amsterdam, pp. 38–70.

Youdim, M.B., and Finberg, J.P. (1987). Monoamine oxidase B inhibition and the "cheese effect". *J. Neural Transm. Suppl.* **25**, 27–33.

Youdim, M.B.H., Finberg, J.P.M., and Tipton, K.F. (1988). Amine oxidases. In: U. Trendlenburg and N. Weiner (Eds.), *Handbook of Experimental Pharmacology.* Springer Verlag, Berlin, pp. 119–192.

Youdim, M.B., Gross, A., and Finberg, J.P. (2001). Rasagiline [N-propargyl-1R(+)-aminoindan], a selective and potent inhibitor of mitochondrial monoamine oxidase B. *Br. J. Pharmacol.* **132**, 500–506.

Youdim, M.B., Edmondson, D., and Tipton, K.F. (2006). The therapeutic potential of monoamine oxidase inhibitors. *Nat. Rev. Neurosci.* **7**, 295–309.

Zaman, M.Z., Abdulamir, A.S., Bakar, F.A., Jinap Selamat, J., and Bakar, J. (2009). A review: microbiological, physicochemical and health impact of high level of biogenic amines in fish sauce. *Am. J. Appl. Sci.* **6**, 1199–1211.

Zimmer, R., Puech, A.J., Philipp, F., and Korn, A. (1990). Interaction between orally administered tyramine and moclobemide. *Acta Psychiatr. Scand. Suppl.* **360**, 78–80.

A NOVEL ANTI-ALZHEIMER'S DISEASE DRUG, LADOSTIGIL: NEUROPROTECTIVE, MULTIMODAL BRAIN-SELECTIVE MONOAMINE OXIDASE AND CHOLINESTERASE INHIBITOR

Orly Weinreb[1], Tamar Amit[1], Orit Bar-Am[1] and Moussa B.H. Youdim[1,2]

[1]Eve Topf Centre of Excellence for Neurodegenerative Diseases and Department of Molecular Pharmacology, Technion-Rappaport Faculty of Medicine, Efron Street, P.O. Box 9697, Haifa, Israel

[2]Department of Biology, Yonsei University, Seoul, South Korea

Abstract

The novel therapeutic strategy in which drug candidates are designed to possess diverse pharmacological properties and act on multiple targets has stimulated the development of the multimodal drug, ladostigil [(*N*-propargyl-(3R) aminoindan-5yl)-ethyl methyl carbamate]. Ladostigil combines neuroprotective effects with monoamine oxidase-A and -B and cholinesterase inhibitory activities in a single molecule, presently in a Phase IIb clinical trial and intended for the treatment of Alzheimer's disease, and dementia comorbid with extrapyramidal disorders and depression. This chapter will discuss the preclinical scientific evidence for the

INTERNATIONAL REVIEW OF NEUROBIOLOGY, VOL. 100
DOI: 10.1016/B978-0-12-386467-3.00010-8

0074-7742/11 $35.00

therapeutic potential use of ladostigil in the clinic and molecular signaling pathways that are considered to be involved in the molecular activities of the drug.

ABBREVIATIONS

Aβ	β-amyloid
ACh	acetylcholine
AChE	acetylcholinesterase
AChEIs	acetylcholinesterase inhibitors
AD	Alzheimer's disease
ALS	amyotrophic lateral sclerosis
APP	amyloid precursor protein
BDNF	brain-derived neurotrophic factor
BuChE	butyrylcholinesterase
ChE	cholinesterase
ChEI	cholinesterase inhibitor
ChAT	choline acetyltransferase
CYP1A2	cytochrome P450 isoenzyme 1A2
CSF	cerebrospinal fluid
DOPAC	dihydroxyphenylacetic acid
ERK	extracellular signal-regulated kinase
GDNF	glial cell line-derived neurotrophic factor
G6PD	glucose-6-phosphate dehydrogenase
GSHPX-P	glutathione peroxidase precursor
HD	Huntington's disease
HVA	homovanillic acid
FDA	Food and Drugs Administration
LB	Lewy body
L-Dopa	L-dihydroxyphenylalanine
NMDA	*N*-methyl-D-aspartate
MAO	monoamine oxidase
MAPK	mitogen-activated protein kinase
MPTP	*N*-methyl-4-phenyl-1,2,3,6-tetrahydropyridine
NQ01	NADPH quinone 1 oxidoreductase
NE	noradrenalin
NO	nitric oxide donor
6-OHDA	6-hydroxydopamine
OS	oxidative stress
PD	Parkinson's disease
Prx-1	peroxiredoxin 1
PKC	protein kinase C

ROS	reactive oxygen species
SIN-1	3-morpholinosydnonimine
STZ	streptozotocin
Syn	synaptotagmin
TCP	tranylcypromine
TDPX1	thioredoxin peroxidase
TK	tyrosine kinase

I. Introduction

Alzheimer's disease (AD) is the most prevalent neurodegenerative disease, affecting millions of people worldwide. It is estimated that in the elderly population, 5% of adults over 65 years are affected by AD (Bullock, 2004). AD can be classified into sporadic AD, which is the most common form, and the familial form which represents a small fraction of all AD cases, where the age of the disease onset is younger than 65 years and patients show autosomal dominant transmission within affected families (Eckert *et al.*, 2003). The autosomal dominant familial forms involve specific mutations in one of three genetic loci, typically associated with the early onset of AD symptoms, amyloid precursor protein (APP), presenilin 1, and presenilin 2. Genetic polymorphism of the apolipoprotein epsilon locus (apoE4 allele) increases the risk of developing AD and is associated with an earlier age of disease onset in sporadic AD (Maccioni *et al.*, 2001). The predominant clinical manifestation of AD is the progressive memory deterioration and other changes in brain function, including disordered behavior and impairment in language, comprehension, and visual–spatial skills (Tsolaki *et al.*, 2001). The neuropathology of AD is characterized by several features, including an extracellular deposition of amyloid β peptide (Aβ)-containing plaques in the cerebral cortical regions, accompanied by the presence of intracellular neurofibrillary tangles composing of a hyperphosphorylated form of the tau (τ) protein and a progressive loss of basal forebrain cholinergic neurons, leading to reductions in cholinergic markers, such as acetylcholine (ACh) levels, choline acetyltransferase (ChAT), and muscarinic and nicotinic ACh receptor binding (Maccioni *et al.*, 2001; Selkoe and Schenk, 2003; Schliebs, 2005). Significant reduction also occurs in serotoninergic and noradrenergic transmission, which might explain the relatively high incidence of depression found in AD patients (Palmer *et al.*, 1988; Newman, 1999).

Initial therapeutic approaches for AD focused on increasing cholinergic transmission in the brain, based on the "cholinergic hypothesis" of memory dysfunction. Among different approaches employed to increase synaptic levels of ACh, blocking the breakdown of ACh by inhibiting acetylcholinesterase (AChE) has proven most successful to date. Inhibiting the enzyme butyrylcholinesterase (BuChE), which is a minor constituent in normal brains, but increased in association with plaques and tangles in the brain of AD patients, may improve cholinergic transmission (Hansen *et al.*, 2006). Moreover, a number of evidence implicated the excitatory neurotransmitter, glutamate as playing an additional role in the pathophysiology of AD (Greenamyre *et al.*, 1988). To date, there are more than 70 compounds in various stages of clinical studies for the treatment of AD and many more candidate molecules are at the preclinical stage of development (www.alzforum.org). The only Food and Drugs Administration (FDA)-approved drugs for the treatment of AD patients are the AChE inhibitors (AChEIs), such as donepezil hydrochloride (donepezil), galantamine hydrochloride (galantamine), rivastigmine tartrate (rivastigmine), and tacrine hydrochloride (tacrine), and the noncompetitive *N*-methyl-D-aspartate (NMDA)-receptor channel antagonist, memantine hydrochloride (memantine) (Kornhuber *et al.*, 1991). AChEIs were found to exert their effects by preventing enzymatic degradation of the neurotransmitter ACh and prolonging the time in which ACh molecules remain in the synaptic gap, resulting in enhanced cholinergic transmission (Lleo *et al.*, 2006).

Indeed, various clinical trials have demonstrated the safety and efficacy of AChEIs in the treatment of AD. Yet, their benefits in AD as symptomatic drugs are thought to be more complex than simply replacement of lost ACh (Francis *et al.*, 1999; Giacobini, 2002; Lopez *et al.*, 2002; Racchi *et al.*, 2004). As previously reviewed (Francis *et al.*, 2005), there is growing preclinical evidence that AChEIs have minor symptomatic activity and may block some of the fundamental neurodegenerative processes involved in AD. Thus, AChEIs, such as tacrine, donepezil, galantamine, and rivastigmine, were reported to preserve neuronal viability and structure in various cell culture models of neurodegenerative diseases (Arias *et al.*, 2004; Takada-Takatori *et al.*, 2006; Akasofu *et al.*, 2008; Bailey and Lahiri, 2010). In addition, there is evidence that several cholinesterase inhibitors (ChEIs) may also affect various neuropathological markers of AD and modulate the cleavage of the nonamyloidogenic APP (Racchi *et al.*, 2004). Therefore, it is suggested that AChEIs, possessing properties of neuroprotection and/or regulation of APP processing, may be more beneficial than those that only inhibit AChE to treat AD and, in particular, to prevent the pathogenesis of the disease. However, the available anti-AD medications for the treatment of AD dementia cases, which include both AChEIs and NMDA-receptor channel antagonists, are efficient to produce modest symptomatic improvements in some of the patients (Clark and Karlawish, 2003; Klafki *et al.*, 2006).

Current novel therapeutic approach suggests that drugs acting at a single target may be insufficient for the treatment of multifactorial neurodegenerative diseases such as AD, Parkinson's disease (PD), Huntington's disease (HD), and amyotrophic lateral sclerosis (ALS), all characterized by the coexistence of multiple etiopathologies. These include among others, oxidative stress (OS) and reactive oxygen species (ROS) formation, protein misfolding and aggregation, mitochondrial dysfunction, inflammation, metal dyshomeostasis, and accumulation at the sites of neurodegeneration (Smith *et al.*, 2000; Rogers and Lahiri, 2004; Joseph *et al.*, 2005). Thus, it seems likely reasonable to conclude that AD therapy will require multiple drug therapy to address the varied pathological aspects of the disease. Indeed, this new approach, using combination therapy, such as donepezil and memantine, has been reported to have significant beneficial effects on cognitive function, activities of daily living, and behavior (Tariot *et al.*, 2004). However, drug therapy of neurodegenerative diseases with pluripotential compounds embracing diverse biological properties with single bioavailability and pharmacokinetic metabolism was suggested to produce higher significant advantage over a cocktail of drugs (Van der Schyf *et al.*, 2006).

Based on this reasoning, the multimodal drug, ladostigil (TV3326; [(*N*-propargyl-(3R)-aminoindan-5yl)-ethyl methyl carbamate]) (Fig. 1A), was designed to possess neuroprotective and neurorestorative activities, thus targeting various pathogenic mechanisms of AD. The underlying principle in the design of ladostigil was to amalgamate the neuroprotective propargyl moiety of the novel anti-Parkinsonian drug and irreversible selective monoamine oxidase (MAO)-B inhibitor, rasagiline (Azilect; *N*-propargyl-1-(*R*)-aminoindan; Teva Pharmaceutical Co, IL, USA) to the carbamate cholinesterase (ChE) inhibitory moiety of the anti-AD drug, rivastigmine (Exelon; (*S*)-3-[1-(dimethylamino) ethyl]phenyl ethylmethylcarbamate; Novartis Pharmaceuticals, NJ) (Fig. 1A). The resulting molecule, ladostigil, is a novel dual AChE–BuChE and brain-selective MAO-A and -B inhibitor intended for the treatment of AD and dementia comorbid with extrapyramidal disorders (Parkinsonism) and depression (presently in a Phase IIb clinical study). This chapter will discuss the preclinical scientific evidence for the therapeutic potential use of ladostigil in the clinic and molecular signaling pathways that are considered to be involved in the molecular activities of the drug.

II. The Underlying Strategy in the Design of Ladostigil

The current notion that AChEIs improve merely AD symptoms and are palliative medication is accompanied with a slight assumption that they might have a disease-modifying effects by reducing the brain, cerebrospinal fluid (CSF), or plasma Aβ concentrations (Sabbagh *et al.*, 2006). However, none of the approved drugs for

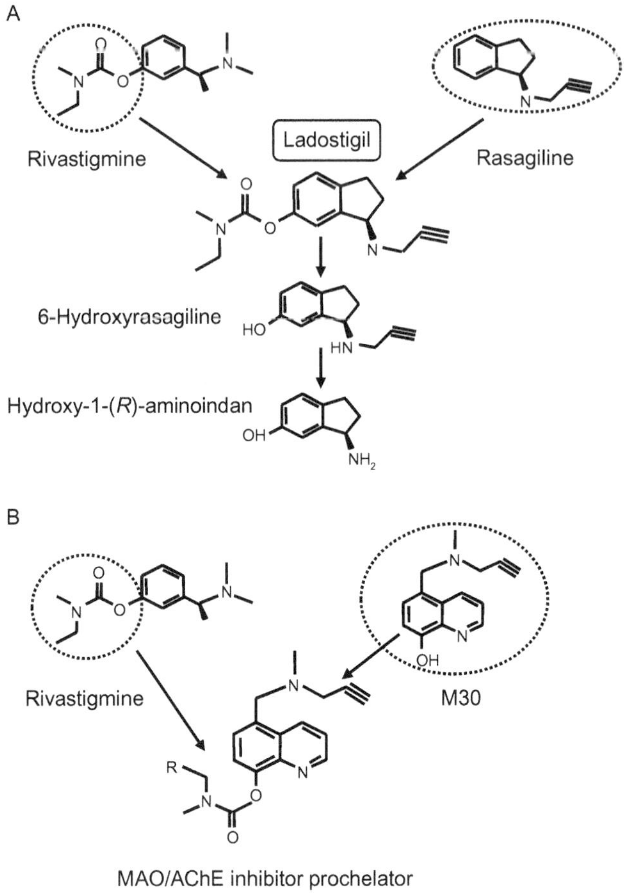

FIG. 1. Strategy for the design of multimodal compounds targeting both ChE and MAO inhibition. (A) Ladostigil amalgamates the propargyl moiety of the anti-Parkinsonian drug and irreversible selective MAO-B inhibitor, rasagiline (Azilect; *N*-propargyl-1-(*R*)-aminoindan; Teva Pharmaceutical Co., IL, USA) with the carbamate ChE inhibitory moiety of the anti-AD drug, rivastigmine (Exelon; (*S*)-3-[1-(dimethylamino) ethyl] phenyl ethylmethylcarbamate; Novartis Pharmaceuticals, NJ). The hydrolysis of the carbamate moiety in ladostigil by pseudo-inhibition of ChE yields the 6-hydroxyrasagiline (6-hydroxy-*N*-propargyl-1(*R*)-aminoindan) and further gives rise to the major metabolite of ladostigil, hydroxy-1-(*R*)-aminoindan. (B) The MAO/AChE inhibitor prochelator (5-]*N*-methyl-*N*-propargyl-aminomethyl[quinolin-8-yl dimethyl carbamate) amalgamates the propargyl moiety of the MAO inhibitor–multifunctional-neuroprotective chelator M30 (5-[*N*-methyl-*N*-propargylaminomethyl]-8-hydroxyquinoline) with the carbamate ChE inhibitory moiety of the anti-AD drug rivastigmine.

the treatment of AD was demonstrated to significantly slow Aβ deposition and hence to retard the disease progression in patients (Sabbagh *et al.*, 2006).

In order to fill this gap, a series of chimeric compounds were synthesized by amalgamating the carbamate ChE inhibitory moiety of rivastigmine into the 6-position of the pharmacophore of the irreversible selective MAO-B inhibitor, rasagiline (Weinstock *et al.*, 2000a,c, 2001; Sterling *et al.*, 2002) (Fig. 1A). The underlying concepts of this design were to preserve the beneficial properties of the parent drugs, rasagiline and rivastigmine, including neuroprotective and neurorestorative activities, AChE/MAO inhibition, induction of cholinergic transmission, and target the essential pathogenic mechanisms in AD and Lewy Body (LB) disease.

A. Rasagiline

Rasagiline (Fig. 1A) is a second generation, selective, and irreversible inhibitor of MAO-B, indicated as monotherapy or adjunct to L-dihydroxyphenylalanine (L-Dopa) for patients with early and late PD (Parkinson Study Group, 2002, 2004, 2005; Olanow *et al.*, 2009). Recent Phase III clinical neuroprotective study, ADAGIO (attenuation of disease progression with Azilect given once daily), has reported that PD patients treated with 1 mg per day of rasagiline showed less functional decline than those whose treatment was delayed, consistent with a possible disease-modifying effect (Olanow *et al.*, 2008, 2009).

Rasagiline is primarily metabolized by hepatic cytochrome P450 isoenzyme 1A2 (CYP1A2)-mediated *N*-dealkylation followed by renal excretion of the parent conjugated compound and its metabolites (Chen and Swope, 2005).

In preclinical studies, rasagiline was shown to possess broad neuroprotective/neurorestorative activities against a variety of neurotoxins *in vivo* and in neuronal cell culture models of neurodegeneration, (as reviewed recently Weinreb *et al.*, 2010).

Structure–activity studies provide evidence that the *N*-propargyl moiety of rasagiline promotes neuronal survival via similar neuroprotective/neurorescue pathways, thus enlightening the importance of this moiety for the novel activities of rasagiline (Bar-Am *et al.*, 2004, 2005). The molecular aspects of the mechanism of action of rasagiline are thoroughly discussed by Weinreb and coauthors in the current book.

B. Rivastigmine

Rivastigmine (Fig. 1A), an established ChEI that inhibits both AChE and BuChE, is indicated for the treatment of mild to moderate dementia associated with AD type and PD and reported to improve cognitive function and activities of daily living of the patients (Weinstock *et al.*, 1994; Machado, 2009). Rivastigmine has also proven effective in decreasing psychiatric symptoms and cognitive deficits in a large double blind, placebo-controlled trial in patients with diffuse LD disease (Farlow, 2003). It was suggested that the continued beneficial effect, observed in rivastigmine-treated patients after drug withdrawal, indicated an effect on disease progression (Leber, 1996; Farlow *et al.*, 2003).

In a preclinical rat model, rivastigmine inhibited ChE and antagonized scopolamine-induced deficits in working and reference memory (Bejar *et al.*, 1999). Previous *in vitro* studies demonstrated that rivastigmine possessed a concentration-dependent neuroprotective effect against cell death induced by various insults (Arias *et al.*, 2005). This effect was obtained at a concentration that differs from its IC_{50} of AChE inhibition, suggesting that the neuroprotective effect of rivastigmine might be through a mechanism, which was unrelated to ChE inhibition (Arias *et al.*, 2005).

III. MAO Inhibition and Antidepressant Activity of Ladostigil

Previous studies have shown that brain MAO-A and -B are inhibited only after chronic, but not acute treatment with ladostigil (Sterling *et al.*, 2000; Weinstock *et al.*, 2000c, 2002b; Sagi *et al.*, 2003; Hubalek *et al.*, 2004) (Table I). This is attributed to the formation of active MAO inhibitory metabolites as a consequence of hydrolysis of the carbamate moiety in ladostigil by ChE, which yields the 6-hydroxyrasagiline derivative (Sterling *et al.*, 2002). 6-

Table I
THE EFFECT OF ACUTE AND CHRONIC TREATMENT OF LADOSTIGIL ON BRAIN MAO AND CHE ACTIVITIES IN RATS.

Enzyme	Enzyme inhibition (%)	
	Acute treatment	Chronic treatment
MAO-A	8.0 ± 4.0	66.0 ± 1.9*
MAO-B	12.0 ± 5.0	71.2 ± 2.1*
ChE	38.2 ± 5.7	36.5 ± 5.8

For measurement of brain MAO-A and -B activities, male Sprague–Dawley rats were orally given water (control) or ladostigil (26 mg/kg) acutely or once daily for 14 days and sacrificed immediately. For measurement of brain ChE activity, rats were orally given water (control) or ladostigil (26 mg/kg) acutely and sacrificed 2 h later, and for chronic paradigm, the rats were administered once daily for 2 weeks and then sacrificed 2 h after the last administration.

Results represent the mean (% enzyme inhibition of control activity) ±SEM ($n = 5$–8).

* $P < 0.01$ significantly different from acute administration. Reproduced from Weinstock *et al.* (2002a).

Hydroxyrasagiline was found to exhibit high affinity towards the activity of MAO-A (K_i = 1.9 μM) and -B (K_i = 17 μM). It is mostly likely that adequate concentration of 6-hydroxyrasagiline and other metabolites of ladostigil significantly inhibit MAO-A and -B and increase levels of striatal dopamine, 5-HT and NE (noradrenalin), which account for the respective anti-Parkinsonian and antidepressant activities of ladostigil, demonstrated in preclinical models (Sagi *et al.*, 2003).

Thus, once daily administration of ladostigil (26 mg/kg) to rats for 2 weeks inhibited brain MAO-A and -B by ~70%, with very little or no effect in the intestine (Weinstock *et al.*, 2000c, 2002b). However, the *S*-isoform of ladostigil, TV3279, has only comparable ChE inhibitory activity but lacks MAO inhibitory activity (Weinstock *et al.*, 2000c). In rats, ladostigil was demonstrated to act like other antidepressants (e.g., amitriptyline and moclobemide, a reversible MAO-A selective inhibitor) in the forced swim test, indicating a potential effect for antidepressant-like activity (Weinstock *et al.*, 2002a, 2003). This antidepressant-like activity of ladostigil was accompanied by increased levels of brain serotonin (Weinstock *et al.*, 2002a).

In rabbits, ladostigil (26 mg/kg) oral administration for 2 weeks inhibited brain MAO by more than 90% but did not inhibit intestinal MAO-A, which comprised ~80% of total intestinal MAO (Weinstock *et al.*, 2002b). This brain selectivity results in minimal potentiation of the pressor to oral tyramine (Weinstock *et al.*, 2002b, 2003). In mice, chronic treatment with ladostigil (52 mg/kg) dependently inhibited MAO-A and -B in the cerebral cortex, brainstem, striatum, and hippocampus of mice (Sagi *et al.*, 2003, 2005). Inhibition of MAO-A and -B reached values of 65% and 90%, respectively (Sagi *et al.*, 2003, 2005). MAO inhibition by ladostigil, like other propargylamines, such as clorgyline, selegiline, and rasagiline, was shown to be irreversible and slow to recover (Sagi *et al.*, 2003).

IV. ChE Inhibitory Activity of Ladostigil

In rats, acute oral administration of ladostigil at doses ranging from 12 to 35 mg/kg was shown to inhibit brain ChE by 25–40% and antagonize scopolamine-induced spatial memory impairments (Weinstock *et al.*, 2003; Weinstock *et al.*, 2000c). The authors suggested that high levels of synaptic ACh induced by ladostigil might be one of the mechanisms involved in the cognitive improvement, by competing with scopolamine for the muscarinic receptors subserving memory (Weinstock *et al.*, 2000c, 2003). However, symptoms arising from access cholinergic stimulation in the periphery (e.g., salivation, diarrhea, and muscle weakness) or the brain (e.g., termer, respiratory depression, or hypothermia) were not observed, until doses in excess of 139 mg/kg of ladostigil were administered orally, which inhibited ChE by 50–60% (Weinstock *et al.*, 2000a). This wide therapeutic ratio is probably due to the relatively small rate of increase in ChE inhibition with increasing drug doses that was shown to occur with ladostigil over the range of 17–69 mg/kg (Weinstock *et al.*, 2000a,c). In addition, the rate of the increase in ChE inhibition with oral administration of ladostigil (doses range 17–69 mg/kg; 10–15% increase in ChE inhibition) is markedly lower, compared to subcutaneous injection of ladostigil (doses range of 8.6–17 mg/kg; 40% increase in ChE inhibition) (Weinstock *et al.*, 2000a, 2001). Ladostigil was found to inhibit both AChE and BuChE *in vitro* (Weinstock *et al.*, 2000a, 2003). The inhibitory effect was ~100 times more potent against AChE than BuChE (Weinstock *et al.*, 2000a, 2003).

Chronic oral treatment of rats with ladostigil (52 mg/kg, for 21 days) was shown to inhibit striatal ChE activity by ~50%, accompanied by inhibition of the hyperactivity behavioral syndrome response induced by L-3,4-dihydroxyphenylalanine (L-dopa) (Sagi *et al.*, 2005). Similar to ladostigil, chronic oral administration of the *S*-isomer of ladostigil, TV3279 (which does not block MAO) (26 mg/kg, for 21 days), inhibited striatal ACh activity and attenuated the hyperactivity syndrome response (Sagi *et al.*, 2005). The inhibitory effects of both ladostigil and TV3279 on these hyperactivity syndromes were possibly mediated by the enhanced functional ACh levels and supported the well-established inhibitory effect of ACh on dopaminergic-mediated stereotypy (Davis and Rosenberg, 1981; Rastogi *et al.*, 1982). Indeed, previous *in vivo* study has shown that 85% and 52% inhibition of ChE by the ChE inhibitors, metrifonate and rivastigmine, respectively, was accompanied by a subsequent significant increase in ACh release in rat cortex (Scali *et al.*, 2002). Accordingly, it was shown that brain ACh levels were induced following ladostigil treatment in rats (G. Pepeu, 2004, personal communication to Teva Pharmaceutical Co., Israel).

Although the role of ACh in the activity of striatal projection neurons is still unclear, previous findings suggested that cholinergic transmission, via muscarinic receptors, ensured the correct processing mechanisms of cortical inputs forward from the cortex to projection cells (Izzo and Bolam, 1988; Calabresi *et al.*, 2000). Thus, the ability of scopolamine to reverse the inhibitory effects of ladostigil and TV3279 plus the nonselective inhibitor, tranylcypromine (TCP) on L-Dopa-induced hyperactivity (Sagi *et al.*, 2005) is in line with the hierarchical structure of the motor loops in the basal ganglia (Calabresi *et al.*, 2000).

V. Neuroprotective Activities of Ladostigil in Preclinical Models of Neurodegeneration

In preclinical studies, ladostigil was shown to have a broad neuroprotective activity against a variety of neurotoxins in *in vivo* and neuronal cell culture models of neurodegeneration (Table II).

Table II
NEUROPROTECTIVE ACTIVITIES OF LADOSTIGIL IN ANIMAL AND NEURONAL CELL MODELS OF NEURODEGENERATION.

Neuroprotective effects	Experimental model	References
Protection against MPTP-induced dopaminergic neurotoxocity	Mice	Sagi *et al.* (2003)
Upregulation of antioxidant activity and mRNA expression of antioxidant enzymes	Aged rats	Weinreb *et al.* (2007a)
Prevention of age-related glial activation and spatial memory deficits	Aged rats	Weinstock *et al.* (2009)
Improvement of cognitive performance	Aged monkeys	Buccafusco *et al.* (2003)
Inhibition of scopolamine-induced impairments in spatial memory	Rats	Weinstock *et al.* (2000c)
Prevention of STZ-induced gliosis, oxidative–nitrative stress, and memory deficits	Rats	Weinstock *et al.* (2001)
Inhibition of behavioral hyperactivity syndromes	Rats	Sagi *et al.* (2005)
Prevention of sodium azide-induced ChAT immunoreactivity reduction	Rats	Luques *et al.* (2007)
Prevention of closed head injury-induced cerebral edema and impaired motor and memory function	Rats	Weinstock *et al.* (2000b)
Prevention of neurotoxicity	Long-term cultured human neuroblastoma SK-N-SH, serum-deprived rat pheochromocytoma PC-12, and human SH-SY5Y neuroblastoma cells	Yogev-Falach *et al.* (2006)
Prevention of H_2O_2,-SIN-1-and glucose–oxygen deprivation-induced OS	PC-12, SH-SY5Y cells	Wadia *et al.* (1998), Weinstock *et al.* (2000c), Maruyama *et al.* (2003), Weinreb *et al.* (2008), Bar-Am *et al.* (2009)

In vivo studies demonstrated that chronic, once daily administration of ladostigil (26 mg/kg) for 14 days, prevented the depletion of striatal dopamine and the reduction in the metabolites dihydroxyphenylacetic acid (DOPAC) and homovanillic acid (HVA) by *N*-methyl-4-phenyl-1,2,3,6-tetrahydropyridine (MPTP) (Sagi *et al.*, 2003). Ladostigil was demonstrated to improve cognitive deficits in aged monkeys (Buccafusco *et al.*, 2003), antagonize spatial memory deficit induced by scopolamine in rats (Weinstock *et al.*, 2000c), and prevent memory impairment induced by intracerebroventricular (ICV) injection of streptozotocin (STZ) in rats (Weinstock *et al.*, 2001; Shoham *et al.*, 2007). In addition, ladostigil (17 mg/kg) restored spatial memory deficits induced in rats by the cytochrome oxidase inhibitor, sodium azide (Luques *et al.*, 2007). Additionally, ladostigil significantly reduced hippocampal cell damage caused by global ischemia in gerbils and cerebral edema induced by closed head injury in mice and induced the recovery of their motor and memory (Weinstock *et al.*, 2000a,c).

Genomic study demonstrated that ladostigil (1 mg/kg) significantly downregulated the hippocampi levels of the familial AD-linked presenilin-1 (PS1) gene (Weinreb *et al.*, 2007b). This effect can be of value in reducing Aβ formation, as PS1 is major component of the γ-secretase complex, which facilitates the generation of Aβ peptide via intramembranous proteolysis of APP (Kern *et al.*, 2006; Vetrivel and Thinakaran, 2006). Further, ladostigil (1 mg/kg) downregulated (~50% of control) casein kinase 1 delta mRNA, associated with pathological hallmarks in several neurodegenerative diseases (Weinreb *et al.*, 2007b).

In vitro studies demonstrated also that ladostigil (1–10 μM) exerts neuroprotective activities, including a prevention of the fall of the mitochondrial membrane potential (ψ) and the initiation of apoptotic cascade (Maruyama *et al.*, 2003), inhibition of ROS production, and cytotoxicity in rat pheochromocytoma PC-12 and human SH-SY5Y neuroblastoma cells, in response to OS induced by hydrogen peroxide (H_2O_2), the nitric oxide donor (NO), 3-morpholinosydnonimine (SIN-1), and glucose–oxygen deprivation (Weinstock *et al.*, 2000c; Maruyama *et al.*, 2003; Weinreb *et al.*, 2008; Bar-Am *et al.*, 2009). In the extremely neurotoxic model of long-term cultured human neuroblastoma SK-N-SH, ladostigil (1–10 μM) was found to possess neuroprotective activity, including inhibition of caspase-3 activation, induction of Bcl-2, and reduction of Bad and Bax gene and protein expression (Tables II and III; Yogev-Falach *et al.*, 2006). Previous studies demonstrated that the propargyl moiety of rasagiline also promoted neuronal survival, mediated by protein kinase C (PKC)- and mitogen-activated protein kinase (MAPK)/extracellular signal-regulated kinase (ERK)-dependent activation associated with Bcl-2 family members (Weinreb *et al.*, 2004), and mitochondrial membrane stabilization (Maruyama *et al.*, 2003; Yogev-Falach *et al.*, 2006), thus suggesting to obtain a crucial role in the neuroprotective activity of ladostigil. Indeed, ladostigil possessed various stimulatory effects on the cell signaling pathways, PKC and MAPK cascades (Yogev-Falach *et al.*, 2002), promoting the phosphorylation of p44 and p42 MAPK, which was abolished by specific inhibitors of MAPK activation (Yogev-Falach *et al.*, 2002) (Table III).

The neuroprotective effect of ladostigil was shown to be associated with upregulation of mRNA expression levels of the neurotrophic factors, brain-derived neurotrophic factor (BDNF), and glial cell line-derived neurotrophic

Table III
THE MOLECULAR MECHANISMS UNDERLYING THE NEUROPROTECTIVE EFFECTS OF LADOSTIGIL, AS DEMONSTRATED IN VARIOUS NEURONAL MODELS OF NEURODEGENERATION.

Molecular effects related to neuroprotective outcome	Experimental systems	References
Upregulation of antioxidant activity and mRNA expression of antioxidant enzymes	SH-SY5Y and SK-N-SH cells;	Bar-Am *et al.* (2009), Maruyama *et al.* (2003), Weinreb *et al.* (2008, 2007b)
Regulation of APP processing	PC-12 and SH-SY5Y cells; SK-N-SH cells	Yogev-Falach *et al.* (2002, 2006)
Activation of PKC	PC-12 and SH-SY5Y cells; SK-N-SH cells	Yogev-Falach *et al.* (2002, 2006)
Activation of ERK1/2	PC-12 cells	Yogev-Falach *et al.* (2002)
Prevention of caspase-3 activation	SK-N-SH cells; SH-SY5Y cells	Yogev-Falach *et al.* (2006), Maruyama *et al.* (2000b)
Regulation of Bcl-2 protein family	SK-N-SH cells	Yogev-Falach *et al.* (2006)
Prevention of the fall in mitochondrial membrane potential (ψ)	SH-SY5Y cells	Maruyama *et al.* (2003)
Induction of neurotrophic factors, GDNF and BDNF	SK-N-SH cells	Weinreb *et al.* (2007b)

factor (GDNF), in neuronal cell cultures (Weinreb *et al.*, 2007a). Upon ligand–receptor binding, BDNF and GDNF stimulate intracellular signaling pathways involved in differentiation and survival, including phospholipase C-γ, phosphatidylinositol 3-kinase (PI3K), and MAPK (Segal and Greenberg, 1996; Murakami *et al.*, 1999; Siegel and Chauhan, 2000). Thus, elevation of neurotrophic factors by ladostigil, which are reported to initiate several cell signaling cascades, indicates an involvement of these neurotrophic factors in the neuroprotective mechanism of action of ladostigil (Table III). Ladostigil was also demonstrated to upregulate hipopcampal brain-specific isoform of calcium sensors synaptotagmin (Syn) family, Syn IV in old rats (Weinreb *et al.*, 2007b).

Similar to ladostigil, its *S*-enantiomer, TV3279, also possesses a neuroprotective activity, demonstrated in various animal and cell culture models of neurodegeneration (e.g., antagonizing the spatial memory impairment induced by scopolamine in rats (Weinstock *et al.*, 2000c); reducing edema, motor function, and memory deficits after closed head injury in mice (Weinstock *et al.*, 2000a,c); and protecting against ischemia-induced cytotoxicity in PC12 cells; Weinstock *et al.*, 2000a), indicating that the neuroprotective activity of ladostigil is independent of MAO inhibition. It is likely that several different mechanisms are in operation, and accumulated data provide evidence that the neuroprotective effects of ladostigil might be associated also with some intrinsic pharmacological action of its propargyl moiety, similar to other propargylamine-containing drugs (Maruyama *et al.*, 2000a,b; Akao *et al.*, 2002; Bar-Am *et al.*, 2005).

VI. Mechanism of Action of Ladostigil

A. Regulation of APP Processing

A significant body of evidence suggests that the accumulation of the Aβ in AD may play a pivotal role, thus a bulk of studies focused on possible drug intervention along the amyloid pathways in AD (Pakaski and Kasa, 2003). Hence, an essential neuroprotective effect, shared by ladostigil, rasagiline, and other propargylamine-derivative compounds, is related to their ability to regulate the processing of APP by the nonamyloidogenic α-secretase pathway (Vetrivel and Thinakaran, 2006; Bar-Am *et al.*, 2010). APP can be processed in at least two alternative, mutually exclusive posttranslational pathways: in the amyloidogenic pathway, APP is cleaved sequentially by β and γ secretases, at the N- and C-terminal of the Aβ domain, respectively, to release Aβ. While, in the nonamyloidogenic pathway, APP is cleaved by α-secretase, within the Aβ sequence, and generates a secreted form of soluble APP (sAPPα), thus precluding the formation of the amyloidogenic Aβ (Vetrivel and Thinakaran, 2006). The precise mechanisms leading to neurodegeneration in AD are not completely clear; however, most studies have focused on the role of Aβ–APP and its products in AD pathogenesis (Zhang and Xu, 2007). Secreted sAPPα has been implicated to possess neurotrophic and neuroprotective properties, and Aβ peptides exert contrasting effects on neuronal survival (Chen and Tang, 2006; Chow *et al.*, 2010). Available evidence has demonstrated that α-secretase activity could be regulated via PKC, calcium ion (Ca^{2+}), tyrosine kinase (TK), MAPK, and hormonal signaling (Mills and Reiner, 1999a,b; Zhang and Xu, 2007). Additional studies have shown that the proteolytic processing of APP is regulated by MAPK pathway involving sequential activation of MAPK and ERK (Hung *et al.*, 1993).

In this context, previous findings demonstrated that ladostigil (*in vivo*, 52 mg/kg/day, i.p. 14 days; *in vitro*, 1–10 μM) markedly suppressed holo-APP protein levels and elevated sAPPα release (Yogev-Falach *et al.*, 2002, 2006; Bar-Am *et al.*, 2004) (Table III), indicating that this drug can be of clinical value toward accelerating the nonamyloidogenic APP processing, thereby reducing the possibility of generation of Aβ. The stimulation of sAPPα release induced by ladostigil was blocked by a hydroxamic acid-based metalloprotease inhibitor, indicating an α-secretase metalloprotease mediation in this effect (Yogev-Falach *et al.*, 2002).

A previous study demonstrated that selective BuChE inhibitors reduced APP and Aβ levels *in vitro* and *in vivo* (Greig *et al.*, 2001). Thus, the mechanism underlying these effects may involve both cholinergic and noncholinergic actions regulating APP synthesis and processing. The observation that ladostigil did not alter APP mRNA levels may suggest that the decrease in APP protein and Aβ levels can be attributed to suppression of APP translation (Yogev-Falach *et al.*, 2006). In addition, using several signaling inhibitors suggested that PKC-, MAPK-, and TK-dependent pathways might be involved in the enhancement of sAPPα release by ladostigil (Yogev-Falach *et al.*, 2002). In accordance to this inhibitory study, it was shown that ladostigil stimulated ERK1- and ERK2-MAPK phosphorylation (Yogev-Falach *et al.*, 2002). *In vivo* studies revealed that ladostigil reduced the levels of holo-APP and upregulated the levels of p-PKC, PKCα, and PKCε in the mice hippocampus (Bar-Am *et al.*, 2004). Moreover, similar to ladostigil, TV3279 regulated APP processing effects, supporting the notion that the mode of action of ladostigil is independent of MAO inhibition (Yogev-Falach *et al.*, 2002, 2003, 2006).

B. Neuroprotective Effects Against OS

Accumulative studies have suggested that OS is an early event AD, occurring prior to cytopathology, and therefore might possess a key role in the disease pathology (Nunomura *et al.*, 2001; Zhu *et al.*, 2004; Mamelak, 2007). H_2O_2, which is one of the main ROS, is produced during the redox process, and is considered as a messenger in intracellular signaling cascades, including cellular metabolism and proliferation (Rhee *et al.*, 1994; Stone and Yang, 2006). It is well acknowledged that H_2O_2 could cause lipid peroxidation and DNA damage, thus inducing apoptosis in many different cell types (Halliwell, 1992, 2001; Zhang *et al.*, 2007). Several lines of evidence suggest that AChE and BuChE activation might be involved in apoptosis associated with H_2O_2 (Xiao *et al.*, 2000; Schallreuter and Elwary, 2007). These links between cholinergic signal and OS provide an additional therapeutic target for AChEIs in AD. Indeed, the AChEIs, tacrine (Xiao *et al.*, 2000) and huperzine A, were demonstrated to significantly protect PC12 cells against H_2O_2 insult (Xiao *et al.*, 1999). Similarly, ladostigil (1 and 5 μM) was found to comprise a significant neuroprotective effect against H_2O_2-induced damage in SH-SY5Y cells (Bar-Am *et al.*, 2009) (Tables II and III).

In addition, the enzyme MAO, which is located in the outer mitochondrial membrane, can produce a significant amount of ROS, in particular, H_2O_2 (Youdim *et al.*, 2006). Consistent with this, rasagiline was found to increase antioxidant enzyme activities in brain dopaminergic system in rats (Carrillo *et al.*, 2000). Other propargylamine derivatives were also shown to possess antioxidant properties assessed by their ability to scavenge peroxynitrite (Dragoni *et al.*, 2006). Ladostigil (1 μM) was demonstrated to exert antioxidant activity through both direct scavenging effect on free radicals overproduced in H_2O_2-treated neuronal cells and an indirect effect by stimulating the expression and activity of cellular antioxidant enzymes, catalase, and glutathione reductase, suggesting their involvement in the cytoprotective effect of ladostigil (Bar-Am *et al.*, 2009). Additionally, the mRNA expression levels of the antioxidant enzymes, catalase, peroxiredoxin 1 (Prx-1), and NADPH quinone 1 oxidoreductase (NQ01) were elevated by ladostigil in H_2O_2-treated SH-SY5Y cells (Weinreb *et al.*, 2008). In support, results obtained from the high-density cytotoxic model of SK-N-SH cells, widely used in neuronal injury studies as a potential source of ROS (Ba *et al.*, 2003), also revealed that ladostigil induced mRNA levels of the same antioxidant enzymes (Bar-Am *et al.*, 2009). Indeed, overexpression of NQ01 was demonstrated to induce neuroprotection against various toxins *in vitro*, suggesting to play an important role in the central nervous system (Stringer *et al.*, 2004; Zafar *et al.*, 2006). Additionally, NQ01 activity was closely colocalized with AD pathology, indicating that increasing NQ01 activity may provide a neuroprotective avenue for the treatment of neurodegenerative diseases (SantaCruz *et al.*, 2004). These findings are in accordance with previous studies showing that ladostigil possesses neuroprotective effects against various *in vitro* insults, including OS damage induced by SIN-1 (Maruyama *et al.*, 2003), serum withdrawal (Yogev-Falach *et al.*, 2003), or glucose–oxygen deprivation (Weinstock *et al.*, 2000b). In the model of SH-SY5Y cells exposed to SIN-1, ladostigil (0.1–10 μM) prevented apoptosis by reducing the decline in the mitochondria membrane potential (ψ) (Maruyama *et al.*, 2003) (Table III).

An additional inspection of the neuroprotective activity of ladostigil against OS is related to its antioxidative effects in aged rats; previous proteomics and genomics analyses identified significant changes in various genes and proteins related to the iron-mediated OS pathway, such as reduction in several antioxidant enzymes and induction of ferritin in the hippocampus of old rats (Weinreb *et al.*, 2007b). Several studies demonstrated age-associated downregulation of gene expression of glutathione-related enzymes and suggested that this decline consequently resulted in a decrease in the activity of these enzymes in the brain (Liu and Choi, 2000; Liu, 2002; Zhu *et al.*, 2006). Indeed, ladostigil was shown to reverse the effect of aging on mRNA expression levels of various genes associated with metabolism and oxidation processes in the hippocampus of old rats, including glutathione peroxidase precursor (GSHPX-P), glutathione *S*-transferase, glutathione synthetase, thioredoxin peroxidase (TDPX1), and glucose-6-phosphate dehydrogenase (G6PD) (Weinreb *et al.*, 2008).

In vivo chronic administration of ladostigil (1 mg/kg) for a week before and after ICV injection of STZ significantly reduced microglia activation and loss of astrocytes, and prevented the increase in nitrotyrosine, a marker of nitrative-OS (Shoham *et al.*, 2007). It was suggested that these actions result from a combination of actions of ladostigil on neuronal and glial cells. The ability to inhibit ChE partially contributes to the effect on episodic memory, as ladostigil inhibits ChE only in the cortex of STZ rat model of AD, and no ChE inhibition occurs at the site of recognition, which depends on hippocampal cholinergic activity. Thus, the prevention of memory deficits by ladostigil may be resulted from its antioxidant activity (Shoham *et al.*, 2007).

C. The Major Metabolite, Hydroxy-1-(*R*)-Aminoindan

Hydroxy-1-(*R*)-aminoindan (Fig. 1A) is the main metabolite of ladostigil (Sterling *et al.*, 2002). Similar to 1-(*R*)-aminoindan, which is the major metabolite of rasagiline (Sterling *et al.*, 1998), hydroxy-1-(*R*)-aminoindan was previously shown to exert neuroprotective properties in high-density cultured SK-N-SH cells and against 6-hydroxydopamine (6-OHDA)-induced neuronal toxicity in PC-12 cells (Bar-Am *et al.*, 2007). Hydroxy-1-(*R*)-aminoindan markedly reduced the levels of the apoptotic-associated phosphorylated protein H2A.X (Ser139) and decreased the cleavage of caspase 9 and caspase 3, while increasing the anti-apoptotic proteins Bcl-2 and Bcl-xl (Bar-Am *et al.*, 2007). These findings suggest that hydroxyl-1-(*R*)-aminoindan may contribute to the overall neuroprotective activity of its parent compound, ladostigil.

VII. Summary and Future Perspectives

Drug combinations, mixing target-acting compounds, provide a practical way to design specific polypharmacology. The challenge of designing polypharmacological drugs has to link multiple *in vitro* activities to *in vivo* models and clinical settings (Keith *et al.*, 2005; Hopkins *et al.*, 2006). The pluripotential "dirty" drug ladostigil was designed to possess the beneficial properties of its parent drugs, rasagiline and rivastigmine, including neuroprotective activities, AChE/MAO inhibition, and induction of cholinergic transmission and cognitive effects, and to address the therapeutic requirements needed to delay the progression of neurodegenerative diseases (such as AD, PD with dementia, and LB disease) with features of dementia, behavioral abnormalities, depression, and extrapyramidal symptoms. Indeed, ladostigil was found to exert antidepressant-like and neuroprotective activities and reverse behavioral deficits in a variety of animal models of neurodegeneration. The pathogenesis of the common sporadic form of AD involves many characteristics, including apoptosis, mitochondrial malfunction, microvascular alterations, insufficient oxidant defense, glial activation, and generation of toxic Aβ peptide, shown to be modulated by ladostigil in preclinical experiments, and thus, clinical therapeutic studies is oriented toward AD and other type of dementias. The putative target genes and proteins regulated by ladostigil are illustrated in Fig. 2. Another unique property of ladostigil, related to

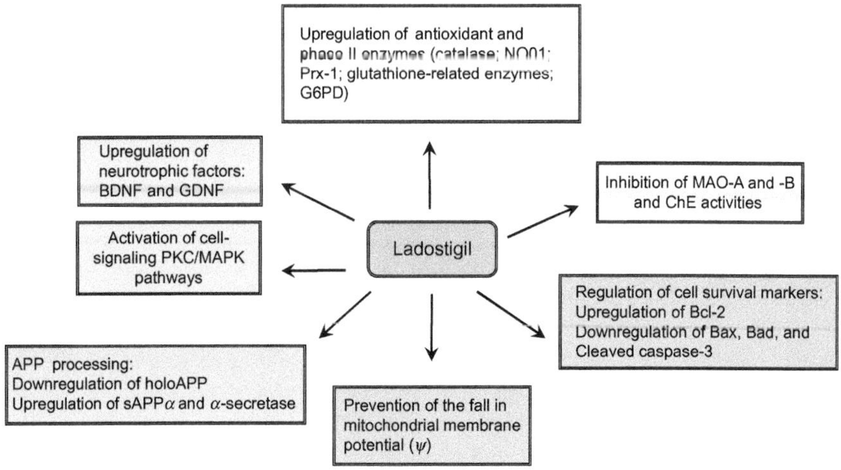

FIG. 2. Schematic overview demonstrating protein and gene targets involved in the neuroprotective activity of ladostigil with respect to the pathological features described for AD. These include brain-selective irreversible MAO-A and -B inhibition, ChE inhibitory activity, regulation of APP processing, regulation of cell survival signaling pathways and neurotrophic factors, and protective activity against OS. Full explanation is discussed in the text. APP, amyloid precursor protein; BDNF, brain-derived neurotrophic factor; ChE, cholinesterase; GDNF, glial cell line-derived neurotrophic factor; G6PD, glucose-6-phosphate dehydrogenase; MAO, monoamine oxidase; MAPK, mitogen-activated protein kinase; NQ01, NADPH quinone 1 oxidoreductase; PKC, protein kinase C; Prx-1, peroxiredoxin 1.

AD, is its higher ability to inhibit both AChE and BuCh (Weinstock *et al.*, 2000a,c). Moreover, ladostigil possesses a brain-selective MAO-A and -B inhibitory activity *in vivo*. The MAO inhibition avoids H_2O_2 generation, thus preventing the Fenton reaction and the formation of neurotoxic free radicals. In addition, the MAO-A inhibition confers potential antidepressant activity. Ladostigil also showed other neuroprotective actions, such as APP processing regulation via MAPK and PKC signaling pathways, mitochondrial membrane potential stabilization, increase in catalase activity and upregulation of the expression levels of antioxidant enzymes.

Based on this multimodal rational, we have recently synthesized a new class of site-activated chelators with dual AChE and MAO inhibitory activities (Zheng *et al.*, 2010). The new compound (5-[*N*-methyl-*N*-propargylaminomethyl] quinolin-8-yl dimethyl carbamate) (Fig. 1B) (Zheng *et al.*, 2010) was shown to be activated by inhibition of AChE to release the multifunctional iron chelating compound M30 (5-[*N*-methyl-*N*-propargylaminomethyl]-8-hydroxyquinoline), demonstrated to be an active nontoxic, brain permeable, and potent neuroprotective chelator and brain-selective MAO-A and -B inhibitor (Gal *et al.*, 2005; Zheng *et al.*, 2005a,b; Avramovich-Tirosh *et al.*, 2007, 2010; Kupershmidt *et al.*, 2009). Currently, simultaneous modulation of multiple targets related to AD etiology by one pluripotential molecule is the most promising drug candidate for the multidysfunctional molecular conditions observed in the incurable AD.

Acknowledgment

The support of Teva Pharmaceutical Co. (Netanya, Israel), Varinel Co. (USA), and Rappaport Family Research Institute, Technion-Israel Institute of Technology is gratefully acknowledged.

References

Akao, Y., Maruyama, W., Shimizu, S., Yi, H., Nakagawa, Y., Shamoto-Nagai, M., Youdim, M.B., Tsujimoto, Y., and Naoi, M. (2002). Mitochondrial permeability transition mediates apoptosis induced by N-methyl(R)salsolinol, an endogenous neurotoxin, and is inhibited by Bcl-2 and rasagiline, N-propargyl-1(R)-aminoindan. *J. Neurochem.* **82**, 913–923.

Akasofu, S., Kimura, M., Kosasa, T., Sawada, K., and Ogura, H. (2008). Study of neuroprotection of donepezil, a therapy for Alzheimer's disease. *Chem. Biol. Interact.* **175**, 222–226.

Arias, E., Ales, E., Gabilan, N.H., Cano-Abad, M.F., Villarroya, M., Garcia, A.G., and Lopez, M.G. (2004). Galantamine prevents apoptosis induced by beta-amyloid and thapsigargin: involvement of nicotinic acetylcholine receptors. *Neuropharmacology* **46**, 103–114.

Arias, E., Gallego-Sandin, S., Villarroya, M., Garcia, A.G., and Lopez, M.G. (2005). Unequal neuroprotection afforded by the acetylcholinesterase inhibitors galantamine, donepezil, and rivastigmine in SH-SY5Y neuroblastoma cells: role of nicotinic receptors. *J. Pharmacol. Exp. Ther.* **315**, 1346–1353.

Avramovich-Tirosh, Y., Amit, T., Bar-Am, O., Zheng, H., Fridkin, M., and Youdim, M.B. (2007). Therapeutic targets and potential of the novel brain- permeable multifunctional iron chelator-monoamine oxidase inhbitor drug, M-30, for the treatment of Alzheimer's disease. *J. Neurochem.* **100**, 490–502.

Avramovich-Tirosh, Y., Bar-Am, O., Amit, T., Youdim, M.B., and Weinreb, O. (2010). Up-regulation of hypoxia-inducible factor (HIF)-1alpha and HIF-target genes in cortical neurons by the novel multifunctional iron chelator anti-Alzheimer drug, M30. *Curr. Alzheimer Res.* **7**, 300–306.

Ba, F., Pang, P.K., and Benishin, C.G. (2003). The establishment of a reliable cytotoxic system with SK-N-SH neuroblastoma cell culture. *J. Neurosci. Methods* **123**, 11–22.

Bailey, J.A., and Lahiri, D.K. (2010). A novel effect of rivastigmine on pre-synaptic proteins and neuronal viability in a neurodegeneration model of fetal rat primary cortical cultures and its implication in Alzheimer's disease. *J. Neurochem.* **112**, 843–853.

Bar-Am, O., Yogev-Falach, M., Amit, T., Sagi, Y., and Youdim, M.B.H. (2004). Regulation of protein kinase C by the anti-Parkinson drug, MAO-B inhibitor, rasagiline and its derivatives, in vivo. *J. Neurochem.* **89**, 1119–1125.

Bar-Am, O., Weinreb, O., Amit, T., and Youdim, M.B. (2005). Regulation of Bcl-2 family proteins, neurotrophic factors, and APP processing in the neurorescue activity of propargylamine. *FASEB J.* **19**, 1899–1901.

Bar-Am, O., Amit, T., and Youdim, M.B. (2007). Aminoindan and hydroxyaminoindan, metabolites of rasagiline and ladostigil, respectively, exert neuroprotective properties in vitro. *J. Neurochem.* **103**, 500–508.

Bar-Am, O., Weinreb, O., Amit, T., and Youdim, M.B. (2009). The novel cholinesterase-monoamine oxidase inhibitor and antioxidant, ladostigil, confers neuroprotection in neuroblastoma cells and aged rats. *J. Mol. Neurosci.* **37**, 135–145.

Bar-Am, O., Amit, T., Weinreb, O., Youdim, M.B., and Mandel, S. (2010). Propargylamine containing compounds as modulators of proteolytic cleavage of amyloid-beta protein precursor: involvement of mitogen activated protein kinase and protein kinase C activation. *J. Alzheimers Dis.* **21**, 361–371.

Bejar, C., Wang, R.H., and Weinstock, M. (1999). Effect of rivastigmine on scopolamine-induced memory impairment in rats. *Eur. J. Pharmacol.* **383**, 231–240.

Buccafusco, J.J., Terry, A.V. Jr., Goren, T., and Blaugrun, E. (2003). Potential cognitive actions of (n-propargly-(3r)-aminoindan-5-yl)-ethyl, methyl carbamate (TV3326), a novel neuroprotective agent, as assessed in old rhesus monkeys in their performance of versions of a delayed matching task. *Neuroscience* **119**, 669–678.

Bullock, R. (2004). Future directions in the treatment of Alzheimer's disease. *Expert Opin. Investig. Drugs* **13**, 303–314.

Calabresi, P., Centonze, D., and Bernardi, G. (2000). Electrophysiology of dopamine in normal and denervated striatal neurons. *Trends Neurosci.* **23**, S57–S63.

Carrillo, M.C., Minami, C., Kitani, K., Maruyama, W., Ohashi, K., Yamamoto, T., Naoi, M., Kanai, S., and Youdim, M.B.H. (2000). Enhancing effect of rasagiline on superoxide dismutase and catalase activities in the dopaminergic system in the rat. *Life Sci.* **67**, 577–585.

Chen, J.J., and Swope, D.M. (2005). Clinical pharmacology of rasagiline: a novel, second-generation propargylamine for the treatment of Parkinson disease. *J. Clin. Pharmacol.* **45**, 878–894.

Chen, Y., and Tang, B.L. (2006). The amyloid precursor protein and postnatal neurogenesis/neuroregeneration. *Biochem. Biophys. Res. Commun.* **341**, 1–5.

Chow, V.W., Mattson, M.P., Wong, P.C., and Gleichmann, M. (2010). An overview of APP processing enzymes and products. *Neuromolecular Med.* **12**, 1–12.

Clark, C.M., and Karlawish, J.H. (2003). Alzheimer disease: current concepts and emerging diagnostic and therapeutic strategies. *Ann. Intern. Med.* **138**, 400–410.

Davis, K.L., and Rosenberg, G.S. (1981). Induction of behavioral supersensitivity to apomorphine by DFP treatment. *Life Sci.* **28**, 1953–1959.

Dragoni, S., Porcari, V., Travagli, M., Castagnolo, D., and Valoti, M. (2006). Antioxidant properties of propargylamine derivatives: assessment of their ability to scavenge peroxynitrite. *J. Pharm. Pharmacol.* **58**, 561–565.

Eckert, A., Keil, U., Marques, C.A., Bonert, A., Frey, C., Schussel, K., and Muller, W.E. (2003). Mitochondrial dysfunction, apoptotic cell death, and Alzheimer's disease. *Biochem. Pharmacol.* **66**, 1627–1634.

Farlow, M.R. (2003). Update on rivastigmine. *Neurologist* **9**, 230–234.

Farlow, M., Potkin, S., Koumaras, B., Veach, J., and Mirski, D. (2003). Analysis of outcome in retrieved dropout patients in a rivastigmine vs placebo, 26-week, Alzheimer disease trial. *Arch. Neurol.* **60**, 843–848.

Francis, P.T., Palmer, A.M., Snape, M., and Wilcock, G.K. (1999). The cholinergic hypothesis of Alzheimer's disease: a review of progress. *J. Neurol. Neurosurg. Psychiatry* **66**, 137–147.

Francis, P.T., Nordberg, A., and Arnold, S.E. (2005). A preclinical view of cholinesterase inhibitors in neuroprotection: do they provide more than symptomatic benefits in Alzheimer's disease? *Trends Pharmacol. Sci.* **26**, 104–111.

Gal, S., Zheng, H., Fridkin, M., and Youdim, M.B. (2005). Novel multifunctional neuroprotective iron chelator-monoamine oxidase inhibitor drugs for neurodegenerative diseases. In vivo selective brain monoamine oxidase inhibition and prevention of MPTP-induced striatal dopamine depletion. *J. Neurochem.* **95**, 79–88.

Giacobini, E. (2002). Long-term stabilizing effect of cholinesterase inhibitors in the therapy of Alzheimer' disease. *J. Neural Transm. Suppl.* **62**, 181–187.

Greenamyre, J.T., Maragos, W.F., Albin, R.L., Penney, J.B., and Young, A.B. (1988). Glutamate transmission and toxicity in Alzheimer's disease. *Prog. Neuropsychopharmacol. Biol. Psychiatry* **12**, 421–430.

Greig, N.H., Utsuki, T., Yu, Q., Zhu, X., Holloway, H.W., Perry, T., Lee, B., Ingram, D.K., and Lahiri, D.K. (2001). A new therapeutic target in Alzheimer's disease treatment: attention to butyrylcholinesterase. *Curr. Med. Res. Opin.* **17**, 159–165.

Halliwell, B. (1992). Reactive oxygen species and the central nervous system. *J. Neurochem.* **59**, 1609–1623.

Halliwell, B. (2001). Role of free radicals in the neurodegenerative diseases: therapeutic implications for antioxidant treatment. *Drugs Aging* **18**, 685–716.

Hansen, R.A., Gartlehner, G., Kaufer, D.J., Lohr, K.N., and Carey, T. (2006). Drug Class Review on Alzheimer's Drugs: Final Report in Drug Class Reviews Oregon Health & Science University, Portland (OR).

Hopkins, A.L., Mason, J.S., and Overington, J.P. (2006). Can we rationally design promiscuous drugs? *Curr. Opin. Struct. Biol.* **16**, 127–136.

Hubalek, F., Binda, C., Li, M., Herzig, Y., Sterling, J., Youdim, M.B.H., Mattevi, A., and Edmondson, D.E. (2004). Inactivation of purified human recombinant monoamine oxidases A and B by rasagiline and its analogues. *J. Med. Chem.* **47**, 1760–1766.

Hung, A.Y., Haass, C., Nitsch, R.M., Qiu, W.Q., Citron, M., Wurtman, R.J., Growdon, J.H., and Selkoe, D.J. (1993). Activation of protein kinase C inhibits cellular production of the amyloid beta-protein. *J. Biol. Chem.* **268**, 22959–22962.

Izzo, P.N., and Bolam, J.P. (1988). Cholinergic synaptic input to different parts of spiny striatonigral neurons in the rat. *J. Comp. Neurol.* **269**, 219–234.

Joseph, J.A., Shukitt-Hale, B., Casadesus, G., and Fisher, D. (2005). Oxidative stress and inflammation in brain aging: nutritional considerations. *Neurochem. Res.* **30**, 927–935.

Keith, C.T., Borisy, A.A., and Stockwell, B.R. (2005). Multicomponent therapeutics for networked systems. *Nat. Rev. Drug Discov.* **4**, 71–78.

Kern, A., Roempp, B., Prager, K., Walter, J., and Behl, C. (2006). Down-regulation of endogenous amyloid precursor protein processing due to cellular aging. *J. Biol. Chem.* **281**, 2405–2413.

Klafki, H.W., Staufenbiel, M., Kornhuber, J., and Wiltfang, J. (2006). Therapeutic approaches to Alzheimer's disease. *Brain* **129**, 2840–2855.

Kornhuber, J., Bormann, J., Hubers, M., Rusche, K., and Riederer, P. (1991). Effects of the 1-amino-adamantanes at the MK-801-binding site of the NMDA-receptor-gated ion channel: a human postmortem brain study. *Eur. J. Pharmacol.* **206**, 297–300.

Kupershmidt, L., Weinreb, O., Amit, T., Mandel, S., Carri, M.T., and Youdim, M.B. (2009). Neuroprotective and neuritogenic activities of novel multimodal iron-chelating drugs in motor-neuron-like NSC-34 cells and transgenic mouse model of amyotrophic lateral sclerosis. *FASEB J.* **23**, 3766–3779.

Leber, P. (1996). Observations and suggestions on antidementia drug development. *Alzheimer Dis. Assoc. Disord.* **10**(Suppl. 1), 31–35.

Liu, R.M. (2002). Down-regulation of gamma-glutamylcysteine synthetase regulatory subunit gene expression in rat brain tissue during aging. *J. Neurosci. Res.* **68**, 344–351.

Liu, R., and Choi, J. (2000). Age-associated decline in gamma-glutamylcysteine synthetase gene expression in rats. *Free Radic. Biol. Med.* **28**, 566–574.

Lleo, A., Greenberg, S.M., and Growdon, J.H. (2006). Current pharmacotherapy for Alzheimer's disease. *Annu. Rev. Med.* **57**, 513–533.

Lopez, O.L., Becker, J.T., Wisniewski, S., Saxton, J., Kaufer, D.I., and DeKosky, S.T. (2002). Cholinesterase inhibitor treatment alters the natural history of Alzheimer's disease. *J. Neurol. Neurosurg. Psychiatry* **72**, 310–314.

Luques, L., Shoham, S., and Weinstock, M. (2007). Chronic brain cytochrome oxidase inhibition selectively alters hippocampal cholinergic innervation and impairs memory: prevention by ladostigil. *Exp. Neurol.* **206**, 209–219.

Maccioni, R.B., Munoz, J.P., and Barbeito, L. (2001). The molecular bases of Alzheimer's disease and other neurodegenerative disorders. *Arch. Med. Res.* **32**, 367–381.

Machado, J.C. (2009). Review: rivastigmine reduces rate of cognitive decline and improves performance in mild to moderate Alzheimer's. *Evid. Based Ment. Health* **12**, 113.

Mamelak, M. (2007). Alzheimer's disease, oxidative stress and gammahydroxybutyrate. *Neurobiol. Aging* **28**, 1340–1360.

Maruyama, W., Akao, Y., Youdim, M.B.H., and Naoi, M. (2000a). Neurotoxins induce apoptosis in dopamine neurons: protection by N-propargylamine-1(R)- and (S)-aminoindan, rasagiline and TV1022. *J. Neural Transm. Suppl.* **60**, 171–186.

Maruyama, W., Naoi, M., and Youdim, M.B.H. (2000b). Propargylamines protect dopamine cells from apoptosis induced by a neurotoxin, N-methyl (R) salsolinol. In: S. Alexander and M.A. Gollins, (Eds.), *Neuromelanin, may Mediate Neurotoxicity via Its Interaction with Redox Active Iron.* Kluwer Academic/ Plenum Publishers, New York, p. 321.

Maruyama, W., Weinstock, M., Youdim, M.B., Nagai, M., and Naoi, M. (2003). Anti-apoptotic action of anti-Alzheimer drug, TV3326 [(N-propargyl)-(3R)-aminoindan-5-yl]-ethyl methyl carbamate, a novel cholinesterase-monoamine oxidase inhibitor. *Neurosci. Lett.* **341**, 233–236.

Mills, J., and Reiner, P.B. (1999a). Regulation of amyloid precursor protein cleavage. *J. Neurochem.* **72**, 443–460.

Mills, J., and Reiner, P.B. (1999b). Mitogen-activated protein kinase is involved in N-methyl-D-aspartate receptor regulation of amyloid precursor protein cleavage. *Neuroscience* **94**, 1333–1338.

Murakami, H., Iwashita, T., Asai, N., Shimono, Y., Iwata, Y., Kawai, K., and Takahashi, M. (1999). Enhanced phosphatidylinositol 3-kinase activity and high phosphorylation state of its downstream signalling molecules mediated by ret with the MEN 2B mutation. *Biochem. Biophys. Res. Commun.* **262**, 68–75.

Newman, S.C. (1999). The prevalence of depression in Alzheimer's disease and vascular dementia in a population sample. *J. Affect. Disord.* **52**, 169–176.

Nunomura, A., Perry, G., Aliev, G., Hirai, K., Takeda, A., Balraj, E.K., Jones, P.K., Ghanbari, H., Wataya, T., Shimohama, S., Chiba, S., Atwood, C.S., *et al.* (2001). Oxidative damage is the earliest event in Alzheimer disease. *J. Neuropathol. Exp. Neurol.* **60**, 759–767.

Olanow, C.W., Hauser, R.A., Jankovic, J., Langston, W., Lang, A., Poewe, W., Tolosa, E., Stocchi, F., Melamed, E., Eyal, E., and Rascol, O. (2008). A randomized, double-blind, placebo-controlled, delayed start study to assess rasagiline as a disease modifying therapy in Parkinson's disease (the ADAGIO study): rationale, design, and baseline characteristics. *Mov. Disord.* **23**, 2194–2201.

Olanow, C.W., Rascol, O., Hauser, R., Feigin, P.D., Jankovic, J., Lang, A., Langston, W., Melamed, E., Poewe, W., Stocchi, F., and Tolosa, E. (2009). A double-blind, delayed-start trial of rasagiline in Parkinson's disease. *N. Engl. J. Med.* **361**, 1268–1278.

Pakaski, M., and Kasa, P. (2003). Role of acetylcholinesterase inhibitors in the metabolism of amyloid precursor protein. *Curr. Drug Targets CNS Neurol. Disord.* **2**, 163–171.

Palmer, A.M., Stratmann, G.C., Procter, A.W., and Bowen, D.M. (1988). Possible neurotransmitter basis of behavioral changes in Alzheimer's disease. *Ann. Neurol.* **23**, 616–620.

Parkinson Study Group, (2002). A controlled trial of rasagiline in early Parkinson disease: the TEMPO Study. *Arch. Neurol.* **59**, 1937–1943.

Parkinson Study Group, (2004). A controlled, randomized, delayed-start study of rasagiline in early Parkinson disease. *Arch. Neurol.* **61**, 561–566.

Parkinson Study Group, (2005). A randomized placebo-controlled trial of rasagiline in levodopa-treated patients with Parkinson disease and motor fluctuations: the PRESTO study. *Arch. Neurol.* **62**, 241–248.

Racchi, M., Mazzucchelli, M., Porrello, E., Lanni, C., and Govoni, S. (2004). Acetylcholinesterase inhibitors: novel activities of old molecules. *Pharmacol. Res.* **50**, 441–451.

Rastogi, S.K., Rastogi, R.B., Singhal, R.L., and Lapierre, Y.D. (1982). Effects of cholinergic and GABAergic drugs on apomorphine-induced gnawing behavior in rats. *Biol. Psychiatry* **17**, 937–946.

Rhee, S.G., Kim, K.H., Chae, H.Z., Yim, M.B., Uchida, K., Netto, L.E., and Stadtman, E.R. (1994). Antioxidant defense mechanisms: a new thiol-specific antioxidant enzyme. *Ann. N. Y. Acad. Sci.* **738**, 86–92.

Rogers, J.T., and Lahiri, D.K. (2004). Metal and inflammatory targets for Alzheimer's disease. *Curr. Drug Targets* **5**, 535–551.

Sabbagh, M.N., Farlow, M.R., Relkin, N., and Beach, T.G. (2006). Do cholinergic therapies have disease-modifying effects in Alzheimer's disease? *Alzheimers Dement.* **2**, 118–125.

Sagi, Y., Weinstock, M., and Youdim, M.B.H. (2003). Attenuation of MPTP-induced dopaminergic neurotoxicity by TV3326, a cholinesterase-monoamine oxidase inhibitor. *J. Neurochem.* **2**, 290–297.

Sagi, Y., Drigues, N., and Youdim, M.B. (2005). The neurochemical and behavioral effects of the novel cholinesterase-monoamine oxidase inhibitor, ladostigil, in response to L-dopa and L-tryptophan, in rats. *Br. J. Pharmacol.* **146**, 553–560.

SantaCruz, K.S., Yazlovitskaya, E., Collins, J., Johnson, J., and DeCarli, C. (2004). Regional NAD(P) H:quinone oxidoreductase activity in Alzheimer's disease. *Neurobiol. Aging* **25**, 63–69.

Scali, C., Casamenti, F., Bellucci, A., Costagli, C., Schmidt, B., and Pepeu, G. (2002). Effect of subchronic administration of metrifonate, rivastigmine and donepezil on brain acetylcholine in aged F344 rats. *J. Neural Transm.* **109**, 1067–1080.

Schallreuter, K.U., and Elwary, S. (2007). Hydrogen peroxide regulates the cholinergic signal in a concentration dependent manner. *Life Sci.* **80**, 2221–2226.

Schliebs, R. (2005). Basal forebrain cholinergic dysfunction in Alzheimer's disease—interrelationship with beta-amyloid, inflammation and neurotrophin signaling. *Neurochem. Res.* **30**, 895–908.

Segal, R.A., and Greenberg, M.E. (1996). Intracellular signaling pathways activated by neurotrophic factors. *Annu. Rev. Neurosci.* **19**, 463–489.

Selkoe, D.J., and Schenk, D. (2003). Alzheimer's disease: molecular understanding predicts amyloid-based therapeutics. *Annu. Rev. Pharmacol. Toxicol.* **43**, 545–584.

Shoham, S., Bejar, C., Kovalev, E., Schorer-Apelbaum, D., and Weinstock, M. (2007). Ladostigil prevents gliosis, oxidative-nitrative stress and memory deficits induced by intracerebroventricular injection of streptozotocin in rats. *Neuropharmacology* **52**, 836–843.

Siegel, G.J., and Chauhan, N.B. (2000). Neurotrophic factors in Alzheimer's and Parkinson's disease brain. *Brain Res. Brain Res. Rev.* **33**, 199–227.

Smith, M.A., Rottkamp, C.A., Nunomura, A., Raina, A.K., and Perry, G. (2000). Oxidative stress in Alzheimer's disease. *Biochim. Biophys. Acta* **1502**, 139–144.

Sterling, J., Veinberg, A., Lerner, D., Goldenberg, W., Levy, R., Youdim, M., and Finberg, J. (1998). (R)(+)-N-propargyl-1-aminoindan (rasagiline) and derivatives: highly selective and potent inhibitors of monoamine oxidase B. *J. Neural Transm. Suppl.* **52**, 301–305.

Sterling, J., Herzig, Y., Youdim, M.B.H., and Weinstock, M. (2000). The Design and Synthesis of Dual Action Compounds for the Treatment of Alzheimer's Disease. Israel Chemical Co., Israel.

Sterling, J., Herzig, Y., Goren, T., Finkelstein, N., Lerner, D., Goldenberg, W., Miskolczi, I., Molnar, S., Rantal, F., Tamas, T., Toth, G., Zagyva, A., *et al.* (2002). Novel dual inhibitors of AChE and MAO derived from hydroxy aminoindan and phenethylamine as potential treatment for Alzheimer's disease. *J. Med. Chem.* **45**, 5260–5279.

Stone, J.R., and Yang, S. (2006). Hydrogen peroxide: a signaling messenger. *Antioxid. Redox Signal.* **8**, 243–270.

Stringer, J.L., Gaikwad, A., Gonzales, B.N., Long, D.J. Jr., Marks, L.M., and Jaiswal, A.K. (2004). Presence and induction of the enzyme NAD(P)H: quinone oxidoreductase 1 in the central nervous system. *J. Comp. Neurol.* **471**, 289–297.

Takada-Takatori, Y., Kume, T., Sugimoto, M., Katsuki, H., Sugimoto, H., and Akaike, A. (2006). Acetylcholinesterase inhibitors used in treatment of Alzheimer's disease prevent glutamate neurotoxicity via nicotinic acetylcholine receptors and phosphatidylinositol 3-kinase cascade. *Neuropharmacology* **51**, 474–486.

Tariot, P.N., Farlow, M.R., Grossberg, G.T., Graham, S.M., McDonald, S., and Gergel, I. (2004). Memantine treatment in patients with moderate to severe Alzheimer disease already receiving donepezil: a randomized controlled trial. *JAMA* **291**, 317–324.

Tsolaki, M., Kokarida, K., Iakovidou, V., Stilopoulos, E., Meimaris, J., and Kazis, A. (2001). Extrapyramidal symptoms and signs in Alzheimer's disease: prevalence and correlation with the first symptom. *Am. J. Alzheimers Dis. Other Demen.* **16**, 268–278.

Van der Schyf, C.J., Gal, S., Geldenhuys, W.J., and Youdim, M.B. (2006). Multifunctional neuroprotective drugs targeting monoamine oxidase inhibition, iron chelation, adenosine receptors, and cholinergic and glutamatergic action for neurodegenerative diseases. *Expert Opin. Investig. Drugs* **15**, 873–886.

Vetrivel, K.S., and Thinakaran, G. (2006). Amyloidogenic processing of beta-amyloid precursor protein in intracellular compartments. *Neurology* **66**, S69–S73.

Wadia, J.S., Chalmers-Redman, R.M.E., Ju, W.J.H., Carlile, G.W., Phillips, J.L., Fraser, A.D., and Tatton, W.G. (1998). Mitochondrial membrane potential and nuclear changes in apoptosis caused by serum and nerve growth factor with drawal: time course and modification by (−)-deprenyl. *J. Neurosci.* **18**, 932–947.

Weinreb, O., Bar-Am, O., Amit, T., Chillag-Talmor, O., and Youdim, M.B.H. (2004). Neuroprotection via pro-survival protein kinase C isoforms associated with Bcl-2 family members. *FASEB J.* **18**, 1471–1473.

Weinreb, O., Amit, T., Bar-Am, O., and Youdim, M.B. (2007a). Induction of neurotrophic factors GDNF and BDNF associated with the mechanism of neurorescue action of rasagiline and ladostigil: new insights and implications for therapy. *Ann. N. Y. Acad. Sci.* **1122**, 155–168.

Weinreb, O., Drigues, N., Sagi, Y., Reznick, A.Z., Amit, T., and Youdim, M.B. (2007b). The application of proteomics and genomics to the study of age-related neurodegeneration and neuroprotection. *Antioxid. Redox Signal.* **9**, 169–179.

Weinreb, O., Bar-Am, O., Amit, T., Drigues, N., Sagi, Y., and Youdim, M.B. (2008). The neuroprotective effect of ladostigil against hydrogen peroxide-mediated cytotoxicity. *Chem. Biol. Interact.* **175**, 318–326.

Weinreb, O., Amit, T., Bar-Am, O., and Youdim, M.B. (2010). Rasagiline: a novel anti-Parkinsonian monoamine oxidase-B inhibitor with neuroprotective activity. *Prog. Neurobiol.* **92**, 330–344.

Weinstock, M., Razin, M., Chorev, M., and Enz, A. (1994). Pharmacological evaluation of phenylcarbamates as CNS-selective acetylcholinesterase inhibitors. *J. Neural Transm.* **43**, 219–225.

Weinstock, M., Goren, T., and Youdim, M.B.H. (2000a). Development of a novel neuroprotective drug (TV3326) for the treatment of Alzheimer's disease, with cholinesterase and monoamine oxidase inhibitory activities. *Drug Dev. Res.* **50**, 216–222.

Weinstock, M., Kirschbaum-Slager, N., Lazarovici, P., Bejar, C., Shoami, E., and Youdim, M.B.H. (2000b). Cell Culture and In Vivo Neuroprotective Effects of Novel Cholinesterase-MAO Inhibitors. *Ann. N. Y. Acad. Sci.* **939**, 148–161.

Weinstock, M., Bejar, C., Wang, R.H., Poltyrev, T., Gross, A., Finberg, J., and Youdim, M.B.H. (2000c). TV3326, a novel neuroprotective drug with cholinesterase and monoamine oxidase inhibitory activities for the treatment of Alzheimer's disease. *J. Neural Transm. Suppl.* **60**, S157–S170.

Weinstock, M., Kirschbaum-Slager, N., Lazarovici, P., Bejar, C., Youdim, M.B.H., and Shoham, S. (2001). Neuroprotective effects of novel cholinesterase inhibitors derived from rasagiline as potential anti-Alzheimer drugs. *Ann. N. Y. Acad. Sci.* **939**, 148–161.

Weinstock, M., Poltyrev, T., Bejar, C., and Youdim, M.B. (2002a). Effect of TV3326, a novel monoamine-oxidase cholinesterase inhibitor, in rat models of anxiety and depression. *Psychopharmacology* **160**, 318–324.

Weinstock, M., Gorodetsky, E., Wang, R.H., Gross, A., Weinreb, O., and Youdim, M.B.H. (2002b). Limited potentiation of blood pressure response to oral tyramine by brain-selective monoamine oxidase A-B inhibitor, TV-3326 in conscious rabbits(1). *Neuropharmacology* **43**, 999–1005.

Weinstock, M., Gorodetsky, E., Poltyrev, T., Gross, A., Sagi, Y., and Youdim, M.B.H. (2003). A novel cholinesterase and brain-selective monoamine oxidase inhibitor for the treatment of dementia comorbid with depression and Parkinson's disease. *Prog. Neuropsychopharmacol. Biol. Psyciatry* **27**, 555–561.

Weinstock, M., Luques, L., Poltyrev, T., Bejar, C., and Shoham, S. (2009). Ladostigil prevents age-related glial activation and spatial memory deficits in rats. *Neurobiol. Aging* **32**, 1069–1078.

Xiao, X.Q., Yang, J.W., and Tang, X.C. (1999). Huperzine A protects rat pheochromocytoma cells against hydrogen peroxide-induced injury. *Neurosci. Lett.* **275**, 73–76.

Xiao, X.Q., Lee, N.T., Carlier, P.R., Pang, Y., and Han, Y.F. (2000). Bis(7)-tacrine, a promising anti-Alzheimer's agent, reduces hydrogen peroxide-induced injury in rat pheochromocytoma cells: comparison with tacrine. *Neurosci. Lett.* **290**, 197–200.

Yogev-Falach, M., Amit, T., Bar-Am, O., Sagi, Y., Weinstock, M., and Youdim, M.B.H. (2002). The involvement of mitogen-activated protein (MAP) kinase in the regulation of amyloid precursor protein processing by novel cholinesterase inhibitors derived from rasagiline. *FASEB J.* **16**, 1674–1676.

Yogev-Falach, M., Amit, T., Bar-AM, O., and Youdim, M.B.H. (2003). The importance of propargylamine moiety in the anti-Parkinson drug rasagiline and its derivatives for MAPK-dependent amyloid precursor protein processing. *FASEB J.* **17**, 2325–2327.

Yogev-Falach, M., Bar-Am, O., Amit, T., Weinreb, O., and Youdim, M.B. (2006). A multifunctional, neuroprotective drug, ladostigil (TV3326), regulates holo-APP translation and processing. *FASEB J.* **20**, 2177–2179.

Youdim, M.B., Edmondson, D., and Tipton, K.F. (2006). The therapeutic potential of monoamine oxidase inhibitors. *Nat. Rev.* **7**, 295–309.

Zafar, K.S., Inayat-Hussain, S.H., Siegel, D., Bao, A., Shieh, B., and Ross, D. (2006). Overexpression of NQO1 protects human SK-N-MC neuroblastoma cells against dopamine-induced cell death. *Toxicol. Lett.* **166**, 261–267.

Zhang, Y.W., and Xu, H. (2007). Molecular and cellular mechanisms for Alzheimer's disease: understanding APP metabolism. *Curr. Mol. Med.* **7**, 687–696.

Zhang, L., Yu, H., Sun, Y., Lin, X., Chen, B., Tan, C., Cao, G., and Wang, Z. (2007). Protective effects of salidroside on hydrogen peroxide-induced apoptosis in SH-SY5Y human neuroblastoma cells. *Eur. J. Pharmacol.* **564**, 18–25.

Zheng, H., Gal, S., Weiner, L.M., Bar-Am, O., Warshawsky, A., Fridkin, M., and Youdim, M.B. (2005a). Novel multifunctional neuroprotective iron chelator-monoamine oxidase inhibitor drugs for neurodegenerative diseases: in vitro studies on antioxidant activity, prevention of lipid peroxide formation and monoamine oxidase inhibition. *J. Neurochem.* **95**, 68–78.

Zheng, H., Weiner, L.M., Bar-Am, O., Epsztejn, S., Cabantchik, Z.I., Warshawsky, A., Youdim, M.B., and Fridkin, M. (2005b). Design, synthesis, and evaluation of novel bifunctional iron-chelators as potential agents for neuroprotection in Alzheimer's, Parkinson's, and other neurodegenerative diseases. *Bioorg. Med. Chem.* **13**, 773–783.

Zheng, H., Youdim, M.B., and Fridkin, M. (2010). Site-activated chelators targeting acetylcholinesterase and monoamine oxidase for Alzheimer's therapy. *ACS Chem. Biol.* **5**, 603–610.

Zhu, X., Raina, A.K., Lee, H.G., Casadesus, G., Smith, M.A., and Perry, G. (2004). Oxidative stress signalling in Alzheimer's disease. *Brain Res.* **1000**, 32–39.

Zhu, Y., Carvey, P.M., and Ling, Z. (2006). Age-related changes in glutathione and glutathione-related enzymes in rat brain. *Brain Res.* **1090**, 35–44.

NOVEL MAO-B INHIBITORS: POTENTIAL THERAPEUTIC USE OF THE SELECTIVE MAO-B INHIBITOR PF9601N IN PARKINSON'S DISEASE

Mercedes Unzeta and Elisenda Sanz

Institut de Neurociències and Departament de Bioquímica i Biologia Molecular, Universitat Autònoma de Barcelona, Cerdanyola del Vallès (Barcelona), Spain

Abstract

Parkinson's disease (PD) is a neurodegenerative disease that is characterized by preferential loss of dopaminergic neurons in the *substantia nigra pars compacta*, leading to declining levels of dopamine in the striatum. In our search for compounds able not only to extend the effects of dopamine by preventing its degradation but also to halt or slow the neurodegenerative process, we designed, synthesized, and biologically tested a series of propargylamines for their potential use as therapeutic agents for PD. Among them, PF9601N, [*N*-(2-propynyl)-2-(5-benzyloxy-indolyl) methylamine], showed high potency and selectivity as a MAO-BI (monoamine oxidase type B inhibitor) and also demonstrated remarkable neuroprotective properties in several *in vivo* and cellular models of PD. In this chapter, we describe the preclinical evidence revealing the novel MAO-BI PF9601N as an interesting candidate for the treatment of PD.

INTERNATIONAL REVIEW OF NEUROBIOLOGY, VOL. 100
DOI: 10.1016/B978-0-12-386467-3.00011-X

I. Introduction

Parkinson's disease (PD) is the second most common neurodegenerative disease after Alzheimer's disease, afflicting more than 6 million people worldwide (Dauer and Przedborski, 2003). This movement disorder has characteristic motor symptoms such as bradykinesia, rigidity, resting tremor, and postural instability (Gálvez-Jiménez, 2007). Most of these motor symptoms result from a depletion of striatal dopamine levels due to the degeneration of dopaminergic neurons in the nigrostriatal pathway (Hirsch *et al.*, 1988; Dauer and Przedborski, 2003). In addition, other nondopaminergic neuronal populations, such as the cholinergic neurons of the *nucleus basalis* of Meynert and the *pedunculopontine nucleus*, the noradrenergic neurons of the *locus coeruleus*, and the serotonergic neurons of the *raphe nuclei*, are affected, leading to several nonmotor symptoms that are also present in PD (Jellinger, 1991).

Currently, treatment for PD is essentially symptomatic, relying on agents that address the dopaminergic deficit and therefore improve the characteristic motor symptoms of the disease (Factor, 2008). However, no current interventions have been shown to slow down the neurodegenerative process. In order to find new neuroprotective agents for PD, a deeper understanding of the mechanisms of neurodegeneration is required. With respect to this point, several factors have been proposed to underlie the nigral degeneration in PD, among them are oxidative stress, mitochondrial dysfunction, and the accumulation of anomalous proteins (Jenner, 2003; Jenner and Olanow, 2006; Esposito *et al.*, 2007; Olanow, 2007; Wang and Takahashi, 2007; Schapira, 2008). In addition, it has also been proposed that these factors lead to neuronal degeneration through apoptotic mechanisms (Mattson, 2006; Olanow, 2007), suggesting the use of antiapoptotic compounds to halt or slow the neurodegeneration present in PD. To date, none of the potential neuroprotective agents assessed in clinical trials was able to significantly slow down the neurodegenerative process (Ahlskog, 2007; Ahlskog and Uitti, 2010), thus pointing to the need for new neuroprotective compounds for the treatment of PD.

In this chapter, we will review the current status of monoamine oxidase type B inhibitors (MAO-BIs) with antiapoptotic properties as potential neuroprotective agents for the treatment of PD and describe the preclinical evidence that allows us to propose the novel MAO-BI PF9601N as a potential candidate for the treatment of PD.

II. MAO-B Inhibitors

MAO-BIs have been used extensively in the treatment of PD due to their ability to extend the effects of endogenous dopamine as well as levodopa, an exogenous source of dopamine (Riederer and Youdim, 1986; The Parkinson

Study Group, 1993). These actions are able to delay the onset of disability in early PD and therefore delay the need for levodopa; they also reduce the effective dose of levodopa required in more advanced PD (The Parkinson Study Group, 1993, 1996; Olanow *et al.*, 1998). Besides these symptomatic benefits, it has also been suggested that the inhibition of the metabolism of dopamine by MAO-BIs may be beneficial, as it reduces the production of neurotoxic metabolites such as hydrogen peroxide and toxic aldehydes that may ultimately contribute to increased neuronal damage (Pizzinat *et al.*, 1999; Kristal *et al.*, 2001).

Several studies have reported that some MAO-BIs, especially the propargylamine-containing MAO-BIs, show neuroprotective properties in different experimental paradigms of PD. Among them, L-deprenyl (or selegiline), rasagiline, and the novel MAO-BI PF9601N were shown in different reports to confer neuroprotection by mechanisms not related to their MAO-B inhibitory effects. While L-deprenyl was the first selective MAO-BI used clinically, rasagiline and PF9601N belong to a second generation of MAO-BIs that, unlike L-deprenyl, do not generate amphetamine derivatives when metabolized (Chen *et al.*, 2007; Valoti, 2007).

A. A Novel MAO-BI: PF9601N

PF9601N [*N*-(2-propynyl)-2-(5-benzyloxy-indolyl) methylamine] (Fig. 1), formerly known as FA-73, is a propargylamine-containing irreversible MAO-BI that was identified in an extensive screen conducted by our group of a series of acetylenic and allenic tryptamine derivatives as potential MAO inhibitors (Balsa *et al.*, 1991; Avila *et al.*, 1993; Balsa *et al.*, 1994; Perez *et al.*, 1996). This compound is more potent and selective than the prototypical MAO-BI L-deprenyl (Table I; Perez *et al.*, 1999), and, unlike L-deprenyl, its metabolism does not generate amphetamine derivatives that may potentially harm the cells (Valoti, 2007). PF9601N extends the effects of exogenous dopamine, which is demonstrated by

Table I

Inhibition Constants (K_i) of PF9601N and L-Deprenyl for the A and B Isoforms of Monoamine Oxidase.

	K_i MAO-B (nM)	K_i MAO-A (nM)
PF9601N	0.75 ± 0.15	800 ± 60
L-Deprenyl	16.8 ± 0.1	376 ± 0.032

Table shows that PF9601N is more potent and selective toward the B isoform than L-deprenyl. Modified from Perez *et al.* (1999).

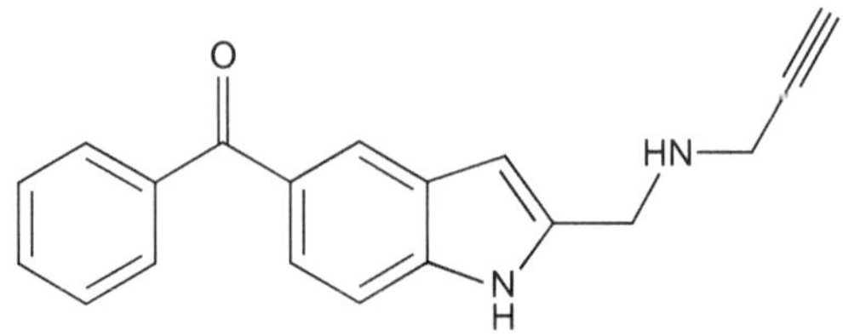

FIG. 1. Chemical structure of PF9601N. The compound [*N*-(2-propynyl)-2-(5-benzyloxy-indolyl) methylamine] PF9601N is a new propargylamine analogue, which turns out to be a very potent and selective monoamine oxidase-B (MAO-B) inhibitor.

its capacity to increase the duration of L-DOPA-induced contralateral turning in an animal model of PD (Prat *et al.*, 2000). In addition, PF9601N might also be able to potentiate endogenous dopamine effects via its reported ability to inhibit dopamine uptake in human and rat striatal synaptosomes (Perez *et al.*, 1999).

In addition to these symptomatic effects, PF9601N is also an effective and potent neuroprotective agent in several *in vivo* models of PD. This compound attenuates the MPTP-induced striatal dopamine depletion in young–adult and old–adult C57BL/6 mice (Perez and Unzeta, 2003) and reduces the loss of tyrosine hydroxylase-positive neurons after nigrostriatal injection of 6-hydroxy-dopamine in rats (Cutillas *et al.*, 2002). Further, PF9601N confers remarkable neuroprotective effects in several cell culture models of PD (Perez and Unzeta, 2003; Sanz *et al.*, 2008, 2009). Like other propargylamine-containing MAO-BIs, the neuroprotective properties of PF9601N appear to be independent of MAO-B inhibition (Perez and Unzeta, 2003; Sanz *et al.*, 2008, 2009); further, the molecular mechanisms involved in this neuroprotective effect appear to be distinct from other propargylamines, such as L-deprenyl or rasagiline; the molecular mechanisms underlying PF9601N neuroprotective actions will be extensively discussed further in this chapter.

B. MAO-BIs and Symptomatic Benefits

Treating PD with dopaminergic symptomatic agents such as levodopa, dopaminergic agonists, and MAO-BIs or catechol-*O*-methyltransferase inhibitors significantly improves the characteristic symptoms of PD, such as bradykinesia and rigidity. Currently, the most effective symptomatic relief is levodopa; this is despite the fact that levopoda was initially considered neurotoxic because of its ability to generate oxidative stress that could potentially contribute to neurodegeneration in PD (Ziv *et al.*, 1997; Melamed *et al.*, 1998; Kostrzewa *et al.*, 2002). Several studies have now shown that levodopa administration does not worsen disease progression (Fahn *et al.*, 2004; Fahn, 2005).

Although there is solid evidence that levodopa does not hasten clinical disease progression, it is widely accepted that levodopa administration has several side effects, such as dyskinesias and motor fluctuations (wearing-off and on–off phenomena) that result in discontinuous dopamine availability in the striatum (Ahlskog and Muenter, 2001; Aminoff, 2006). For this reason, the current suggested practices for the treatment of PD are to withhold levodopa administration until the PD symptoms limit patient function (Hauser, 2010) or to provide adjunct treatment combined with levodopa to treat PD with motor complications (Pahwa and Lyons, 2009; Stowe *et al.*, 2010).

In this context, several agents have been shown to delay the onset of the disabling symptoms in early PD and hence delay the need for treatment with levodopa (Chen and Swope, 2007), or to reduce the off-time and improve scores on the Unified Parkinson's Disease Rating Scale (UPDRS) when coadministered with levodopa (Stowe *et al.*, 2010). Among them, MAO-BIs have not only been proven to delay the progression of the disease (The Parkinson Study Group, 1993, 2002, 2004; Macleod *et al.*, 2005; Caslake *et al.*, 2009) but also to ameliorate the motor complications in levodopa-treated PD patients (Marconi *et al.*, 1992; Waters *et al.*, 2004; The Parkinson Study Group, 2005; Rascol *et al.*, 2005; Stowe *et al.*, 2010).

Several clinical trials initially performed to assess the potential neuroprotective actions of these compounds have shown that MAO-BIs improve PD symptoms as measured by the UPDRS. These effects are thought to be mediated by enhanced endogenous dopamine availability in early PD (The Parkinson Study Group, 1993, 1996, 2004; Ives *et al.*, 2004) and extension of the effects of exogenous dopamine (levopoda) in late PD (Heinonen *et al.*, 1989; The Parkinson Study Group, 2005; Rascol *et al.*, 2005).

The above-mentioned findings support the conclusion that treatment with MAO-BIs results in better clinical progression in PD (Schapira and Obeso, 2006) and highlight the value of these compounds in the symptomatic treatment of PD.

C. MAO-BIs and Neuroprotection

Despite the recent advances in our knowledge of the pathophysiology and the environmental and genetic factors that influence neurodegeneration in PD, slowing or halting the neurodegenerative process has not been accomplished, and neuroprotection is thus still considered an unmet need in PD. Neuroprotection is defined as an intervention able to influence the etiology or the pathogenesis underlying neurodegenerative diseases, thus preventing or delaying the onset or the progression of the disease (Shoulson, 1998). Therefore, neuroprotective compounds must display long-term benefits that are not related to symptomatic effects. In this regard, progress has been made in the field with the publication of the ADAGIO study, a delayed-start clinical trial to assess the disease-modifying effects of the MAO-BI rasagiline (Olanow *et al.*, 2009). The study found a potential disease-modifying effect of rasagiline (Sampaio and Ferreira, 2010); however, some concerns were raised about the interpretation of this study due to the unavoidable need to rely on complex trial designs to demonstrate the disease-modifying effect (Ahlskog and Uitti, 2010; Schwarzschild, 2010). These concerns and the relatively small benefit provided by rasagiline in this study emphasize the need for new compounds that not only provide symptomatic benefits but that also significantly modify the neurodegenerative process.

III. Apoptosis in PD

Several lines of evidence support the hypothesis that neurodegeneration in PD occurs via apoptosis (Vila and Przedborski, 2003; Mattson, 2006; Olanow, 2007; Burke, 2008). Apoptosis, which can proceed through three distinct pathways, is one of the mechanisms through which programmed cell death is executed. These pathways are the intrinsic pathway, which originates in the mitochondria (Spierings *et al.*, 2005), the extrinsic pathway, which is mediated by death receptors (Thorburn, 2004), and the endoplasmic reticulum (ER) stress pathway, which is initiated by the accumulation of unfolded or misfolded proteins in the ER (Rao *et al.*, 2004; Szegezdi *et al.*, 2006). The mediators of these pathways are specific components of the apoptotic machinery, such as the caspase family of cysteine proteases essential for the initiation and execution of the apoptotic process (Kumar, 2007), the proteins of the Bcl-2 family with both proapoptotic and antiapoptotic members (Youle and Strasser, 2008), and transcription factors such as p53 (Culmsee and Mattson, 2005).

The first evidence of apoptosis in PD came from several postmortem studies where investigators were able to show the presence of apoptotic neurons in the *substantia nigra* (SN) of PD brains (Mochizuki *et al.*, 1996; Anglade *et al.*, 1997; Tatton *et al.*, 1998) using the *TdT-mediated X-dUTP nick end labeling* and *in situ end labeling* techniques. Unfortunately, these approaches have several drawbacks (Charriaut-Marlangue and Ben-Ari, 1995; Labat-Moleur *et al.*, 1998) and further research was needed to support the observation of apoptosis in PD. Posterior studies showing the presence of proapoptotic proteins in nigral neurons in postmortem PD brains as well as clear evidence of apoptosis in different experimental models of PD have confirmed the role of apoptosis in the death of SN dopaminergic neurons in PD.

Consistent with the above observations, several proapoptotic proteins have been identified in PD brains, including caspase-3 (a main executioner caspase in the apoptotic process), which showed activation and increased immunoreactivity in several studies (Hartmann *et al.*, 2000; Mogi *et al.*, 2000; Tatton, 2000). Other caspases (such as caspase-8 and -9) that are known for their role in initiating the two main apoptotic pathways were also activated in PD brains (Hartmann *et al.*, 2001; Viswanath *et al.*, 2001). Increased levels of Bax and Bcl-xL, proapoptotic and antiapoptotic members of the Bcl-2 family, respectively, were also detected (Tatton, 2000; Hartmann *et al.*, 2002). Other features of apoptotic cell death such as the nuclear accumulation of the glycolytic enzyme GAPDH (Tatton, 2000) and increased levels of the death receptor TNF-α R1 (Mogi *et al.*, 2000) have also been observed.

More recently, the proapoptotic transcription factor p53 and the ER stress pathway have also been implicated in the apoptosis observed in PD (Wang and Takahashi, 2007; Matus *et al.*, 2008; Alves da Costa and Checler, 2011). In postmortem tissue, increased levels of phospho-p53 (a posttranslational modification that increases p53 stability) were reported (Nair *et al.*, 2006), and the presence of several mediators of the ER stress pathway in degenerating neurons was also established (Hoozemans *et al.*, 2007; Slodzinski *et al.*, 2009).

In addition to the compelling postmortem data, evidence obtained from different experimental models of PD highlights the importance of apoptosis in the neurodegenative process. These models mimic both the histological and biochemical traits of PD (Shimohama *et al.*, 2003; Terzioglu and Galter, 2008) and are therefore extremely helpful in defining factors that may be involved in the nigral degeneration. These experimental models reveal not only the prominent role of apoptosis in neuronal cell death in PD, but also that blockage of the apoptotic process is sufficient to prevent neurodegeneration in most of these experimental approaches (Zhou *et al.*, 2000; Blum *et al.*, 2001; Nicotra and Parvez, 2002; Eberhardt and Schulz, 2003; Vila and Przedborski, 2003; Yamada *et al.*, 2004), thus highlighting the need for relevant antiapoptotic compounds able to halt neurodegeneration in PD.

A. Apoptosis and Propargyl-Containing MAO-BIs

Numerous preclinical studies have suggested the usefulness of MAO-BIs as neuroprotective agents in PD (Jenner, 2004; Tabakman *et al.*, 2004; Fernandez and Chen, 2007). However, it seems unlikely that the neuroprotective activities are exclusively related to the ability of MAO-BIs to decrease the production of free radicals and toxic aldehydes via inhibition of the enzymatic oxidation of dopamine. With regard to this, several studies suggest that the neuroprotective properties of MAO-BIs are independent of their capability to inhibit MAO-B and are actually related to the antiapoptotic properties of the propargyl group present in these molecules (Tatton *et al.*, 2003; Weinreb *et al.*, 2006; Naoi *et al.*, 2007).

Propargilamines are molecules containing a propargyl moiety that typically inhibits MAO-B, including the well-characterized compounds rasagiline and L-deprenyl. These compounds have antiapoptotic properties that are independent of their ability to inhibit MAO-B, and some studies suggest that these propargylamine-containing compounds prevent apoptosis by such diverse mechanisms as: (1) stabilizing the mitochondrial membrane potential, (2) blocking the opening of the permeability transition pore (PTP), (3) inducing the expression of antioxidant and antiapoptotic proteins of the Bcl-2 family, (4) increasing the expression of trophic factors, (5) regulating APP processing, or (6) inhibiting the nuclear translocation of the proapoptotic factor GAPDH (Jenner, 2004; Bar-Am *et al.*, 2005; Yi *et al.*, 2006). Unfortunately, none of these compounds has been convincingly shown to be neuroprotective when assessed in clinical trials.

PF9601N is a novel MAO-BI with a propargyl moiety that has demonstrated neuroprotective activity in different experimental models of PD. As with other propargylamine-containing MAO-BIs, PF9601N has also been shown to have neuroprotective effects independent of its capacity to inhibit MAO-B (Perez and Unzeta, 2003). Similar to other propargylamines, PF9601N has significant antioxidant properties (Sanz *et al.*, 2004; Bellik *et al.*, 2010) as well as remarkable effects in maintaining the mitochondrial membrane potential and preventing the opening of the PTP (Battaglia *et al.*, 2006). Nonetheless, further research from our group has showed that the neuroprotective effects of PF9601N may not be due to these or other mechanisms typically attributed to propargylamines, such as the induction of the expression of antiapoptotic proteins of the Bcl-2 family or the inhibition of the nuclear translocation of GAPDH. For example, in the MPP^+ model, where PF9601N displayed a prominent antiapoptotic effect, PF9601N was unable to induce the expression of the antiapoptotic proteins of the Bcl-2 family (Sanz *et al.*, 2008). Further, in this experimental paradigm, apoptosis is not mediated by nuclear translocation of GAPDH (Kodama *et al.*, 2005), suggesting that the neuroprotective effects of PF9601N were not related to potential actions at this level.

B. Antiapoptotic Mechanisms of PF9601N

Among all the factors proposed to underlie the pathogenesis of PD, mitochondrial dysfunction has been a long-term candidate. Evidence of decreased activity in complex I of the mitochondrial respiratory chain in PD brains (Schapira *et al.*, 1990) and the fact that complex I inhibitors cause a selective degeneration of nigral neurons contribute to the idea that mitochondrial dysfunction may be a causative factor in the pathogenesis of PD (Langston *et al.*, 1984; Betarbet *et al.*, 2000). It has also been proposed recently that ER stress may also be one of the factors responsible for the neuronal cell death in several neurodegenerative diseases, including PD (Lindholm *et al.*, 2006; Hoozemans *et al.*, 2007; Wang and Takahashi, 2007). For these reasons, our studies were designed to discern the molecular mechanisms underlying PF9601N neuroprotective effects using two well-established models of PD: the mitochondrial dysfunction model using the nigrotoxin MPP^+ (Sanz *et al.*, 2008) and the ER stress model using the ER–Golgi disruptor Brefeldin A (Sanz *et al.*, 2009).

In our first study, the results show how PF9601N confers neuroprotection in an experimental model where a complex I inhibitor (MPP^+) is used to mimic the respiratory chain dysfunction suggested to be responsible for the pathogenesis of PD (Table IIA) (Sanz *et al.*, 2008). In this experimental model, PF9601N showed a prominent antiapoptotic effect by preventing the stabilization of the proapoptotic transcription factor p53 and its subsequent transcriptional activity. This inhibitory effect on p53 activation resulted in reduced induction of p53-responsive proapoptotic proteins such as caspase-2 and the proapoptotic Bcl-2 family protein PUMA-α (Fig. 2A). These results are significant because p53 plays a key role in cellular stress responses and has been widely suggested to be one of the main proapoptotic mediators in several neurodegenerative diseases, including PD (Culmsee and Mattson, 2005; Levy *et al.*, 2009; Alves da Costa and Checler, 2010). These findings and others implicate p53 as a key mediator of apoptosis in several experimental models of PD (Trimmer *et al.*, 1996; Duan *et al.*, 2002; Nair *et al.*, 2006; Bretaud *et al.*, 2007; Perier *et al.*, 2007; Karunakaran *et al.*, 2008). p53 activation has also been reported in human PD brains (Nair *et al.*, 2006).

Table II

Percentage of Cell Viability (mean ± SEM) After MPP^+ (A) or Brefeldin A (B) Exposure in Cells Pretreated with the MAO-B Inhibitors PF9601N or l-Deprenyl.

	% Cell viability (30 h)	% Cell viability (48 h)
(A)		
MPP^+ 5 mM	24.23 ± 1.52^a	2.15 ± 0.16^c
PF9601N (10 μM)+MPP^+	69.48 ± 3.30^b	50.94 ± 2.07^d
L-Deprenyl (10 μM)+MPP^+	27.13 ± 1.03^a	2.81 ± 0.33^c
(B)		
Brefeldin A (BrA) 50 nM	54.65 ± 0.79^a	37.38 ± 1.47^c
PF9601N (10 μM)+BrA	93.39 ± 1.08^b	88.73 ± 1.92^b
L-Deprenyl (10 μM)+BrA	54.49 ± 4.65^a	39.78 ± 1.17^c

SH-SY5Y cells were treated with PF9601N (10 μM) or L-deprenyl (10 μM) for 30 min before the addition of MPP^+ (5 mM) or Brefeldin A (50 nM) to the cell culture. After 30 and 48 h, cell viability was determined using the Calcein-AM (A) or the Alamar Blue (B) assay. Values not sharing a superscript are significantly different ($P < 0.05$) to the others. Modified from Sanz *et al.* (2008) and Sanz *et al.* (2009).

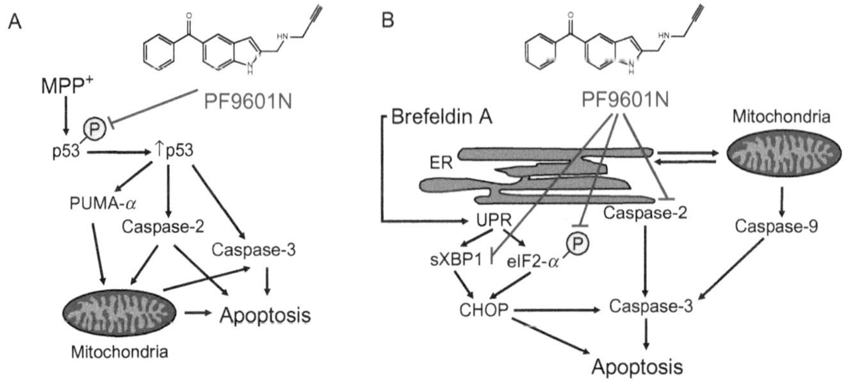

FIG. 2. PF9601N confers neuroprotection through different mechanisms in two cellular models of PD. (A) In the MPP^+ model, PF9601N confers neuroprotection by preventing the stabilization of the proapoptotic transcription factor p53. Stabilization of this transcription factor by phosphorylation results in its accumulation resulting in increased PUMA-α levels, caspase-2, and caspase-3 activation. Both caspase-2 activation and increased levels of PUMA-α result in mitochondrial proapoptotic signals, which, in conjunction with caspase-3 activity, can lead to cell death. (B) Brefeldin A treatment activates the unfolded protein response (UPR), as indicated by XBP1 splicing and eIF2-α phosphorylation. This activation results in the expression of the proapoptotic factor CHOP. PF9601N pretreatment blocks XBP1 splicing and reduces eIF2-α phosphorylation, which prevents CHOP induction. Brefeldin A exposure also induces the activation of the initiator caspases, caspase-2 and −9, which are significantly reduced in cultures treated with PF9601N. The increase in CHOP levels and the activation of initiator caspases may be responsible for caspase-3 activation and ultimately lead to cell death.

In a separate study, we have also described how PF9601N prevents ER stress-induced cell death (Table IIB) (Sanz *et al.*, 2009), demonstrating its efficacy in interfering with cell death induced by misfolding and aberrant accumulation of proteins. The ability of PF9601N to prevent the cell death induced by the ER stressor Brefeldin A is especially relevant because no other propargylamine has been previously shown to prevent ER stress-induced apoptosis. Our data indicate that PF9601N is able to prevent ER stress-induced cell death by partially blocking the unfolded protein response and consequently preventing the expression of the ER stress-related apoptotic mediator GADD153/CHOP. Further, under these same experimental conditions, PF9601N is also able to prevent the activation of the initiator caspases 2 and 9. Both effects may contribute to the neuroprotective effect of PF9601N in the cell death induced by this ER stressor (Fig. 2B).

In this context, it should be noted that ER stress is one of the factors recently suggested to underlie the pathogenesis of several other neurodegenerative diseases, including Alzheimer's disease (Nakagawa *et al.*, 2000; Hoozemans *et al.*, 2005), amyotrophic lateral sclerosis (Ilieva *et al.*, 2007), retinitis pigmentosa (Ryoo *et al.*, 2007), prionic diseases (Hetz *et al.*, 2003), polyglutamine diseases (Nishitoh *et al.*, 2002; Reijonen *et al.*, 2008), and several other acute cerebral disorders such as ischemic (DeGracia and Montie, 2004) or traumatic (Larner *et al.*, 2004, 2006 processes. The ability of PF9601N to block ER stress-induced cell death suggests that this compound may have benefit in the treatment of other neurodegenerative diseases. Nonetheless, further studies will be required in order to confirm this potential neuroprotective activity in these other diseases.

Surprisingly, the propargylamine-containing MAO-BI L-deprenyl did not show neuroprotective effects in either of the above-mentioned experimental models, which suggests that the neuroprotective effect of PF9601N is independent of its ability to inhibit MAO-B and may also be independent of the antiapoptotic properties of the propargylamine moiety of the molecule.

In summary, our data demonstrate that PF9601N blocks apoptotic pathways that are suggested to underlie neurodegeneration in PD; these apoptotic pathways do not appear to be targeted by other propargilamines.

IV. Clinical Potential of PF9601N

In 1995, the UK Parkinson Study Group reported an increased mortality linked to L-deprenyl and levodopa coadministration, raising concerns about the safety of L-deprenyl and other MAO-BIs (Lees, 1995). Importantly, PF9601N has been shown to be safer than L-deprenyl in oral acute toxicity studies and well-tolerated when used in animal models (Prodesfarma, 1997). In addition, PF9601N is a MAO-B suicide-type irreversible inhibitor that is more potent and selective than L-deprenyl (Table I) (Perez *et al.*, 1999), making PF9601N pharmacologically safer than L-deprenyl, as it has a lower risk of inhibiting peripheral MAO-A. In the periphery, MAO-A plays a prominent role in the deactivation of circulating catecholamines and diet-derived vasopressors, such as tyramine. Ingestion of tyramine-rich foods when peripheral MAO-A is inhibited may result in severe hypertensive crisis. Therefore, the high selectivity of PF9601N toward MAO-B makes it unlikely that PF9601N could lead to this adverse effect.

Regarding the metabolic profile and bioavailability of PF9601N, it has been shown that this compound does not affect its own metabolism and has a lower clearance rate than L-deprenyl (Dragoni *et al.*, 2007), which has low oral bioavailability as a result of extensive first-pass hepatic metabolism (Haberle *et al.*, 2002). Both PF9601N and L-deprenyl undergo extensive hepatic metabolism via different cytochrome p450 enzymes in the CYP system. L-Deprenyl is *N*-dealkylated mainly through CYP2B6 and CYP2C19 and to a lesser extend through CYP1A2 and CYP3A4 (Hidestrand *et al.*, 2001; Kamada *et al.*, 2002) to L-metamphetamine and L-amphetamine. In contrast, PF9601N is metabolized through CYP3A to the

N-dealkylated metabolite FA-72 (Dragoni *et al.*, 2007). It has been shown that L-deprenyl inhibits cytochrome CYP2B6 and CYP2C19 (Kamada *et al.*, 2002) and reduces CYP microsomal content as well as pNPH (*p*-nitrophenol hydroxylase) activity (Valoti *et al.*, 2000). This effect may have important clinical consequences, as CYP inactivation usually leads to delayed metabolic clearance and increased concentration of the drug, which may result in prolonged and more severe effects and undesired interactions between different compounds (Lin and Lu, 1998). In contrast, PF9601N does not affect its own metabolism, suggesting a lower risk of adverse effects (Dragoni *et al.*, 2007); the favorable metabolic profile of PF9601N supports its potential clinical use.

V. Concluding Remarks

All of the compounds that have been proposed so far as neuroprotectors, including MAO-BIs (with or without propargylamine moiety), dopaminergic agonists, glutamatergic antagonists, antioxidants, antiapoptotic compounds, anti-inflammatories, immunomodulators, and trophic factors, have failed or given inconclusive results when assessed in clinical trials for neuroprotection (Waldmeier *et al.*, 2006; Ahlskog, 2007; Hung and Schwarzschild, 2007; Kieburtz and Ravina, 2007; Schapira, 2008).

In particular, although numerous studies have proposed antiapoptotic agents as promising candidates for neuroprotection, several clinical trials have resulted, at best, in inconclusive results. It has been suggested that exposure (dosage or duration of treatment) to these antiapoptotic agents or the clinical trial design was not appropriate to reveal the neuroprotective effect of these compounds (Li and Beg, 2000). Another explanation is that the dopaminergic neurons that would have been rescued by apoptosis inhibition were either already dead or in an advanced state of degeneration when neuroprotective treatment was initiated, making their rescue impossible. Other explanations suggest that when apoptosis is blocked, cell death may proceed through other cell death mechanisms, such as necrosis or autophagy (Lum *et al.*, 2005). Besides the possible compensation through other cell death mechanisms, it is also possible that the different apoptotic pathways are highly redundant; this pathway redundancy may result in the activation of other apoptotic mechanisms when a currently active one is blocked. Another possibility is that these potential neuroprotective compounds preserve cell viability but are unable to maintain cell functionality. It must also be taken into account that the current experimental models of PD may not accurately reflect the progressive neurodegeneration present in PD, or that apoptosis may not play as prominent a role as originally thought.

Among all these possible explanations, the most accepted explanation for the failure of antiapoptotic compounds to confer neuroprotection in clinical trials is that these agents target apoptotic pathways that may not be significantly relevant for neurodegeneration in PD, and thus, these antiapoptotic compounds are only functional in the neurodegeneration present in the different experimental models. Current thinking suggests that apoptosis does indeed play an active role in the neurodegeneration seen in PD, and this fact emphasizes the need for more research on other neuroprotective compounds that act through mechanisms different from the ones previously described and more relevant to the neurodegeneration observed in PD.

It is important to note that PF9601N has been demonstrated to have neuroprotective properties that target apoptotic pathways suggested to be involved in PD pathogenesis, and these pathways are distinct from those previously described for other propargylamines.

Because of these observations, PF9601N emerges as a promising new therapeutic agent for PD, and further studies to verify the potential usefulness of PF9601N in the treatment of PD are warranted.

References

Ahlskog, J.E. (2007). I can't get no satisfaction: still no neuroprotection for Parkinson disease. *Neurology* **69**, 1476–1477.

Ahlskog, J.E., and Muenter, M.D. (2001). Frequency of levodopa-related dyskinesias and motor fluctuations as estimated from the cumulative literature. *Mov. Disord.* **16**, 448–458.

Ahlskog, J.E., and Uitti, R.J. (2010). Rasagiline, Parkinson neuroprotection, and delayed-start trials: still no satisfaction? *Neurology* **74**, 1143–1148.

Alves da Costa, C., and Checler, F. (2011). Apoptosis in Parkinson's disease: is p53 the missing link between genetic and sporadic Parkinsonism? *Cell. Signal* **6**, 963–968.

Aminoff, M.J. (2006). Treatment should not be initiated too soon in Parkinson's disease. *Ann. Neurol.* **59**, 562–564, (discussion 564–565).

Anglade, P., Vyas, S., Javoy-Agid, F., Herrero, M.T., Michel, P.P., Marquez, J., Mouatt-Prigent, A., Ruberg, M., Hirsch, E.C., and Agid, Y. (1997). Apoptosis and autophagy in nigral neurons of patients with Parkinson's disease. *Histol. Histopathol.* **12**, 25–31.

Avila, M., Balsa, M.D., Fernandez-Alvarez, E., Tipton, K.F., and Unzeta, M. (1993). The effect of side chain substitution at positions 2 and 3 of the heterocyclic ring of N-acetylenic analogues of tryptamine as monoamine oxidase inhibitors. *Biochem. Pharmacol.* **45**, 2231–2237.

Balsa, D., Fernandez-Alverez, E., Tipton, K.F., and Unzeta, M. (1991). Monoamine oxidase inhibitory potencies and selectivities of 2-[N-(2-propynyl)-aminomethyl]-1-methyl indole derivatives. *Biochem. Soc. Trans.* **19**, 215–218.

Balsa, D., Perez, V., Fernandez-Alvarez, E., and Unzeta, M. (1994). Kinetic behaviour of some acetylenic indolalkylamine derivatives and their corresponding parent amines. *J. Neural Transm. Suppl.* **41**, 281–285.

Bar-Am, O., Weinreb, O., Amit, T., and Youdim, M.B. (2005). Regulation of Bcl-2 family proteins, neurotrophic factors, and APP processing in the neurorescue activity of propargylamine. *FASEB J.* **19**, 1899–1901.

Battaglia, V., Sanz, E., Salvi, M., Unzeta, M., and Toninello, A. (2006). Protective effect of N-(2-propynyl)-2-(5-benzyloxy-indolyl) methylamine (PF9601N) on mitochondrial permeability transition. *Cell. Mol. Life Sci.* **63**, 1440–1448.

Bellik, L., Dragoni, S., Pessina, F., Sanz, E., Unzeta, M., and Valoti, M. (2010). Antioxidant properties of PF9601N, a novel MAO-B inhibitor: assessment of its ability to interact with reactive nitrogen species. *Acta Biochim. Pol.* **57**, 235–239.

Betarbet, R., Sherer, T.B., MacKenzie, G., Garcia-Osuna, M., Panov, A.V., and Greenamyre, J.T. (2000). Chronic systemic pesticide exposure reproduces features of Parkinson's disease. *Nat. Neurosci.* **3**, 1301–1306.

Blum, D., Torch, S., Lambeng, N., Nissou, M., Benabid, A.L., Sadoul, R., and Verna, J.M. (2001). Molecular pathways involved in the neurotoxicity of 6-OHDA, dopamine and MPTP: contribution to the apoptotic theory in Parkinson's disease. *Prog. Neurobiol.* **65**, 135–172.

Bretaud, S., Allen, C., Ingham, P.W., and Bandmann, O. (2007). p53-dependent neuronal cell death in a DJ-1-deficient zebrafish model of Parkinson's disease. *J. Neurochem.* **100**, 1626–1635.

Burke, R.E. (2008). Programmed cell death and new discoveries in the genetics of parkinsonism. *J. Neurochem.* **104**, 875–890.

Caslake, R., Macleod, A., Ives, N., Stowe, R., and Counsell, C. (2009). Monoamine oxidase B inhibitors versus other dopaminergic agents in early Parkinson's disease. *Cochrane Database Syst. Rev.* **4**, CD006661.

Charriaut-Marlangue, C., and Ben-Ari, Y. (1995). A cautionary note on the use of the TUNEL stain to determine apoptosis. *Neuroreport* **7**, 61–64.

Chen, J.J., and Swope, D.M. (2007). Pharmacotherapy for Parkinson's disease. *Pharmacotherapy* **27**, 161S–173S.

Chen, J.J., Swope, D.M., and Dashtipour, K. (2007). Comprehensive review of rasagiline, a second-generation monoamine oxidase inhibitor, for the treatment of Parkinson's disease. *Clin. Ther.* **29**, 1825–1849.

Culmsee, C., and Mattson, M.P. (2005). p53 in neuronal apoptosis. *Biochem. Biophys. Res. Commun.* **331**, 761–777.

Cutillas, B., Ambrosio, S., and Unzeta, M. (2002). Neuroprotective effect of the monoamine oxidase inhibitor PF 9601N [N-(2-propynyl)-2-(5-benzyloxy-indolyl) methylamine] on rat nigral neurons after 6-hydroxydopamine-striatal lesion. *Neurosci. Lett.* **329**, 165–168.

Dauer, W., and Przedborski, S. (2003). Parkinson's disease: mechanisms and models. *Neuron* **39**, 889–909.

DeGracia, D.J., and Montie, H.L. (2004). Cerebral ischemia and the unfolded protein response. *J. Neurochem.* **91**, 1–8.

Dragoni, S., Materozzi, G., Pessina, F., Frosini, M., Marco, J.L., Unzeta, M., Sgaragli, G., and Valoti, M. (2007). CYP-dependent metabolism of PF9601N, a new monoamine oxidase-B inhibitor, by C57BL/6 mouse and human liver microsomes. *J. Pharm. Pharm. Sci.* **10**, 473–485.

Duan, W., Zhu, X., Ladenheim, B., Yu, Q.S., Guo, Z., Oyler, J., Cutler, R.G., Cadet, J.L., Greig, N.H., and Mattson, M.P. (2002). p53 inhibitors preserve dopamine neurons and motor function in experimental parkinsonism. *Ann. Neurol.* **52**, 597–606.

Eberhardt, O., and Schulz, J.B. (2003). Apoptotic mechanisms and antiapoptotic therapy in the MPTP model of Parkinson's disease. *Toxicol. Lett.* **139**, 135–151.

Esposito, E., Di Matteo, V., and Di Giovanni, G. (2007). Death in the substantia nigra: a motor tragedy. *Expert Rev. Neurother.* **7**, 677–697.

Factor, S.A. (2008). Current status of symptomatic medical therapy in Parkinson's disease. *Neurotherapeutics* **5**, 164–180.

Fahn, S. (2005). Does levodopa slow or hasten the rate of progression of Parkinson's disease? *J. Neurol.* **252**(Suppl. 4), IV37–IV42.

Fahn, S., Oakes, D., Shoulson, I., Kieburtz, K., Rudolph, A., Lang, A., Olanow, C.W., Tanner, C., and Marek, K. (2004). Levodopa and the progression of Parkinson's disease. *N. Engl. J. Med.* **351**, 2498–2508.

Fernandez, H.H., and Chen, J.J. (2007). Monamine oxidase inhibitors: current and emerging agents for Parkinson disease. *Clin. Neuropharmacol.* **30**, 150–168.

Gálvez-Jiménez, N. (2007). Parkinson's disease. In: S. Gilman (Ed.), *Neurobiology of Disease*. Elsevier Academic Press, Maryland Heights, MO, USA, pp. 51–67.

Haberle, D., Szoko, E., and Magyar, K. (2002). The influence of metabolism on the MAO-B inhibitory potency of selegiline. *Curr. Med. Chem.* **9**, 47–51.

Hartmann, A., Hunot, S., Michel, P.P., Muriel, M.P., Vyas, S., Faucheux, B.A., Mouatt-Prigent, A., Turmel, H., Srinivasan, A., Ruberg, M., Evan, G.I., Agid, Y., *et al.* (2000). Caspase-3: a vulnerability factor and final effector in apoptotic death of dopaminergic neurons in Parkinson's disease. *Proc. Natl. Acad. Sci. USA* **97**, 2875–2880.

Hartmann, A., Troadec, J.D., Hunot, S., Kikly, K., Faucheux, B.A., Mouatt-Prigent, A., Ruberg, M., Agid, Y., and Hirsch, E.C. (2001). Caspase-8 is an effector in apoptotic death of dopaminergic neurons in Parkinson's disease, but pathway inhibition results in neuronal necrosis. *J. Neurosci.* **21**, 2247–2255.

Hartmann, A., Mouatt-Prigent, A., Vila, M., Abbas, N., Perier, C., Faucheux, B.A., Vyas, S., and Hirsch, E.C. (2002). Increased expression and redistribution of the antiapoptotic molecule Bcl-xL in Parkinson's disease. *Neurobiol. Dis.* **10**, 28–32.

Hauser, R.A. (2010). Early pharmacologic treatment in Parkinson's disease. *Am. J. Manag. Care* **16** (Suppl Implications), S100–S107.

Heinonen, E.H., Rinne, U.K., and Tuominen, J. (1989). Selegiline in the treatment of daily fluctuations in disability of parkinsonian patients with long-term levodopa treatment. *Acta Neurol. Scand. Suppl.* **126**, 113–118.

Hetz, C., Russelakis-Carneiro, M., Maundrell, K., Castilla, J., and Soto, C. (2003). Caspase-12 and endoplasmic reticulum stress mediate neurotoxicity of pathological prion protein. *EMBO J.* **22**, 5435–5445.

Hidestrand, M., Oscarson, M., Salonen, J.S., Nyman, L., Pelkonen, O., Turpeinen, M., and Ingelman-Sundberg, M. (2001). CYP2B6 and CYP2C19 as the major enzymes responsible for the metabolism of selegiline, a drug used in the treatment of Parkinson's disease, as revealed from experiments with recombinant enzymes. *Drug Metab. Dispos.* **29**, 1480–1484.

Hirsch, E., Graybiel, A.M., and Agid, Y.A. (1988). Melanized dopaminergic neurons are differentially susceptible to degeneration in Parkinson's disease. *Nature* **334**, 345–348.

Hoozemans, J.J., Veerhuis, R., Van Haastert, E.S., Rozemuller, J.M., Baas, F., Eikelenboom, P., and Scheper, W. (2005). The unfolded protein response is activated in Alzheimer's disease. *Acta Neuropathol.* **110**, 165–172.

Hoozemans, J.J., van Haastert, E.S., Eikelenboom, P., de Vos, R.A., Rozemuller, J.M., and Scheper, W. (2007). Activation of the unfolded protein response in Parkinson's disease. *Biochem. Biophys. Res. Commun.* **354**, 707–711.

Hung, A.Y., and Schwarzschild, M.A. (2007). Clinical trials for neuroprotection in Parkinson's disease: overcoming angst and futility? *Curr. Opin. Neurol.* **20**, 477–483.

Ilieva, E.V., Ayala, V., Jove, M., Dalfo, E., Cacabelos, D., Povedano, M., Bellmunt, M.J., Ferrer, I., Pamplona, R., and Portero-Otin, M. (2007). Oxidative and endoplasmic reticulum stress interplay in sporadic amyotrophic lateral sclerosis. *Brain* **130**, 3111–3123.

Ives, N.J., Stowe, R.L., Marro, J., Counsell, C., Macleod, A., Clarke, C.E., Gray, R., and Wheatley, K. (2004). Monoamine oxidase type B inhibitors in early Parkinson's disease: meta-analysis of 17 randomised trials involving 3525 patients. *BMJ* **329**, 593.

Jellinger, K.A. (1991). Pathology of Parkinson's disease. Changes other than the nigrostriatal pathway. *Mol. Chem. Neuropathol.* **14**, 153–197.

Jenner, P. (2003). Oxidative stress in Parkinson's disease. *Ann. Neurol.* **53**(Suppl. 3), S26–S36; (discussion S36–S38).
Jenner, P. (2004). Preclinical evidence for neuroprotection with monoamine oxidase-B inhibitors in Parkinson's disease. *Neurology* **63**, S13–S22.
Jenner, P., and Olanow, C.W. (2006). The pathogenesis of cell death in Parkinson's disease. *Neurology* **66**, S24–S36.
Kamada, T., Chow, T., Hiroi, T., Imaoka, S., Morimoto, K., Ohde, H., and Funae, Y. (2002). Metabolism of selegiline hydrochloride, a selective monoamine b-type inhibitor, in human liver microsomes. *Drug Metab. Pharmacokinet.* **17**, 199–206.
Karunakaran, S., Saeed, U., Mishra, M., Valli, R.K., Joshi, S.D., Meka, D.P., Seth, P., and Ravindranath, V. (2008). Selective activation of p38 mitogen-activated protein kinase in dopaminergic neurons of substantia nigra leads to nuclear translocation of p53 in 1-methyl-4-phenyl-1,2,3,6-tetrahydropyridine-treated mice. *J. Neurosci.* **28**, 12500–12509.
Kieburtz, K., and Ravina, B. (2007). Why hasn't neuroprotection worked in Parkinson's disease? *Nat. Clin. Pract. Neurol.* **3**, 240–241.
Kodama, R., Kondo, T., Yokote, H., Jing, X., Sawada, T., Hironishi, M., and Sakaguchi, K. (2005). Nuclear localization of glyceraldehyde-3-phosphate dehydrogenase is not involved in the initiation of apoptosis induced by 1-methyl-4-phenyl-pyridium iodide (MPP+). *Genes Cells* **10**, 1211–1219.
Kostrzewa, R.M., Kostrzewa, J.P., and Brus, R. (2002). Neuroprotective and neurotoxic roles of levodopa (L-DOPA) in neurodegenerative disorders relating to Parkinson's disease. *Amino Acids* **23**, 57–63.
Kristal, B.S., Conway, A.D., Brown, A.M., Jain, J.C., Ulluci, P.A., Li, S.W., and Burke, W.J. (2001). Selective dopaminergic vulnerability: 3,4-dihydroxyphenylacetaldehyde targets mitochondria. *Free Radic. Biol. Med.* **30**, 924–931.
Kumar, S. (2007). Caspase function in programmed cell death. *Cell Death Differ.* **14**, 32–43.
Labat-Moleur, F., Guillermet, C., Lorimier, P., Robert, C., Lantuejoul, S., Brambilla, E., and Negoescu, A. (1998). TUNEL apoptotic cell detection in tissue sections: critical evaluation and improvement. *J. Histochem. Cytochem.* **46**, 327–334.
Langston, J.W., Forno, L.S., Rebert, C.S., and Irwin, I. (1984). Selective nigral toxicity after systemic administration of 1-methyl-4-phenyl-1,2,5,6-tetrahydropyrine (MPTP) in the squirrel monkey. *Brain Res.* **292**, 390–394.
Larner, S.F., Hayes, R.L., McKinsey, D.M., Pike, B.R., and Wang, K.K. (2004). Increased expression and processing of caspase-12 after traumatic brain injury in rats. *J. Neurochem.* **88**, 78–90.
Larner, S.F., Hayes, R.L., and Wang, K.K. (2006). Unfolded protein response after neurotrauma. *J. Neurotrauma* **23**, 807–829.
Lees, A.J. (1995). Comparison of therapeutic effects and mortality data of levodopa and levodopa combined with selegiline in patients with early, mild Parkinson's disease. Parkinson's Disease Research Group of the United Kingdom. *BMJ* **311**, 1602–1607.
Levy, O.A., Malagelada, C., and Greene, L.A. (2009). Cell death pathways in Parkinson's disease: proximal triggers, distal effectors, and final steps. *Apoptosis* **14**, 478–500.
Li, M., and Beg, A.A. (2000). Induction of necrotic-like cell death by tumor necrosis factor alpha and caspase inhibitors: novel mechanism for killing virus-infected cells. *J. Virol.* **74**, 7470–7477.
Lin, J.H., and Lu, A.Y. (1998). Inhibition and induction of cytochrome P450 and the clinical implications. *Clin. Pharmacokinet.* **35**, 361–390.
Lindholm, D., Wootz, H., and Korhonen, L. (2006). ER stress and neurodegenerative diseases. *Cell Death Differ.* **13**, 385–392.
Lum, J.J., Bauer, D.E., Kong, M., Harris, M.H., Li, C., Lindsten, T., and Thompson, C.B. (2005). Growth factor regulation of autophagy and cell survival in the absence of apoptosis. *Cell* **120**, 237–248.

Macleod, A.D., Counsell, C.E., Ives, N., and Stowe, R. (2005). Monoamine oxidase B inhibitors for early Parkinson's disease. *Cochrane Database Syst. Rev.* **3**, CD004898.

Marconi, R., Bonnet, A.M., Vidailhet, M., and Agid, Y. (1992). The IMAO-B MDL 72.974 A in Parkinson's disease. *J. Neurol. Neurosurg. Psychiatry* **55**, 1096–1097.

Mattson, M.P. (2006). Neuronal life-and-death signaling, apoptosis, and neurodegenerative disorders. *Antioxid. Redox Signal.* **8**, 1997–2006.

Matus, S., Lisbona, F., Torres, M., Leon, C., Thielen, P., and Hetz, C. (2008). The stress rheostat: an interplay between the unfolded protein response (UPR) and autophagy in neurodegeneration. *Curr. Mol. Med.* **8**, 157–172.

Melamed, E., Offen, D., Shirvan, A., Djaldetti, R., Barzilai, A., and Ziv, I. (1998). Levodopa toxicity and apoptosis. *Ann. Neurol.* **44**, S149–S154.

Mochizuki, H., Goto, K., Mori, H., and Mizuno, Y. (1996). Histochemical detection of apoptosis in Parkinson's disease. *J. Neurol. Sci.* **137**, 120–123.

Mogi, M., Togari, A., Kondo, T., Mizuno, Y., Komure, O., Kuno, S., Ichinose, H., and Nagatsu, T. (2000). Caspase activities and tumor necrosis factor receptor R1 (p55) level are elevated in the substantia nigra from parkinsonian brain. *J. Neural Transm.* **107**, 335–341.

Nair, V.D., McNaught, K.S., Gonzalez-Maeso, J., Sealfon, S.C., and Olanow, C.W. (2006). p53 mediates nontranscriptional cell death in dopaminergic cells in response to proteasome inhibition. *J. Biol. Chem.* **281**, 39550–39560.

Nakagawa, T., Zhu, H., Morishima, N., Li, E., Xu, J., Yankner, B.A., and Yuan, J. (2000). Caspase-12 mediates endoplasmic-reticulum-specific apoptosis and cytotoxicity by amyloid-beta. *Nature* **403**, 98–103.

Naoi, M., Maruyama, W., Yi, H., Akao, Y., Yamaoka, Y., and Shamoto-Nagai, M. (2007). Neuroprotection by propargylamines in Parkinson's disease intracellular mechanism underlying the anti-apoptotic function and search for clinical markers. *J. Neural Transm. Suppl.* **72**, 121–131.

Nicotra, A., and Parvez, S. (2002). Apoptotic molecules and MPTP-induced cell death. *Neurotoxicol. Teratol.* **24**, 599–605.

Nishitoh, H., Matsuzawa, A., Tobiume, K., Saegusa, K., Takeda, K., Inoue, K., Hori, S., Kakizuka, A., and Ichijo, H. (2002). ASK1 is essential for endoplasmic reticulum stress-induced neuronal cell death triggered by expanded polyglutamine repeats. *Genes Dev.* **16**, 1345–1355.

Olanow, C.W. (2007). The pathogenesis of cell death in Parkinson's disease—2007. *Mov. Disord.* **22** (Suppl. 17), S335–S342.

Olanow, C.W., Myllyla, V.V., Sotaniemi, K.A., Larsen, J.P., Palhagen, S., Przuntek, H., Heinonen, E. H., Kilkku, O., Lammintausta, R., Maki-Ikola, O., and Rinne, U.K. (1998). Effect of selegiline on mortality in patients with Parkinson's disease: a meta-analysis. *Neurology* **51**, 825–830.

Olanow, C.W., Rascol, O., Hauser, R., Feigin, P.D., Jankovic, J., Lang, A., Langston, W., Melamed, E., Poewe, W., Stocchi, F., and Tolosa, E. (2009). A double-blind, delayed-start trial of rasagiline in Parkinson's disease. *N. Engl. J. Med.* **361**, 1268–1278.

Pahwa, R., and Lyons, K.E. (2009). Levodopa-related wearing-off in Parkinson's disease: identification and management. *Curr. Med. Res. Opin.* **25**, 841–849.

Perez, V., and Unzeta, M. (2003). PF 9601N [N-(2-propynyl)-2-(5-benzyloxy-indolyl) methylamine], a new MAO-B inhibitor, attenuates MPTP-induced depletion of striatal dopamine levels in C57/BL6 mice. *Neurochem. Int.* **42**, 221–229.

Perez, V., Marco, J.L., Fernandez-Alvarez, E., and Unzeta, M. (1996). Kinetic studies of N-allenic analogues of tryptamine as monoamine oxidase inhibitors. *J. Pharm. Pharmacol.* **48**, 718–722.

Perez, V., Marco, J.L., Fernandez-Alvarez, E., and Unzeta, M. (1999). Relevance of benzyloxy group in 2-indolyl methylamines in the selective MAO-B inhibition. *Br. J. Pharmacol.* **127**, 869–876.

Perier, C., Bove, J., Wu, D.C., Dehay, B., Choi, D.K., Jackson-Lewis, V., Rathke-Hartlieb, S., Bouillet, P., Strasser, A., Schulz, J.B., Przedborski, S., and Vila, M. (2007). Two molecular pathways initiate mitochondria-dependent dopaminergic neurodegeneration in experimental Parkinson's disease. *Proc. Natl. Acad. Sci. USA* **104**, 8161–8166.

Pizzinat, N., Copin, N., Vindis, C., Parini, A., and Cambon, C. (1999). Reactive oxygen species production by monoamine oxidases in intact cells. *Naunyn Schmiedebergs Arch. Pharmacol.* **359**, 428–431.

Prat, G., Perez, V., Rubi, A., Casas, M., and Unzeta, M. (2000). The novel type B MAO inhibitor PF9601N enhances the duration of L-DOPA-induced contralateral turning in 6-hydroxydopamine lesioned rats. *J. Neural Transm.* **107**, 409–417.

Prodesfarma (1997). PF9601N un nuevo farmaco inhibidor de la MAO-B para el tratamiento de la Enfermedad de Parkinson. Resúmenes de los estudios realizados. Centro de Investigación Prodesfarma.

Rao, R.V., Ellerby, H.M., and Bredesen, D.E. (2004). Coupling endoplasmic reticulum stress to the cell death program. *Cell Death Differ.* **11**, 372–380.

Rascol, O., Brooks, D.J., Melamed, E., Oertel, W., Poewe, W., Stocchi, F., and Tolosa, E. (2005). Rasagiline as an adjunct to levodopa in patients with Parkinson's disease and motor fluctuations (LARGO, Lasting effect in Adjunct therapy with Rasagiline Given Once daily, study): a randomised, double-blind, parallel-group trial. *Lancet* **365**, 947–954.

Reijonen, S., Putkonen, N., Norremolle, A., Lindholm, D., and Korhonen, L. (2008). Inhibition of endoplasmic reticulum stress counteracts neuronal cell death and protein aggregation caused by N-terminal mutant huntingtin proteins. *Exp. Cell Res.* **314**, 950–960.

Riederer, P., and Youdim, M.B. (1986). Monoamine oxidase activity and monoamine metabolism in brains of parkinsonian patients treated with l-deprenyl. *J. Neurochem.* **46**, 1359–1365.

Ryoo, H.D., Domingos, P.M., Kang, M.J., and Steller, H. (2007). Unfolded protein response in a Drosophila model for retinal degeneration. *EMBO J.* **26**, 242–252.

Sampaio, C., and Ferreira, J.J. (2010). Parkinson disease: ADAGIO trial hints that rasagiline slows disease progression. *Nat. Rev. Neurol.* **6**, 126–128.

Sanz, E., Romera, M., Bellik, L., Marco, J.I., and Unzeta, M. (2004). Indolalkylamines derivatives as antioxidant and neuroprotective agents in an experimental model of Parkinson's disease. *Med. Sci. Monit.* **10**, BR477–BR484.

Sanz, E., Quintana, A., Battaglia, V., Toninello, A., Hidalgo, J., Ambrosio, S., Valoti, M., Marco, J.L., Tipton, K.F., and Unzeta, M. (2008). Anti-apoptotic effect of Mao-B inhibitor PF9601N [N-(2-propynyl)-2-(5-benzyloxy-indolyl) methylamine] is mediated by p53 pathway inhibition in MPP (+)-treated SH-SY5Y human dopaminergic cells. *J. Neurochem.* **105**, 2404–2417.

Sanz, E., Quintana, A., Hidalgo, J., Marco, J.L., and Unzeta, M. (2009). PF9601N [N-(2-propynyl)-2-(5-benzyloxy-indolyl) methylamine] confers MAO-B independent neuroprotection in ER stress-induced cell death. *Mol. Cell. Neurosci.* **41**, 19–31.

Schapira, A.H. (2008). Progress in neuroprotection in Parkinson's disease. *Eur. J. Neurol.* **15**(Suppl. 1), 5–13.

Schapira, A.H., and Obeso, J. (2006). Timing of treatment initiation in Parkinson's disease: a need for reappraisal? *Ann. Neurol.* **59**, 559–562.

Schapira, A.H., Cooper, J.M., Dexter, D., Clark, J.B., Jenner, P., and Marsden, C.D. (1990). Mitochondrial complex I deficiency in Parkinson's disease. *J. Neurochem.* **54**, 823–827.

Schwarzschild, M.A. (2010). Rasagiline in Parkinson's disease. *N. Engl. J. Med.* **362**, 658(author reply 658–659).

Shimohama, S., Sawada, H., Kitamura, Y., and Taniguchi, T. (2003). Disease model: Parkinson's disease. *Trends Mol. Med.* **9**, 360–365.

Shoulson, I. (1998). Where do we stand on neuroprotection? Where do we go from here? *Mov. Disord.* **13**(Suppl. 1), 46–48.

Slodzinski, H., Moran, L.B., Michael, G.J., Wang, B., Novoselov, S., Cheetham, M.E., Pearce, R.K., and Graeber, M.B. (2009). Homocysteine-induced endoplasmic reticulum protein (herp) is up-regulated in parkinsonian substantia nigra and present in the core of Lewy bodies. *Clin. Neuropathol.* **28**, 333–343.

Spierings, D., McStay, G., Saleh, M., Bender, C., Chipuk, J., Maurer, U., and Green, D.R. (2005). Connected to death: the (unexpurgated) mitochondrial pathway of apoptosis. *Science* **310**, 66–67.

Stowe, R., Ives, N., Clarke, C.E., Deane, K., Wheatley, K., Gray, R., Handley, K., and Furmston, A. (2010). Evaluation of the efficacy and safety of adjuvant treatment to levodopa therapy in Parkinson s disease patients with motor complications. *Cochrane Database Syst. Rev.* **7**, CD007166.

Szegezdi, E., Logue, S.E., Gorman, A.M., and Samali, A. (2006). Mediators of endoplasmic reticulum stress-induced apoptosis. *EMBO Rep.* **7**, 880–885.

Tabakman, R., Lecht, S., and Lazarovici, P. (2004). Neuroprotection by monoamine oxidase B inhibitors: a therapeutic strategy for Parkinson's disease? *Bioessays* **26**, 80–90.

Tatton, N.A. (2000). Increased caspase 3 and Bax immunoreactivity accompany nuclear GAPDH translocation and neuronal apoptosis in Parkinson's disease. *Exp. Neurol.* **166**, 29–43.

Tatton, N.A., Maclean-Fraser, A., Tatton, W.G., Perl, D.P., and Olanow, C.W. (1998). A fluorescent double-labeling method to detect and confirm apoptotic nuclei in Parkinson's disease. *Ann. Neurol.* **44**, S142–S148.

Tatton, W., Chalmers-Redman, R., and Tatton, N. (2003). Neuroprotection by deprenyl and other propargylamines: glyceraldehyde-3-phosphate dehydrogenase rather than monoamine oxidase B. *J. Neural Transm.* **110**, 509–515.

Terzioglu, M., and Galter, D. (2008). Parkinson's disease: genetic versus toxin-induced rodent models. *FEBS J.* **275**, 1384–1391.

The Parkinson Study Group (1993). Effects of tocopherol and deprenyl on the progression of disability in early Parkinson's disease. The Parkinson Study Group. *N. Engl. J. Med.* **328**, 176–183.

The Parkinson Study Group (1996). Effect of lazabemide on the progression of disability in early Parkinson's disease. The Parkinson Study Group. *Ann. Neurol.* **40**, 99–107.

The Parkinson Study Group (2002). A controlled trial of rasagiline in early Parkinson disease: the TEMPO Study. *Arch. Neurol.* **59**, 1937–1943.

The Parkinson Study Group (2004). A controlled, randomized, delayed-start study of rasagiline in early Parkinson disease. *Arch. Neurol.* **61**, 561–566.

The Parkinson Study Group (2005). A randomized placebo-controlled trial of rasagiline in levodopa-treated patients with Parkinson disease and motor fluctuations: the PRESTO study. *Arch. Neurol.* **62**, 241–248.

Thorburn, A. (2004). Death receptor-induced cell killing. *Cell. Signal.* **16**, 139–144.

Trimmer, P.A., Smith, T.S., Jung, A.B., and Bennett, J.P. Jr. (1996). Dopamine neurons from transgenic mice with a knockout of the p53 gene resist MPTP neurotoxicity. *Neurodegeneration* **5**, 233–239.

Valoti, M. (2007). CYP-dependent metabolism of PF9601N, a new monoamine oxidase-B inhibitor, by C57BL/6 mouse and human liver microsomes. *J. Pharm. Pharm. Sci.* **10**, 473–485.

Valoti, M., Fusi, F., Frosini, M., Pessina, F., Tipton, K.F., and Sgaragli, G.P. (2000). Cytochrome P450-dependent N-dealkylation of L-deprenyl in C57BL mouse liver microsomes: effects of in vivo pretreatment with ethanol, phenobarbital, beta-naphthoflavone and L-deprenyl. *Eur. J. Pharmacol.* **391**, 199–206.

Vila, M., and Przedborski, S. (2003). Targeting programmed cell death in neurodegenerative diseases. *Nat. Rev. Neurosci.* **4**, 365–375.

Viswanath, V., Wu, Y., Boonplueang, R., Chen, S., Stevenson, F.F., Yantiri, F., Yang, L., Beal, M.F., and Andersen, J.K. (2001). Caspase-9 activation results in downstream caspase-8 activation and bid cleavage in 1-methyl-4-phenyl-1,2,3,6-tetrahydropyridine-induced Parkinson's disease. *J. Neurosci.* **21**, 9519–9528.

Waldmeier, P., Bozyczko-Coyne, D., Williams, M., and Vaught, J.L. (2006). Recent clinical failures in Parkinson's disease with apoptosis inhibitors underline the need for a paradigm shift in drug discovery for neurodegenerative diseases. *Biochem. Pharmacol.* **72**, 1197–1206.

Wang, H.Q., and Takahashi, R. (2007). Expanding insights on the involvement of endoplasmic reticulum stress in Parkinson's disease. *Antioxid. Redox Signal.* **9**, 553–561.
Waters, C.H., Sethi, K.D., Hauser, R.A., Molho, E., and Bertoni, J.M. (2004). Zydis selegiline reduces off time in Parkinson's disease patients with motor fluctuations: a 3-month, randomized, placebo-controlled study. *Mov. Disord.* **19**, 426–432.
Weinreb, O., Amit, T., Bar-Am, O., Sagi, Y., Mandel, S., and Youdim, M.B. (2006). Involvement of multiple survival signal transduction pathways in the neuroprotective, neurorescue and APP processing activity of rasagiline and its propargyl moiety. *J. Neural Transm. Suppl.* **70**, 457–465.
Yamada, M., Iwatsubo, T., Mizuno, Y., and Mochizuki, H. (2004). Overexpression of alpha-synuclein in rat substantia nigra results in loss of dopaminergic neurons, phosphorylation of alpha-synuclein and activation of caspase-9: resemblance to pathogenetic changes in Parkinson's disease. *J. Neurochem.* **91**, 451–461.
Yi, H., Maruyama, W., Akao, Y., Takahashi, T., Iwasa, K., Youdim, M.B., and Naoi, M. (2006). N-Propargylamine protects SH-SY5Y cells from apoptosis induced by an endogenous neurotoxin, N-methyl(R)salsolinol, through stabilization of mitochondrial membrane and induction of anti-apoptotic Bcl-2. *J. Neural Transm.* **113**, 21–32.
Youle, R.J., and Strasser, A. (2008). The BCL-2 protein family: opposing activities that mediate cell death. *Nat. Rev. Mol. Cell Biol.* **9**, 47–59.
Zhou, W., Hurlbert, M.S., Schaack, J., Prasad, K.N., and Freed, C.R. (2000). Overexpression of human alpha-synuclein causes dopamine neuron death in rat primary culture and immortalized mesencephalon-derived cells. *Brain Res.* **866**, 33–43.
Ziv, I., Zilkha-Falb, R., Offen, D., Shirvan, A., Barzilai, A., and Melamed, E. (1997). Levodopa induces apoptosis in cultured neuronal cells—a possible accelerator of nigrostriatal degeneration in Parkinson's disease? *Mov. Disord.* **12**, 17–23.

Subject Index

A

B

C

N

S

CONTENTS OF RECENT VOLUMES

Volume 39

Volume 40

Volume 41

Volume 42

Volume 43

Volume 44

Volume 45

Volume 47

Volume 48

Volume 49

Volume 50

Volume 51

Volume 52

Volume 53

Volume 54

Volume 55

Volume 60

Volume 61

Volume 62

Volume 63

Volume 64

Volume 65

Volume 66

Volume 67

Volume 68

Volume 69

Volume 70

Volume 71

Volume 72

Volume 73

Volume 74

Volume 75

Volume 76

Volume 77

Volume 78

Volume 79

Volume 79

Volume 81

Volume 82

Volume 83

Volume 84

Volume 85

Volume 86

Volume 87

Volume 88

Volume 89

Volume 90

Volume 91

Volume 94

Volume 95

Volume 96

Volume 97

Volume 98

Volume 99

FSC
www.fsc.org
MIX
Papier | Fördert gute Waldnutzung
FSC® C083411

Kolibri 360 GmbH
Ferdinand-Jühlke-Straße 7
99095 Erfurt, Deutschland
produktsicherheit@kolibri360.de